Peter Gendolla, Jörgen Schäfer (eds.)
The Aesthetics of Net Literature

The series "Medienumbrüche | Media Upheavals" is edited by Peter Gendolla.

Peter Gendolla, Jörgen Schäfer (eds.)

The Aesthetics of Net Literature

Writing, Reading and Playing in Programmable Media

[transcript]

Medienumbrüche | Media Upheavals | Volume 16

The Collaborative Research Center 615 at the University of Siegen with funding by the Deutsche Forschungsgemeinschaft (German Research Foundation) produced this book.

Bibliographic information published by Die Deutsche Bibliothek
Die Deutsche Bibliothek lists this publication in the Deutsche National-bibliografie; detailed bibliographic data are available in the Internet at http://dnb.ddb.de

Cover by: Kordula Röckenhaus, Bielefeld; Maik Pluschke, modulbüro, Siegen
Layout & Typeset by: Maik Pluschke, modulbüro, Siegen
Printed by: Majuskel Medienproduktion GmbH, Wetzlar
ISBN 978-3-89942-493-5

Distributed in North America by:

Transaction Publishers
New Brunswick (U.S.A.) and London (U.K.)

Transaction Publishers	Tel.: (732) 445-2280
Rutgers University	Fax: (732) 445-3138
35 Berrue Circle	for orders (U.S. only):
Piscataway, NJ 08854	toll free 888-999-6778

Contents

Playable Literature

Stories and Games

Writing and Programming

Practical Examples

Preface

In this volume we are dealing with ruptures and upheavals in literary communication through computerized networked media. As a result of the programmed manipulations of signs and the networking of computers and their users forms of literature have developed that can only be described as results of feedback processes between man and machine. No longer can we define computer nets as mere channels for messages generated by more or less brilliant creators. Unlike processes of linguistic signs in print media with their comparably stabilized textual meanings computers can process signifiers according to programmed instructions that cannot be controlled any longer in a universal way by "authors," "lectors," or "readers." We therefore have to ask in which of these literature projects indeed a new, a different *literary* quality is developing. The question will be in what way the fundamental openness of networked communications can be successfully amalgamated with the aesthetic demands of closure, perfection or coherence of literary creations as they at one point were formulated for works that were written by hand or that were printed so that after their publication any change of the signifying chains was excluded.

The contributions of this volume are grouped around this question beginning with terminological and conceptual questions discussing terms like "digital literature," "net literature," "ergodic literature," or "hyperfiction." *Peter Gendolla* and *Jörgen Schäfer* are trying to argue why we should talk of "net literature" regarding the above-mentioned literary communications and where in this case the specific cognitive interest might lie. The term is focusing on the decisive difference to traditional literary texts that lies in the recursive processes between humans and machines. Consequently we have to look at the *non-hierarchical* exchange-processes between *all* participating elements—between "authors," "works," and "readers"—in the analysis of literary processes in computer nets. Not only individual roles of activities in the literary system are redistributed here; we also experience characteristic hybridizations encompassing the whole system: In many cases we cannot decide without doubt which textual elements can be attributed to the intentions of an "author," to calculations of a computer or to the (re-)actions of a "reader."

Additionally, Gendolla and Schäfer are discussing the question in what way the specific *aesthetic* difference characterizing literature in traditional print-media as "language *art*" can be also identified under the current altered medial conditions. Their proposal maintains that literature, in the sense of "networked" trial-activities, can aesthetically reflect the far-reaching (man-machine-)communications that occur in networked computer systems, thereby making invisible processes visible and communicable. For it seems that only in current computer-aided and networked media literary forms emerge that cannot any longer be described adequately as aesthetic re-internalizations of perceptions since in networked processes these are immediately followed by externalization into activities of the recipient.

Conversely, *Roberto Simanowski* favors the term "digital literature" on the grounds of a media*technological* concept of "digitality," thereby differing from the concerns of Gendolla and Schäfer. His considerations are centered on the question of the genuineness and relevance of "digital" media for "literary texts." With this Simanowski not only is reaching out for a widening of the traditional hermeneutical approach in literary studies trying to interpret the meaning of signs. He rather is looking for a hermeneutics of intermedial, interactive and performative signs that also considers the meaning of the *"behavior"* of the sign on the screen as in the intermedial connections of words with visual signs or as reactions to the activities of readers, machines, or—as in "biopoetry"—bacteria.

Frank Furtwängler is dealing with one of the subjects with the most theoretical impact concerning our topic: the idea of "ergodic" literature that Espen Aarseth has developed in his publication *Cybertext: Perspectives on Ergodic Literature* as well as in numerous essays on literature in computer-aided media and lately especially regarding computer games. To be precise, Furtwängler here is criticizing the misleading adaptation of the "ergodicity"-concept into literary and game studies through which Aarseth wanted to re-determine the function of the recipient of "cybertexts" on the grounds of a changed materiality of the medium. This concept is derived from physics and there precisely characterizes the *unpredictability* of random processes. In computer games, however, consisting of scripted sequences and similarly non-chaotic structures the attentiveness of the player is directed to a high degree by the technical system. The adaptation of the theory of ergodicity, according to Furtwängler, therefore would only make sense if we were talking about leaving the player out of the system altogether. Conversely, he is advising game studies to model itself on the concepts of literary anthropology as developed for instance by Wolfgang Iser.

Philippe Bootz, by introducing an elaborate procedural model of literary communication between man and machine, also aims into that direction. He is discussing literary forms as processes of programming aesthetic rules that are independent of the surface aesthetics of the screen. In this model, writing and reading are two functions within a system: Authors and readers produce mental representations that he calls *texte-écrit* (text-as-written) and *texte-lu* (text-as-read). Between these cognitive processes, however, an "author"-, a "text"-

and a "reader"-domain is placed so that a procedural text can only be explained as the working together of three functions or processes: The author creates a *texte-auteur* (author-text) to which everything belongs which he has encoded in a significant form. He therefore can control the structure of the text but not all parameters of execution like the varieties of "programming" or reader activities. This transformed notion of the text therefore does not only encompass the *texte-auteur* but also the autonomous technological processes. The reader, on his end, observes a *texte-à-voir* (text-to-be-seen), i.e. a temporary, multimedial status that is bound to time and space and that includes the interface.

Mela Kocher attempts to structure the range of literature in computer-aided media. She introduces a "ludoliterary circle" that is able to cover similarities as well as differences between adventure games, hyperfictions, interactive movies, society simulations, cyberdrama and role-playing games by means of the parameters interactivity, perspective and narrative mode.

The fact that all these novel literary forms can in no way be contemplated without a long literary tradition or concrete historical predecessors is substantiated by *Jörgen Schäfer* with numerous examples of combinatorial, hypertextual and collaborative texts from German literary history since baroque times. Therewith he is providing us with a historical basis in order to work out the common features and differences that with computers have entered literary texts.

Friedrich W. Block is describing comical and diabolical aspects in current media art. He is using Peter Dittmer's *Die Amme* as an example for an installation that stages machines as grotesque caricatures of a robot capable of communicating. Block furthermore shows with the help of Cornelia Sollfrank's network-happening *female extension*, in which the artist identified the productions of a net-art generator as the artistic work of women, submitting them (without success!) to an international art competition, how both gender-politics of the art system as well as habitual man-machine symbols are stymied.

As might have become clear already in the conceptual contributions, the analysis of the new forms, procedures and objects that are manifest in literary processes in the net can rely on theories and methods of game studies. For example, the "ludologists" like Espen Aarseth, Jesper Juul or Gonzalo Frasca mainly focus on working out the ludic character of net literature or computer games—quite in the spirit of the double meaning of *ludus* and *paidia*, i.e. game and play. The scholarly work on these genres in the meantime has established its own transmedial discourse regarding its topics not only as mere derivatives of narratives or plays. Thus *Markku Eskelinen* in his characteristic provocative attitude demands of literary scholars to turn the tables and—if they cannot part with their traditional terminology—instead of persistently searching for narrative elements in computer games to define all narratives as "games of interpretation."

In six steps he is dealing with Aarseth's concepts of cybertext and game studies that challenge literary theory. Among other things he demands giving up the fear of variety and multiplicity, thereby renouncing the idea of literary entities or totalities and instead trying one's luck with the proposals of individual parts and phases. One possibility of such literary works seems to be on its way in the agenda of "playable media," "instrumental texts," and "textual instruments" developed by *Noah Wardrip-Fruin* in analyzing those works with which admittedly we are playing (and are producing for playing) but that cannot be called games in the more narrow sense of the word. Here we are rather dealing with textual and literary structures in which elements of play are used as means of interaction. In all his projects—like *Screen, Regime Change* and *News Reader*—Wardrip-Fruin is concerned with the artistic investigation of the specificities of (electronic) media—be they the World Wide Web or even the virtual reality environment of CAVE. This has led to the construction of texts as instruments, and we all know that instruments can be played: They constitute the frame for the activities of the user allowing the production of interesting variations of the original material. This approach avoids the reliance on formal definitions of games. The aim is neither to reach a certain goal nor to come to a specific result. We are rather talking about the playability of the texts, e.g. the question *how* they are played.

The challenge mentioned above is that of ludology to traditional narratology. It has to ask whether and in what way the logics of playing and that of the reading of narrative texts that seem at first glance to contradict each other (keywords would be: playing as acting vs. reading as interpretation) now can be brought together productively for the analysis of net literature—especially regarding the combination of man and machine.

Marie-Laure Ryan is pursuing the observation that even though "digital" forms of texts are reaching an academic public with avant-garde literary experiments and a mass public with computer games they have not yet, according to her thesis, found a cultured public that primarily reads for fun. Contrary to Eskelinen, in her contribution she is looking for possibilities to overcome this gap by strengthening the narrativity of digital texts. She is looking at three possibilities of interactive narratives: embedded stories as they appear in mystery games, emergent stories such as *The Sims*, and texts with a prescripted, but variable story, represented by Michael Mateas and Andrew Stern's interactive drama *Façade*.

Fotis Jannidis is equally critically analyzing both the "pure" ludologists and the model suggested by Ryan. In a sort of "close reading" of two computer games he is developing a differentiated position: Compared to "prototypes" of narration the organization of event-sequences in games establishes phenomena of liminality that as such, however, can be nevertheless described and analyzed from a narratological point of view.

Additional contributions are concerned with the relationship of writing and programming; they deal with literary texts created by computer programs or "literary machines" as well as with the consequences this has for concepts like "author," "work," or "reading." *Jean-Pierre Balpe,* for example, is defining generative literature as the production of texts that continually change since they are based on a specific dictionary, on a set of rules and the use of algorithms. He shows that texts like his generative novel *Trajectoires* that were produced by a computer and not written by an author demand a specific type of "engrammation" and reading. The point of departure of *Loss Pequeño Glazier* is that the mutation or modulation of words as a rule leads to orthographical relationships between variants. Sometimes, however, they also engender hardly tangible interactions, which he is trying to confirm in poems that use sequences and gaps in signifying chains. From these observations he is developing his concept of the reading of codes as poetic material.

Finally, the last three texts allow for variations of the perspective regarding our subject by firstly giving an insight into quite different practical realms like the didactics of literature as well as into the work of an internationally renowned author. *Laura Borràs Castanyer* shows how computer-aided media can be made useful for an innovative teaching of literature at the Universitat Oberta de Catalunya in Barcelona, a completely virtual distance learning university. She is trying to find out where the chances and dangers of E-Learning/Teaching can be located. *Susanne Berkenheger* introduces her latest play: the dramatic chat-distillation *i'm dying, honey: dramatized proceedings from super-space.* It was generated in chats and consists to 90 percent of original dialogues in the chat-room. The main character in the play is playing with chunks of communication as if she were a computer program and the chatters speculate wildly about her identity, become impatient, curse and carry on, thereby co-writing a play that in the meantime has been performed at several German theaters and that in the end will find its way back into the net as an audio-installation.

Thomas Kamphusmann's contribution at first glance seems out of place. From the perspective of information management, i.e. the analysis of the flow of communication in business, his contribution is questioning the possibilities but most of all the limits of improving or augmenting the exchange of information within business communication. In order to optimize these processes he is pleading for the inclusion of supplementary structures like aesthetic or literary forms into the development of the more abstract models put into operation.

This book is based on a three-day conference *Netzliteratur: Umbrüche in der literarischen Kommunikation [Net Literature: Upheavals in Literary Communication]* that took place between November 25-27, 2004 at Siegen University. We want to thank the Volkswagen Foundation and especially Dr. Vera Szöllösi-Brenig for the generous financial support for this international symposium that allowed our guests to travel on in some cases somewhat arduous paths from the US, France, Finland, Spain, Switzerland and Germany to Siegen in southern Westphalia.

Our special thanks go to Brigitte Pichon and Dorian Rudnytsky for translating several of the texts into English and for giving the necessary finishing touches to most of the others. Patricia Tomaszek completed the essential task of unifying quotations and bibliographic information. We are also indebted to her for proof-reading the manuscript. Maria Beissel finalized the typesetting of this volume. We are also grateful to Georg Rademacher for his critical advice.

Siegen, November 2006
Peter Gendolla and Jörgen Schäfer

Translated by Brigitte Pichon and Dorian Rudnytsky

Approaches and Terminologies

Peter Gendolla and Jörgen Schäfer

Playing With Signs

Towards an Aesthetic Theory of Net Literature

Abstract:

Considering upheavals in literary communications through computerized, networked media we suggest describing communications in computer-nets as open, multiple recursive processes between "authors" and "readers" whose practices of writing and reading are transformed by "autonomous" programs, "agents" etc. This, however, poses several questions: What are the common features that allow us to speak of an operational field called "literature" under historically changed medial conditions? Can the specific aesthetic difference that in traditional media characterizes literature as language art be also pinned down under these changed medial conditions?

As a rule, events are characterized as upheavals if they have far-reaching and long-lasting consequences, both negative and positive. This entails natural or cultural catastrophes, massive destruction of things, circumstances, structures, or systems and their elements just as much as it does their completely new construction, i.e. the long-term establishment of other, historically not yet existing constellations. Both meanings only refer to the two poles of one single process: the radical capsizing or the sudden end of conditions and orders of things. Typically, their partial or complete dissolution is followed by new sets of laws, different circumstances, structures or newly stabilized conditions.

Literature as a specific organ of perception, as a "high-voltage" sixth sense for socio-cultural change of any kind, takes part in such upheavals on several levels. It always has sensed the slightest tremors, the hairline cracks, the amplifying reverberations of conditions that announce or launch the catastrophic breakdown—even if never directly, but by more or less labyrinthine detours albeit in a very lucid manner. A good example would be *Der Mann ohne Eigenschaften [The Man Without Qualities]* by Robert Musil heralding the end of the Austro-Hungarian Empire, but also *Der Process [The Trial]* by Kafka that registered the bureaucratically automated genocide. During the upheaval itself no other instance of observation is able to record and comment on the events as does literature. Even though, compared to the world of objects, literature is a wholly arbitrary medium composed of letters, yet its signs (and maybe for this very reason) are able to amalgamate the perception of all our senses in a specific, synaesthetic way. This is possible only here in the literary interface as a sensitive area of contact between outer and inner worlds. We only have to remember the titles accompanying the great socio-cultural disruptions from the 19th to the 20th century: *Manhattan Transfer*, *USA*, *Ulysses*, or *Berlin Alexanderplatz*. These novels find images and stories for a new urban perception that had never existed in such density; they find possibilities to express the stream of unmitigated impressions of disjointed, freely combinable series of voices and glances in radio, cinema and telecommunications that were assimilated and made available by the new technical media.

But literature not only records and comments on the ruptures of social, cultural and—last but not at all least—medial conditions. It also participates constructively in their modeling as we can see already at the beginning of the 20th century when literature had not only registered or commented on the first great medial ruptures of audio-visions. It had also, both positively and negatively, designed, expelled or stabilized identities, self-images, goals of action, and possible ways of behavior for individuals, groups or even whole collectives. The whisper of literature not only reflects the stream of daily images and sounds. It also orders, frames and perspectivizes, thereby not only tying together multiple sensory perceptions but also furnishing them with possible meaning. A series of literary texts on the so-called Jewish world conspiracy for example equipped the bio-politics of fascism with quite practical goals and the "Protokolle der Weisen von Zion" ('Protocols of the Elders of Zion') transferred them into quite realistic actions (cf. Eco). And without the dissolution of traditional role models in women's literature from Virginia Woolf to Margaret Atwood the development of the questions regarding gender, family, or career in at least Western cultures would have been quite improbable.

In the 21st century another position is assigned to the participation of literature in socio-cultural transformations; literature acts on a different level, so to speak, or, to be more exact, it acts within a hitherto inaccessible realm. Within the context of the developments in media technology, the so-called second rupture of media—i.e. in short, the digitalization of all technical media—literature operates not only on the two levels of registering and commenting. Its third domain, namely that of designing or of conceptualizing, today with the help of networked computers has widened in such a way that it surpasses the familiar interplay of authors, media and public. Computers and nets are therefore not only more comfortable typewriters and faster channels of distribution; they also creatively compose together with their users that specific interplay of signs that we call literature.

Tele-Communication: Language, Texts, Nets

We are talking of a literature then that is now not taking place in books any more but in computers and of a literature, to name but a few of the equally well reasoned terms—depending on the theoretical perspective—that is traded under the name of "digital," "electronic," "ergodic," or "net" literature.[1] In our thesis of upheavals it may have become quite obvious that we refer to far-reaching and wide-ranging changes in literary communication regarding both literature in print-media and also literature in computer-aided media. Yet, we still need a theoretical frame enabling us to describe this specific phenomenon as a literary one. To be more precise, two questions arise that within the presented argument inevitably lead to a third one: Wherein, despite all differences, lie the common features that allow us to talk of a sphere of objects we can continue to call literature even across historically replaced constellations of media? And what difference is produced by the various media of production, distribution, and reception of literature, or, to put it otherwise: Wherein lies the distinguishing feature between the chains of letters of a text fixated in print and the "flickering signifiers" (Katherine Hayles) of computer-aided media? And even if it should be possible to find plausible arguments for these two questions, they will lead directly to a third one: Why do we, referring to literature in computer-aided media, speak of "net literature" and do not elect one of the earlier mentioned alternatives?

Let us at first remain in a quite general area, since in answering these questions once again the tension between the letters of literature, i.e. the realm of linguistic signs, and that of media technology has to be considered. In his latest publications on the "inner economy" of media, Hartmut Winkler has made an interesting suggestion to describe this relationship. His point of departure is language as a basic technology on the basis of which the linguistic praxis—as a fluid discourse—is connected inexorably with its material inscriptions or components that are manifested on the one hand in concrete media products like texts, films, images etc., and on the other in a more general sense in the machines and infrastructures of media technology itself. Already this first step in his media-theoretical argument connects the semiotic with the medial or better, with the media-technological process. Winkler finds two definitions for writing and—more generally—signs, which at first glance are surprising, but that are able to perspectivize the relationship between the symbolic and the technical realm in a manner that is helpful for our project.

On the one hand, he is adopting Derrida's concept of the deferral of signs, or simpler, its change from the context of production to the context of reception. On the other hand, he stresses the aspect of the repeatability of signs by labeling them "Maschinen der Wiederholung" ('machines of repetition') (Winkler, *Diskursökonomie* 25).

> Der enge Rahmen von Sender und Empfänger ist damit überschritten; beide sind nur Teil einer unendlichen Kette von Wiederholungsakten, die sie nicht überschauen oder kontrollieren können; Zeichen werden von Kontext zu Kontext verschickt oder übertragen, Zeichen sind insofern immer Telekommunikation. Dass sie auf Kontexte verweisen, die im Moment ihrer Aktualisierung nicht zur Verfügung stehen, macht ihren eigentümlich fremden Charakter aus. (98)

> Thereby the narrow frame of sender and receiver is transgressed; both are only a part of an infinite chain of acts of repetition, which cannot be grasped or controlled; from context to context signs are sent or transferred, signs then insofar are [always] telecommunication. Their peculiar, strange character is constituted by the fact that they refer to contexts that are not present in the moment they are actualized.

One could of course intuitively object that by using the expression "insofar" he is hiding the fact that the term telecommunication, contrary to its common usage, here is generalized to an extent that makes it useless. But in our context the aim is not to understand the elaborate and very abstract derivation which lets Winkler arrive at this definition of the sign. The crucial element is firstly that semiosis is always already technical, i.e. that it skillfully connects in the sense of *techné* (i.e. operation or procedure); secondly that it always takes place in media and that these media invariably organize distance communications, irregardless whether the distance is bridged by the human voice, the physical transport of a book, or the immaterial sending by wires; thirdly that already in the process of the creation of signs the logic of transferal or transmission begins to work.[2]

In order to conceptualize a communicative model it follows that media not only are added as whatever kinds of objects, appliances or infrastructures after the conclusion of the process of sign-creation in order to establish neutral channels of transport between sender and receiver, but that they are already also participating in this very process:

> Dies bedeutet, dass die technische Leistung der Medien, räumlich-zeitliche Abgründe zu überbrücken, in den Kern des Semiotischen vorrückt. Zeichen also werden keineswegs zuerst konstituiert und dann (sekundär) verschickt. Das Zeichen selbst ist die Klammer, die die unterschiedlichen Kontexte zusammenzieht, und die technischen Medien

> exekutieren nur, was als Kontextwechsel im Zeichen immer schon angelegt ist. (Winkler, *Diskursökonomie* 168)
>
> This means that the technical achievement of media to bridge spatial-temporal chasms now advances into the core of semiosis. I.e. signs in no way are first constituted and then (secondarily) sent. The sign itself is the bracket drawing together the different contexts whereas the technical media only execute those elements that are always already existent as contextual change within the sign.

But first this point has to be re-concretized, and we have to make this turn if we want to use Winkler's model for our purposes. How can we concretely conceptualize this continuity between the semiotic mechanism of a change of context and its implementation through technical media? And wherein do the different modalities of these implementations differ, once we compare different media of storage and transmission?

We therefore suggest to expand the ideas attained with the help of our brief outline of Winkler's media theory with some of the categories of Espen Aarseth's cybertext theory in order to ultimately attain a model of levels for the man-machine-dynamics in net literature, which, admittedly initially will be somewhat coarse. Such a model begins with the two traditional instances of language that cyclically refer to each other: the paradigmatic system of "langue" and its expression, "parole." Ordinarily—and this is important—these expressions materialize in syntagmatic chains. In the special case of a literary text in the medium book, the specific verbal externalizations of an author that he derives from "langue" are fixed in a storage medium and then transported or transmitted to the reader in various ways.

The essential point of Aarseth's model consists in the fact that he, in a very concrete and not only metaphorical sense, considers such texts to be machines for the production and reception of signs; however, in another place he also regards them as machines for the transmission of signs, hereby following Winkler.[3] This machine consists of a technical medium, a user, and a set of signs composed of two levels, the so-called "scriptons," i.e. the character strings which the reader encounters on the user interface and the "textons," the character strings that hide in the textual memory.

Such a text then not only consists of a syntagmatic chain, but is in itself composed of two levels, which are again—and this is the main point—related to each other by the so-called "traversal function." It is described in more detail by seven variables (dynamics, determinability, transience, perspective, access, linking, user function). In our context the most important variable is the user/

reader function: The literature that Aarseth calls "ergodic" demands more of its users or "readers" than the mere interpretation of the read text. In addition it requires non-trivial efforts in order to navigate the text. These activities can be explorative (when the user is able to decide how to navigate, like in hyperfictions), or they can be configurative (when the user can vary or add "textons" or traversal functions). This model then can provide a first approach to relate verbal expressions—i.e. also literary texts—with technical media.[4] However, by regarding a text as a machine does not yet bring us to the computer, since the outline is explicitly usable for all texts. Wherein, then, lies the specificity of the universal medium computer, and—in a further step—wherein lies the specificity of networked individual computers for literary communication?

With computer usage literary communications for the first time have acquired a programmable medium that does not only store and pass on its input. Rather, programming activities (protocols, browsers, word processing, electronic tools of whichever kind) enter the process thereby producing an output that is contingent for the "reader." Unlike in print media, in literary communication recursive processes between human beings (programmers, authors, readers) and machines are built in.

But even the idea that we are dealing with feedback between these two actants is too simple. Both the human- and the machine-related share in these processes in itself is characterized internally by feedbacks between various levels. Regarding the computer, on the one hand on the hardware level we have to differentiate between input and output devices, as well as the CPU (processor, bus and RAM). However, we also have to add the various levels of software. If several individual computers are connected into a net, these effects that already are executed offline still amplify, since the mediality of a computer is owed to the structural correspondence of the ('space *between* the individual digital machines and the space *within* these machines') (Winkler, "Medium Computer"). We might even venture to say with Winkler that the computer has only become a medium because within it circulate signs, so that it is controlled by the logic of telecommunication. In this realm as well signifiers are transmitted, stored, processed and permuted.

On the other hand, neurobiological theories regarding creative processes exist maintaining that we also have to regard the products of the human brain as effects of feedback loops between nerves within networks of the brain. For example Wolf Singer describes the emergence of creative processes as an outcome of the human capacity to generate symbolic representations of internally coded relations in abstract form and to transmit these to other brains:

> Die bereits für die einzelnen Gehirne charakteristischen rekursiven Prozesse weiten sich aus und beziehen die Gehirne der kommunikationsfähigen Artgenossen mit ein. Diese Iteration von Perzeption, Reflexion, Rekombination, Abstraktion, Kommunikation und Perzeption, die sich als unendliche Reihe fortsetzen kann, ist in der Lage, neue Systeme von fast beliebiger Komplexität hervorzubringen. (221)

> The recursive processes already characteristic for individual brains open out and enfold the brains of members of the same species that are able to communicate. This iteration of perception, reflection, recombination, abstraction, communication and perception, which can be continued in a never-ending chain, is able to bring forth new systems of almost any complexity.

Of course we do not intend to put the functioning of human brains on the same level with computers in an offensively simplified way. The decisive question is, however, to what extent we can talk of an expanded *ars combinatoria* regarding creative processes between human beings, whose internal life is structured symbolically, and machines that process symbols.

It seems that we are dealing with recursive loops on and between different levels and actants. Following Winkler once more, the point of this recursion lies in the fact that it allows relating repetition and variation to each other in a very specific way. In fact, recursions permit the

> . . . Wiederanwendung einer Verarbeitungsvorschrift auf eine Variable, die bereits Ergebnis derselben Verarbeitungsvorschrift ist. Der Variablenwert ändert sich mit jedem Durchlauf der Schleife, und Effekt der Wiederholung ist gerade nicht die Herstellung von Identität sondern einer vordefinierten Variation. Rekursion ist insofern nicht einfache, sondern erweiterte Reproduktion; und Rekursion verschränkt Wiederholung und Variation mit dem Ziel, ein Neues hervorzubringen, ein Ergebnis, das in dieser Form nicht vorvollzogen werden kann. (*Diskursökonomie* 173)

> . . . renewed application of a processing rule to a variable that already is the result of the same processing rule. The value of the variable is changing with each loop and the effect of the repetition is exactly not the creation of an identity but of a predefined variation. Recursion then is not a simple reproduction but an expanded one, and recursion

> links repetition and variation in order to bring forth something new, i.e. a result that could not be predefined in this manner.

These are the merely sketched reasons for centering our thoughts for the analysis of these recursive processes between humans and machines on the notion of the net. On purpose we are using here a wide notion of the net, encompassing more than computer networks so that we can include in it the interplay of humans and machines outlined above. The philosopher Hartmut Böhme for example (and we could just as well name Manuel Castells or others) has recently defined "nets" in the following way:

> Netze sind biologische oder anthropogen artifizielle Organisationsformen zur Produktion, Distribution und Kommunikation von materiellen und symbolischen Objekten. . . . Netze bilden komplexe zeiträumliche dynamische Systeme. . . . Sie tun dies nach stabilen Prinzipien, doch in instabilen Gleichgewichten, selbstgenerativ, selbststeuernd, selbsterweiternd, also autopoietisch und evolutionär. (19)

> Nets are biological or artificial anthropogenetic forms of organization to produce, distribute and communicate material and symbolic objects. . . . Nets create complex, dynamic systems of time-space. . . . They do this in a self-generating, self-controlling and self-expanding way according to stable principles, however, with instable balances, i.e. in an autopoietic and evolutionary manner.

Relating this to our field of literature this means that when we speak of "net literature" we do not speak of literary texts on the World Wide Web only, even though this may be somewhat irritating considering the general usage of the term "net" that directly thinks of the global internet.[5]

For those traditional models of literary communication that have stood the test of time for the analysis of literary communication in book culture this would then mean that the strict and clean division between the acts of production, distribution and reception on the level of different users now is collapsing, or at least is becoming less complex. This was the system that bound the well-known triad "author," "work," "reader." And as long as we have not found any better terminology, we will necessarily have to hold on to this triad, even if only as a horizon—if necessary, we will have to use quotation marks! If we want to understand the specific characteristics of literary communication in computerized, networked media, as a point of departure we will have to take distributed generating systems and reciprocal connections between "authors," "works," and "readers." For an author like Jean-Pierre Balpe this means that he has to work on all the described levels—he calls them "niveaux d'engrammation" ('levels of engrammation'). This differentiates him from the "author" of a book, who only has to bring the text to paper, but who can then leave the whole process of production and distribution to the publisher, the printer and the bookstores or publishing trade:

> Faire écrire du texte par un ordinateur implique donc une relation à l'ensemble des composantes de la structure, donc, d'une façon ou d'une autre, une relation à la programmation que celle-ci soit directe ou indirecte. L'écrivain se trouve là face à des niveaux d'engrammations techniques très différents qui vont de la conception abstraite de la notion de texte à la définition d'un style en passant par la représentation de connaissances dans les univers sur lesquels ses textes doivent se construire. (Balpe, "Littérature numérique")
>
> To have a computer write a text implies a connection between all components of the structure. One way or another this entails also a connection to the programming process, whether this is a direct or an indirect one. The writer thus finds himself facing very different technical levels of engrammation that extend from an abstract text-concept to the definition of a style, also traversing a representation of those levels of world-knowledge on which its texts are based.[6]

If in the specific man-machine-dynamics we can see the new and interesting elements of literary processes within, with, or between computers we may provisionally differentiate net literature according to the following parameters of communication:

- Communications of humans with humans, i.e. several authors cooperate via nets and become co-authors of cooperative texts.
- Communications of humans with machines, i.e. programmed processing of signs from which letters, poems, scripts, or narratives "automatically" emerge (text-generators or "literary machines"); this also enables the reader to choose between different possibilities of ramifications defined at certain nodal points.
- Summing up the two other levels brings us to the last possibility, namely communications of a man-machine-man-machine-etc. approach, tending towards an interminable cooperation of several authors, editors, designers, censors or whomever. In these processes between "authors" and "read-

ers"—i.e. before, during and after the production, transmission and reception of "communication"—technical media are present in diverse forms and on various levels. They already contribute to the intentions, or better the strategic efforts of authors, and even more extremely in their realization. This in-between, or rather this intersubjectivity, gains a quite different dimension with the help of a certain autonomy that is granted the automatic transcriptions of the intended and realized texts.

Quite in our line of thought Philippe Bootz has worked out a model of procedural function in the communication between humans and machines. Writing and reading are seen as two functions within a system: authors and readers bring forth mental representations that are called "texte-écrit" ('text-as-written') and "texte lu" ('text-as-read'). Between these cognitive processes an "author"-, a "text"-, and a "reader"-domain are placed. A procedural text therefore can only be explained as a cooperation of three functions or processes that are autonomous:

- *Writing function:* The author creates a "texte-auteur" ('author-text') that includes everything since he has encoded it in a form that is significant for him. Therefore, he can indeed control the structure of the text, but not all parameters of their implementation like different types of programming or readers' activities.
- *Generating function:* The changed notion of text not only comprises the "texte-auteur" but also autonomous technical processes. Bootz therefore defines the textual domain in which the text is generated and which in the end is read by the reader as "subsystem with a principal function: the generating of entrances and exits, in connection with its substances of works" ("Reader/Readers" 104).[7]
- *Reading function:* Finally, the reader observes a "texte-à-voir" ('text-to-be-seen'), a transitory status that is bound to time and space and that includes the interface.

However, also these proposals can refer to any form of a text. They do not yet reflect a specific literaricity. If in computerized media that alternately "read" and "write" each other on different human and mechanical levels multiple encoding and decoding steps are necessary between the writing and the reading of texts, then we can identify literature as a reflection or as an aesthetic perception of the transitions and disturbances between these levels.

The Aesthetics of Net Literature

So far, our considerations of the determining technical factors and of the formal possibilities of an altered textuality in electronic space have proven to be insufficient for a critical study of net literature since they might just as well describe all communication in computer networks. This also seems to be the blind spot in Aarseth's theory of "ergodic literature," since in analyzing structures and functional differences of the media in literary communication it looses track of a specific literaricity. We therefore have to add the question whether in changed medial conditions we can also discover a specific aesthetic difference like fictionality[8] and metafictionality characterizing literature as language art in traditional media. Literature, with its very special usage of words, indeed differs from everyday language and writing in two ways.

For one, it de-automates through distancing, exhibiting irony, and by using effects of alienation, i.e. through playing with utilitarian processes of communication, thereby creating the above mentioned aesthetic difference that always is a difference in perception by observing itself, thereby simultaneously distancing itself from becoming merely utilitarian. It seems that literature internally negates itself as a unified entity thereby enabling itself to capture its readers. This differentiates its specific texts from other texts like newscasts, cooking recipes, business letters etc. This means that literature precisely does not allow for a simple distancing from every day concerns, institutional constraints or too personal obsessions. To be precise, it distances itself from itself and from its own story, i.e. from the fictitious worlds it creates with the same amount of energy that for example a more trivial travel-, adventure-, or love-story uses in ever-new variations in order to create their imagined worlds. This means that it asserts a power, an aesthetic imagination or imaginativeness surpassing all dialectics between the constraints of every-day life and our small flights of fancy. This power is always able to map out more than any book can fulfill.

Secondly, literature changes with and through the respective media through which it is processed and experienced. Above, we have already made some comments about this. It is important to us that only through the interplay or the interaction of both components, i.e. the text and its medium, the concrete literary form emerges. For, as we have seen, media are always necessary to give a form to the communicated aspect. We therefore might say that the medium always inscribes itself into the contents. If with regard to computerized and networked media we then want to speak of net literature we have to ask the question whether it conserves this aesthetic difference described above regarding communication within the net itself as much as regarding traditional literary texts. Up to date literary studies that bear in mind the interplay of texts with their media therefore have to consider to what extent—i.e. in what forms and with which operations—this literature, in the sense of networked experimental activities, aesthetically reflects the far-reaching (man-machine-)communications, thereby making invisible processes visible and communicable.

Castells has shown extensively that computer nets create both technical connections between spatially distanced computers and also social connections between their users. Therefore, it should have become clear by now that the forms and contents of these communications in computer-aided networks are not merely transmitted neutrally. During the recursive circulation of the signals, signs, and symbols defined phases, sections, or parallel processes are automated more and more. This is true for the simple corrections of texts as well as for the organization of complete procedures of repair, operation or processes activated with the help of expert systems or software agents that do not need interceptions by human agents, who might not even be able to intervene or control the procedure without the knowledge of the respective program codes. Despite that, traditional sociology—and this is true from Parsons to Luhmann—continues to define communication as a simple relation of sender–channel–receiver and here the medium as "channel" is attributed no more function than simple transfers containing more or less disruptions and noise pollutions. Our position clearly is a different one: the channel is not merely neutral but generates unpredictable events.

It is just this development that constitutes the motivation and basis, the original object of perception of net literature insofar as we can continue to recognize it as literature, as literary *aisthesis,* aesthetic perception, and perception of perceptions. We then see the current pivotal and most momentous disruption in media systems in the growing autonomy of the technical medium within the processes of communication between man and machine and we see these processes as quite conflicting and at odds with each other: By implementing natural and especially cultural processes in networks they are made calculable; but at the same time they are thereby right away again eluding the desired control.

Net literature inscribes its narratives onto this open flank of technically supported processes of socio-cultural differentiation. Media technology then precisely does not widen—as some of its engineers maintain—but withdraws individual control; it hinders the transfer of control from individuals or groups to systems functioning without friction as much as possible thereby enabling a transfer necessary for cultural evolution. This transfer to systems becomes necessary because these systems are the agents that guarantee the survival of society and not foremost that of individuals. Aesthetic processes in their turn comment on and reflect this disinterest in individual concerns, be they imaginary self-positionings or physical needs. They have visions of short- or long-term consequences and effects of this withdrawal of control from or its handing over to the individual. They activate or intensify, they condense or focus those perceptions that take place in processes of socialization, be they the most sensitive feelings or the most brutal, traumatizing pain. Literature is—as are all the arts—engaged in individual aesthetization rather than social "anesthetization."

This engagement would be quite practical and could be directly observed and analyzed on the net. For example, if a client of a bank is relieved of the fussy banking transaction by a smart agent in internet banking, or if virtual travel agencies choose the holiday destination and accurately program the itinerary including even pit stops or if the traveler even can be relieved of writing a love letter to his Dulcinea at <http://www.liebste.de/liebesbrief2.html> or <http://www.writeexpress.com/love.htm>.[9] Here, or in the Chats, MUDs and MOOs introduced by Uwe Wirth in his article on "talkative correspondence" as a new form of initiating real love stories, we can see the development of new original literary forms within which Werther's fever is repeated on an electronic level as "modem-fever."

We could also show this for a literary contemplation of violence. Computer games can allow us to observe the changes of perspective, the role playing, i.e. the aesthetic difference or distance to these social feedbacks that hardly can be blocked off, that even feed on themselves and that then culminate in concrete acts of violence when we weigh the high-tech control of killings in Iraq against the high-speed training at the Joystick.

And we maybe should also name the quite unstoppable automation of our movement in space represented by the growing number of systems of orientation, GPS, the proximity sensors, the above mentioned travel agencies to which net literature in its turn reacts with its specific "discovery of slowness." We might name experiments in space, games of communication with the webcams in banks and the monitors in airports, as well as with special programs of exchange between text and body (cf. Gendolla).

Net literature now can be divided into different aspects, some of which are commented on in the contributions of this book. Looking at Michel Chaouli's article on the possibilities of computers and nets, which has generally challenged the notion that the interactive potential of the net can generate this *other* aesthetic state that we normally call fiction, we may have to ask: What about the great narrative, the conclusive, indissoluble work that can't be supplemented or shortened that until now has kept us sitting in our reading- (or viewing-)chair if the machine is constantly interrupting our processes of imagination? A similar question is posed by Florian Cramer: What about the aesthetic moment (*kairos*, epiphany), the moment of a feeling of truth that was made conceivable by the poetic avant-garde with their effective simultaneous double illumination of literary idea and realizing method if the poets on the net do not even understand the methods through which they work so that the codes play with them and not vice versa? What happens if the good will—or let us say the intention—to write the great "Hyper"-novel of the "Lost-in-Cyberspace-Generation" is located clearly before one's mind's eye but when in the process of writing this beautiful idea other authors always intervene or if a transformation grammar or a hidden Markov algorithm carries out a spelling reform of the third kind?

The furthest reaching dimension of our whole subject probably is playing. Literature always has had an affinity with playing if we regard the many word-games of antiquity, or the lipograms of women in the baroque era or even the methods suggested by Oulipo, even though without computer. With electronic devices and nets this playful literature now enters other dimensions and not only economic ones in the sense of "The novel after the videogame"—in this realm, film continues to be much more successful. No, our thesis is that in today's computer-aided and networked media literary forms are developing that cannot be described any longer solely in the sense of internalization or sublimation of processes of disciplining or civilizing, as mentioned above, because in networked processes internalization is immediately followed by *externalization,* which cannot be instantly understood in Marcuse's sense as a repressive de-sublimation. On the contrary: "authors," "readers," and their "media" are not playing as silent machines but are playing with, against or into one another in a highly reflexive way. Such practical integrations of traditional literary subjects, forms and methods into the activities of "authors" and "readers" that are connected with the narrative or performative staging of their own technical aids, automations and translations in a new way provoke the critical examination of some traditional aesthetic categories in computer-aided networked literature. First of all we will have to describe the present and long-ranging intentions of the contributing writers and the activities of the participating programs in detail. Secondly the specific emergence of poetical and narrative elements has to be considered, and thirdly we have to illustrate the performative contexts of their implementation. And in all probability—this would be a fourth premise—the new forms of net literature, their methods and "works" could be integrated within the notion of "games" as "playing" or vice versa. We cannot go into this in more detail. All we can do here is to hint at a direction, which then would have to be examined in individual analyses.

We are therefore coming back to our idea of upheaval, which implies that in current computer-aided and networked communications the quality of interaction between man and machine is once again effectively changing and that net literature registers, comments on and above all annotates this change—even if in an experimental way. We would like to repeat the claim that the automated or programmed part of communications directed at practical solutions in our present societies are inexorably growing for reasons of differentiation and complexity and that the generation, circulation and storage of socio-economically or culturally necessary sign-processes is more and more turned over to agents of communication so that this results in hardly separable amalgamations of software with "human" ideas and of manager-brains with mechanized processes of understanding, converting and deciding.

For the programming of software agents we can turn to the projects of AI research by Luc Steels. In the *Talking Heads* project, for example, the experiment between "human" and so-called "artificial" intelligence has indeed already moved forward quite far: Here programmed agents get into simulated situations that force them into the development of a "dialogue" with rudimentary syntax and grammar, the semantics of which we, as observers, are able to decode:

> Die Agenten spielen ein Sprachspiel, das wir Ratespiel nennen. Dabei übernimmt ein Agent den Part des Sprechers, ein anderer den des Zuhörers. Die Agenten wechseln sich in diesem Rollenspiel ab, sodass am Ende jeder beide Fähigkeiten entwickelt hat. Das Spiel kann so erweitert werden, dass ein menschlicher Mitspieler eine der beiden Rollen übernimmt und an die Stelle des Künstlichen Agenten tritt. (Steels, "Kognitive Roboter" 181)

> The agents are playing a language game, which we are calling guessing game. Here an agent takes over the part of a speaker, and another the part of a listener. The agents take turns in this role-playing so that in the end each has developed both capacities. We can develop the game to the extent that a human player takes over one of the two roles, taking the place of one of the artificial agents.

Of course such experiments trigger off the primordial fears of the replacement of the species by their own technical creations. Recently, Bill Joy with his theses of the end of humanity, of a post-human era in societies of clones with manipulated genes, mixed beings, cyborgs, autonomous robots who finally can go about their business peacefully has set off a debate of pros and cons that reverberated throughout all media. As if the species had not already replaced itself with its first scream and its first carving of a line into a branch by articulating or inventing signs through which it could invent itself away from its present state into an open future. Of course, the species believed it was talking about something that already had to be there, something that gave reason and coherence to things, from the stones to the angels. But this is only the undoubtedly necessary illusion of signs, their insurance system, their anchoring in the past so that the design would not become arbitrary, random, and unbearably open. In our present time, however, with the possibilities of media, or better technologies of simulation this illusion is dissolving more and more. The fact that we rigidly hold on to it probably rather proves the fear of individuals and of societies to have to take over the responsibilities for their own designs instead of leaving them forever to other, "higher," metaphysical authorities. At the present time, under the socio-technological conditions of a species that could through its media to all intents and purposes become aware of its activities, the fact that it is holding on to such metaphysical illusions of a different, higher will cannot lead to any other than a grotesque fundamentalism with all its atrocious consequences. The creation of artificial intelligence indeed is nothing but the exploration of one's own intelligence, even though it is a practical exploration that makes one's own possibilities visible and communicable and explicitly not its replacement or its end. For Steels there is still a "weiter Weg, bis wir eine der menschlichen Intelligenz vergleichbare künstliche Intelligenz erreichen" ('long way until we have reached an artificial intelligence that is comparable to human intelligence') (182). But it is this experiment that allows us to see the beginnings how intelligence might have developed and what we might be able to launch with its liberation:

> Indem eine breite Öffentlichkeit hier erstmals direkt in die Interaktion mit Künstlichen Agenten involviert war, konnten alle Beteiligten für sich selbst entscheiden, inwieweit das, was die Agenten tun, sich tatsächlich mit unseren Vorstellungen von kognitiven Phänomenen wie Lernen, Wahrnehmen, Sprechen, Kommunizieren deckt und ob die Kohabitation zwischen Künstlichen Agenten und Menschen in der Praxis überhaupt durchführbar ist. (Steels 182)

> By involving a wide public for the first time in the interaction with artificial agents, all participants could decide for themselves in how much the activities of the agents match with our ideas of cognitive phenomena like learning, perceiving, speaking, or communicating and whether the cohabitation between artificial agents and human beings in praxis can even be accomplished.

This suggests cohabitation, the living together with media-technological duplications instead of technophobia and aggression, which is nothing but a denial of our own capacities, nothing but self-blinding regarding the notion that somebody else, outside of ourselves, is controlling and directing our own activities.

It is this, and not more or less than this permanent and fatal self-illusion that the projects of net literature are trying to bring to the screens, monitors, or interfaces and they thereby are initiating a playing with the possibilities of self-imposed determinations as well as surprising realms of freedom.

The linkage or coupling of our perceptions with sensor technology and of our activities with technical motoric functions or the not yet experienced net of interdependencies of neural patterns and computer programs does not simply run along by itself effectively. Net projects attempt to open up this linkage to cognition in another way. Like already one hundred years ago, in classical avant-garde like Futurism, Dada or Surrealism, practical experiences are also at present transformed into aesthetic perception, and technical configurations into more or less artistic designs. The inability to perceive electronic processes that remain unclear but that (almost) with the speed of light increasingly take part in the scopes of political, economical or social activities here are made perceptible in—also technically—artistic ways. Its means or methods are those that for a long time already literature has articulated: it disturbs, irritates, and breaks expectations and thereby de-automates etc., i.e. it suggests alternatives in perception by inventing new shapes, coherences, "gestalts" that have not yet been perceived in this specific way.

Special projects for instance make the delineated close connection or the control of realms of perception or action perceivable with the help of programs and protocols; they playfully show the dependence of human behavior on automated rules and illustrate them aesthetically as irritations. For example in Susanne Berkenheger's *Die Schwimmmeisterin [The Bubble Bath]* users at first glance in groping their way into virtual space believe that they can choose freely among the directives, requests, commands or prohibitions within the windows of the monitor, thereby quickly making decisions. However, the interaction between them and the machine decides even faster than their eyes can see. A sort of multiple being made of human impulses, movement and programmed activities sets off quite unexpected images and texts; it makes windows disappear and open up new ones, sometimes obeying the users' will, often, however, only following the chance decisions of the machine. The group *Netzaktivismus* tries to illustrate the artificial processes behind it, the scripts that generate the texts, the relationship of authors' intentions and their programs by visualizing the source codes. This type of illumination of computer processes was continued for example by Vuk Cosic, a member of the *ASCII Art Ensemble*. Dragan Espenschied and Alvar C.H. Freude tried to demonstrate the possibilities of control in their art project *insert_coin* that was accompanied by a research program. The images of traditional audiovisual media at present are massively digitalized and the analogue archives of images are brought into archives on the net. Thereby with a little bit of programming the digital can be made visible as a shape. The *ASCII Art Ensemble* is carrying out an about-turn, a reverse transfer of moving images in film into "net-based moving ASCII":

> Erklärtes Ziel des 1998 gegründeten ASCII Art Ensembles (eine Gruppe mit Mitgliedern in Amsterdam, Ljubljana und Berlin) ist die "Rückübertragung" bewegter Filmbilder in "netz-basiertes bewegtes ASCII." Hier ist es nicht, wie bei Jodi, der Sourcecode, der zum Bild wird, sondern hier werden (bewegte) Bilder durch ASCII-Zeichen dargestellt. Das Verfahren erinnert an frühe, grafiklose und 24-nadelige Stadien der Druckertechnologie, als Bilder nur durch im Computer vorhandene ASCII-Zeichen dargestellt werden konnten und dementsprechend unentzifferbar waren. Das ASCII Art Ensemble hat bereits ein Javascript und einen Java Player für bewegte ASCII-Bilder entwickelt. Nun wird noch an einem schnellen Konverter gearbeitet, der bewegtes ASCII in Echtzeit im Netz unterstützt. Hehres Endziel ist die Entwicklung eines RealPlayer G2 Plug-Ins, das besagtes neues Dateiformat unterstützt und für eine weite Verbreitung sorgen könnte. Bislang entwickelt worden sind u.a. die *ASCII to Speech history of art for the blind,* die in ASCII-Zeichen gewandelte Bilder aus der Kunstgeschichte Zeichen für Zeichen vorliest. . . . Auch existiert bereits eine *History of Moving Image,* die in sieben Clips eine Übersicht über die Stilentwicklung und die Distributionsmedien des bewegten Bildes gibt, sowie *Deep ASCII,* eine ASCII-Version des Films *Deep Throat,* die auf einer Pong Arcade läuft. Hier sind nicht die pornografischen Bilder, sondern nur deren unentzifferbare ASCII-Versionen zu sehen. (Arns 239f.)

> The declared aim of the ASCII Art Ensemble, founded in 1998 (with members in Amsterdam, Ljubljana, and Berlin) is the transfer of moving images in film back into "net-based moving ASCII." It is not, as in Jodi's case, the source code that becomes the image; here (moving) images are represented with ASCII signs. The method reminds us of early states of 24-pin needle-printer technology without graphics, when images could only be shown with the help of ASCII signs available in the computer, thereby becoming quite indecipherable. ASCII Art Ensemble has already developed a JavaScript and a Java Player for moving ASCII-images. Now they are working on a fast converter supporting moving ASCII on the internet in real time. The noble final aim is the development of a RealPlayer G2 PlugIn that supports the mentioned new format and could thereby take care of a wide circulation. So far,

> the *ASCII to Speech History of Art for the Blind* was developed, reading sign by sign images from art history that were transformed into ASCII-signs. . . . There is also a *History of Moving Image* offering an overview of the development of styles and media of distribution of the moving image in seven clips as well as *Deep ASCII,* an ASCII-version of *Deep Throat* running on a pong arcade. One does not see the pornographic images but only their indecipherable ASCII-versions.

Finally, the media art project *]insofern[* attempts to visualize the aforementioned coupling of the central nervous system or certain neuronal patterns with computer programs:

> Die von Cod.Act entwickelte Maschine ist ein mobiler Scanner, der es erlaubt, ein künstliches menschliches Gehirn zu erforschen. Der Zuschauer steht isoliert in einer Schleuse, allein mit dem Gehirn. Er ist eingeladen zu einer Reise voller Klänge und Bilder, während er in eine bewegte, transluzide Welt eintaucht. Sobald der Lichtstrahl des Scanners die Nervenmasse durchdringt, erwacht diese zum Leben: sie bewegt sich, schwillt an und sendet Töne aus. Fragmentarisch offenbart sie die in ihrem Gewebe eingeschlossenen Informationen, lässt sie zu Zeugen des sensoriellen Kontakts mit der Außenwelt werden. Mittels einer Simultanübertragung verlässt das individuelle Erlebnis den geschützten Raum der Schleuse und gelangt hinaus zum Publikum. Der Vorgang lässt sich erinnern, er ist Gedächtnis geworden und verbreitet sich im öffentlichen Raum. (Cod.Act)

> The machine developed by Cod.Act is a mobile Scanner allowing investigating an artificial human brain. The viewer stands isolated in a "lock," alone with the brain. S/he is invited to a journey filled with sounds and images while being submerged in a translucid world. As soon as the light-ray of the scanner penetrates the mass of nerves, it awakens to life: it is moving, swelling, and sending out sounds. It reveals the information contained in its tissue in fragments, making them into witnesses of the sensorial contacts with the outer world. With the help of simultaneous transmission this sensorial experience is leaving the protected room of the lock and reaches the outside public. The incidence can be memorized; it has become memory, thereby spreading into the public realm.

These projects may mean nothing to many lovers of familiar art and literature; in the best-case scenario they seem to be intelligent games with the complex technologies of the computer or an interested exploration of public possibilities of cooperation between networked cultures. The actual dimensions of this upheaval in media history may only be visible in fragments, more as the disappearance of beloved shapes of medial dimensions than as the discernible outlines of new ones. Networked computer technologies currently are only offering scaffolds that help to construct something. The forms, structures and shapes that can be erected with these scaffolds are less apparent than in the gothic cathedrals or the skyscrapers of the 20th century. Net literature does not only register or comment on the socio-cultural dislocations and shifts connected to these upheavals. The fact that it is directly—and not only later on—connected to these processes, and above all the fact that it is a direct contributor to the process of designing the 21st century allows it to also discern some possible shapes, so that perceiving the obscure, highly accelerated recursive processes becomes a bit more easy by opening up the realm of sensual perception, thereby possibly allowing more premeditated decisions.

Translated by Brigitte Pichon and Dorian Rudnytsky

Notes

1. Cf. Roberto Simanowski's article in this book.
2. Ludwig Jäger uses similar arguments by stressing the fundamental importance of language as anthropological archimedium regarding cognitive processes. Language for him is a "semiological form of processing," which actually allows the creation of inner, mental episodes only within the networks of linguistic signs (cf. Jäger). These episodes then are externalized via other media (cf. Schäfer).
3. Aarseth's main definitions can be found in two areas. In order to define a text as a machine he says: "As the *cyber* prefix indicates, the text is seen as a machine—not metaphorically but as a mechanical device for the production and consumption of verbal signs. Just as a film is useless without a projector and a screen, so a text must consist of a material medium as well as a collection of words. The machine, of course, is not complete without a third party, the (human) operator, and it is within this triad that the text takes place. The boundaries between these three elements are not clear but fluid and transgressive, and each part can be defined only in terms of the other

two. Furthermore, the utilitarian possibilities of each element combine with those of the two others to produce a large number of actual text types. . . . Cybertext . . . is the wide range (or perspective) of possible textualities seen as a typology of machines, as various kinds of literary communication systems where the utilitarian differences among the mechanical parts play a defining role in determining the aesthetic process." (Aarseth 21f.) Further on, he sketches the aforementioned parameters of his typology: "A text, then, is any object with the primary function to relay verbal information. Two observations follow from this definition: (1) a text cannot operate independently of some material medium, and this influences its behavior, and (2) a text is not equal to the information it transmits. *Information* is here understood as a string of signs, which may (but does not have to) make sense to a given observer. It is useful to distinguish between strings as they appear to readers and strings as they exist in the text, since these may not always be the same. For want of better terms, I call the former *scriptons* and the latter *textons*. . . . In addition to textons and scriptons, a text consists of what I call a traversal function—the mechanism by which scriptons are revealed or generated from textons and presented to the user of the text." (Aarseth 62)

4. Concerning the relationship between language and technical media, Winkler even goes a step further, thereby surpassing Aarseth who presupposes a more or less autonomous development of media technology. Winkler, on the other hand, presupposes an erratic transition, or better turn from "soft," symbolic to "hard" technological procedures—and back! This is not the place to discuss advantages or disadvantages of this idea. However, it could be helpful for a further discussion of a media history of literature to view discursive articulations and media technology as two cyclically connected forms of enunciation instead of ideas that follow a deterministic approach regarding technology. Winkler maintains "dass es ein Kontinuum gibt zwischen der Abrundung/Freistellung einzelner Äußerungen oder Texte (die nur auf Basis dieser Abrundung zirkulieren können), der Systembildung im Umschlag von Diskurs in den Code (die oben Verdichtung genannt wurde), der Herausbildung von Institutionen und der Einschreibung/Abrundung/Einkapslung spezifisch ‚technischer' Infrastrukturen. . . . Der Übergang von der Enunziation_1 zur Enunziation_2 nämlich hat seine Besonderheit darin, *dass die Enunziation_2 den Raum des Symbolischen verlässt.* Die Enunziation_2 argumentiert nicht im Symbolischen sondern im Faktischen" ('that there is a continuum between the rounding off/release of individual articulations or texts (which can only circulate on the basis of this release), the creation of systems within the change of discourses into

codes (which was called condensation above), the forming of institutions and the inscription/rounding off/encapsulation of specifically technological infrastructures. . . . The transition of enunciation_1 to enunciation_2 is specifically remarkable *because enunciation_2 leaves the realm of the symbolic.* Enunciation_2 argues not in the symbolic but in the factual realm') (Winkler, *Diskursökonomie* 145f.).

5. However, how do we deal in this context with texts that are stored as linear chains of signs in connected computers, but that could just as well be stored and distributed as print media? This type of literature for heuristic purposes might rather be called "literature on the internet," in order to differentiate it from "net literature." We make this differentiation on the following grounds: Computers are disperse machines that are integrated only within a specific frame of discursive practice into specific constellations (cf. Schröter). As a universal medium the computer potentially can simulate all media. This is why in the literary system at first the orientation was focussed on the functions and semantics of book culture. Widely discussed in the nineties was the metaphor of a "universal library." It stood for the provisional limitation of a freely programmable universal machine to the function of a medium for downloading (also literary) texts that seemed to realize the utopia of a universal archive on the internet. This brings forth specific extensions and upheavals of the traditional literary scene, or rather the "social system literature," from which in the 18th century the four activities production, mediation, reception and processing emerged (cf. Schmidt). The mediation of literature in particular has been expanded through the World Wide Web. The transfer of texts through electronic media consists in a replacement of the channel of distribution with a completely differently structured system. It reaches from text databases like the "Gutenberg Project" <http://www.gutenberg.org>; <http://gutenberg.spiegel.de>, that makes classic texts available in a free online edition to the innumerable websites of authors that cannot be systematized—mostly hobby writers who cannot interest any publishers—to e-book or book-on-demand-offers of commercial publishers.
6. Cf. also his contribution to this book: "Principles and Processes of Generative Literature: Questions to Literature."
7. Cf. Bootz' article in this book: "The Problem of Form: *Transitoire Observable,* a Laboratory for Emergent Programmed Art."
8. We do not talk of fictitiousness in the sense of things or events that are simply invented. We remain insistent on Käte Hamburger's idea of a difference between the fictional and the fictitious. From this very difference she has deduced the specific "Logic of Literature." Cf. *The Logic of Literature.*

9. For example: "Dear Yinka, I am still on my knees begging for love, not to a stranger but to one I know . . . you. I've been in love and know it's fun. Each night and day I hope and pray that you'll be mine forever more. I know it's hard to trust a man—giving your heart, body and soul, but know it's me that's on my knees swallowing my pride, begging you, please . . . once again I say please; be mine. Love always, Lanre."
"Dear Chris, I'm so in love; I get this warm sensual feeling every time I think of you. When I sleep at night, all I can dream about is you. I wish I could hold you in my arms. I dream of the day when I could kiss you passionately on your soft sweet lips and listen to your heart beating because it's sweet music to my ears. All I can say is let me be the one you love; let me be the one whose love you need. Love always, Secret Admirer."
On March 21, 2005 an automat writes: "At the moment I am imagining you you're standing in front of me dressed only in heels and coat. No woman I've ever met was as beautiful as you. I hardly can wait till I can sip champagne from your navel. I also want to kiss your shoulders, knees, and lips, especially if you've eaten pudding earlier. I have never experienced a relationship that was as exceptional as ours. Just forget for a moment modesty and telephone and fulfill my dreams. I also would do everything for you—I would even walk through town with you as a dog on a leash." We can't get rid of the suspicion that this machine was fed with texts by followers of Peter Weibel—some contemporaries may remember quite well when he was pulled through the streets of Vienna by Valie Export. Cf. Valie Export, "Aus der Mappe der Hundigkeit."

Works Cited

Aarseth, Espen. *Cybertext: Perspectives on Ergodic Literature.* Baltimore: Johns Hopkins UP, 1997.

Arns, Inke. "'Unformatierter ASCII-Text sieht ziemlich gut aus': Die Geburt der Netzkunst aus dem Geiste des Unfalls." *Kunstforum International: Der gerissene Faden. Nichtlineare Techniken in der Kunst* 155 (2001): 236-241.

Balpe, Jean-Pierre. "Littérature numérique, contraintes et ouvertures de l'écran: Du stylo à l'ordinateur ou du livre à l'écran." *Dichtung Digital* 6.3 (2004). 27 June 2006 <http://www.dichtung-digital.com/2004/3-Balpe.htm>.

Berkenheger, Susanne. *Die Schwimmmeisterin.* 2002. 27 June 2006 <http://www.schwimmmeisterin.de>.

———. *The Bubble Bath.* 2003. 27 June 2006 <http://www.thebubblebath.de>.

Böhme, Hartmut. "Netzwerke: Zur Theorie und Geschichte einer Konstruktion." *Netzwerke: Eine Kulturtechnik der Moderne.* Ed. Jürgen Barkhoff, Hartmut Böhme, and Jeanne Riou. Cologne: Böhlau, 2004. 17-36.

Bootz, Philippe. "Reader/Readers." *p0es1s: Ästhetik digitaler Poesie/The Aesthetics of Digital Literature.* Ed. Friedrich W. Block, Christiane Heibach, and Karin Wenz. Ostfildern: Cantz, 2004. 93-121.

Chaouli, Michel. "How Interactive Can Fiction Be?" *Critical Inquiry* 31.3 (2005): 559-617.

Cod.Act. *]insofern[.* 2003. 27 June 2006 <http://db.swr.de/imkp/IMKP.detail?p_lw=g&p_kwid=586>.

Cosic, Vuk. *Unreal.* 1999. 27 June 2006 <http://www.ljudmila.org/~vuk/ascii/unreal/>.

Cramer, Florian. "Warum es zuwenig interessante Netzdichtung gibt." *Netzliteratur.net.* 2000. 27 June 2006 <http://www.netzliteratur.net/cramer/karlsruher_thesen.html>.

Eco, Umberto. "Fictional Protocols." *Six Walks in the Fictional Woods.* Cambridge, MA: Harvard UP, 1994. 117-142.

Espenschied, Dragan, and Alvar C.H. Freude. *insert_coin.* 2001. 27 June 2006 <http://odem.org/insert_coin/>.

Export, Valie. "Aus der Mappe der Hundigkeit." 1969. 27 June 2006 <http://www.medienkunstnetz.de/werke/mappe-der-hundigkeit>.

Gendolla, Peter. "Auf 50 Meter genau: Die neueren Ephemeriden." *Ephemeres. Medicale Innovationen 1900/200.* Ed. Ralf Schnell and Georg Stanitzek. Bielefeld: Transcript, 2005. 217-235.

Hamburger, Käte. *The Logic of Literature.* Trans. Marilynn J. Rose. Bloomington, IN: Indiana UP, 1973.

Jäger, Ludwig. "Zeichen/Spuren: Skizze zum Problem der Sprachzeichenmedialität." *Schnittstelle: Medien und kulturelle Kommunikation.* Ed. Georg Stanitzek and Wilhelm Voßkamp. Cologne: DuMont, 2001. 17-31.

Joy, Bill. "Why the future doesn't need us: Our most powerful 21st century technologies—robotics, genetic engineering, and nanotech—are threatening to make humans an endangered species." *Wired Magazine* 8.4 (2000). 27 June 2006 <http://www.wired.com/wired/archive/8.04/joy.html>.

Schäfer, Jörgen. "Sprachzeichenprozesse: Überlegungen zur Codierung von Literatur in 'alten' und 'neuen' Medien." *Analog/digital—Opposition oder Kontinuum? Zur Theorie und Geschichte einer Unterscheidung.* Ed. Jens Schröter and Alexander Böhnke. Bielefeld: Transcript, 2004. 143-168.

Schmidt, Siegfried J. *Die Selbstorganisation des Sozialsystems Literatur im 18. Jahrhundert.* Frankfurt am Main: Suhrkamp, 1989.

Schröter, Jens. *Das Netz und die Virtuelle Realität: Zur Selbstprogrammierung der Gesellschaft durch die universelle Maschine.* Bielefeld: Transcript, 2004.

Singer, Wolf. "Neurobiologische Anmerkungen zum Wesen und zur Notwendigkeit von Kunst." *Der Beobachter im Gehirn: Essays zur Hirnforschung.* Frankfurt am Main: Suhrkamp, 2002. 211-234.

Steels, Luc. "Kognitive Roboter und Teleportation: Artefakte reagieren auf ihre Umwelt und erfinden sich eine Sprache." *Weltwissen/Wissenswelt: Das globale Netz von Text und Bild.* Ed. Christa Maar, Ernst Pöppel, and Hans U. Obrist. Cologne: DuMont, 2000. 175-184.

Winkler, Hartmut. *Diskursökonomie: Versuch über die innere Ökonomie der Medien.* Frankfurt am Main: Suhrkamp, 2004.

———. "Medium Computer: Zehn populäre Thesen zum Thema und warum sie möglicherweise falsch sind." *Understanding New Media,* Heinz-Nixdorf-Forum, Paderborn. 19 Feb. 2003. 27 June 2006 <http://www.uni-paderborn.de/~winkler/compmed2.html>.

Wirth, Uwe. "Schwatzhafter Schriftverkehr: Chatten in den Zeiten des Modemfiebers." *Praxis Internet.* Ed. Stefan Münker and Alexander Roesler. Frankfurt am Main: Suhrkamp, 2002. 208-231.

Roberto Simanowski

Holopoetry, Biopoetry and Digital Literature

Close Reading and Terminological Debates

Abstract:
What are the characteristics of digital literature? Does it already emerge when it is transferred from one medium to the next? What is the relevance of genuineness in media, what that of the medium itself? How much text does a hyper-medial work have to contain in order to belong to the realm of digital literature and not to that of digital arts? How does the role of the author change when reader, machine or bacteria take his/her place? This article connects terminological questions with case studies of some interesting examples.

Last year a book was published for which I had been waiting for a long time: *Close Reading New Media: Analyzing Electronic Literature*, edited by Jan Van Looy and Jan Baetens. After years of terminological and theoretical debates this book is aiming at providing the case studies long overdue. It does not insist on asking whether we should talk of digital, electronic, interactive, or ergodic, of hypertext-, net-, cyber-, or code-literature but rather asks how we can *read* this literature.[1]

According to the editors there are three reasons for the rarity of such hermeneutic endeavors in comparison to terminological or theoretical works: "First of all there is the basic conviction that critical attention does not matter, or even that it is not appropriate to works belonging to a medium which has as one of its primary principles the absence of—literally—fixed shapes and—literally—fixed meanings" (Looy and Baetens 7).

The fact that electronic literature principally is open indeed seems challenging for literary studies whose main orientation is determining signs. But this is not problematic for the first time; concrete poetry as well demanded of literary studies an adaptation of its methodical tools by assigning predicative processes to the non-linguistic realm. The editors of the book affirm that we have to develop a dialogue with form and I can only confirm this. If form consists in elusiveness or variableness then we must interpret this very characteristic. "Second, there is the idea, which is not entirely false, that hyperfiction is born on the margins of a medium, the computer, which is still considered a number cruncher rather than a literary device" (8). In this case the editors refer to Espen Aarseth: "The emerging new media technologies . . . should be studied for what they can tell us about the principles and evolution of human communication" (8). We might add that this technology should also be looked at regarding the principles of the *aesthetic* communication this involves.

Thirdly—and this is the true problem—"we often hear the argument that hyperfiction has not yet produced enough interesting works to justify a turn towards a more literal and literally tackling of the material" (8). The editors correctly object to this argument maintaining with a quote from Marie-Laure Ryan that even in the case of lacking quality working out a close analysis would be worth while since "[f]or the literary scholar, the importance of the electronic movement is twofold: it problematizes familiar notions, and it challenges the limits of language" (8). I would like to support this argument underlining that even works of mediocre quality—and maybe precisely these—can illuminate to what extent electronic literature can handle its material in a meaningful and convincing way and where the traps of it may lie.

The book itself, however, gives us the impression that the editors themselves do not really believe that hyperfiction has produced sufficiently interesting works. The three contributions to classic hyperfictions—Stephanie Strickland's *True North*, Shelley Jackson's *Patchwork Girl* und Marjorie Luesebrink's *Califia*—are followed by analyses of Geoff Ryman's novel *253* and Raymond Federman's novel *Eating Books*, which were already published as books, thereby providing evidence for the fact that they hardly need the electronic medium. The last three contributions deal with a site promoting a film, with the "Narrative of an Interface" and with *Hypertextual Consciousness*. This book, finally promising concrete case studies after all these theoretical considerations seems barely convinced of its own interests. The meager result of this beginning hardly lets us hope for a continuation.

Let us therefore describe the situation more positively. The range of electronic literature is vast. The transfer of a novel from print can be as much part of it as the website for a film playing with interesting text-graphic-elements. But is it true that already the transformation from one medium to the next creates electronic literature? How much text does a hypermedial work need in order to pertain to electronic *literature* and not to electronic *art*? This is the moment, however, in which after all terminological considerations are necessary. How quickly we get to a point where we didn't want to be: We have to talk about terminology and concepts. The matter at hand as well as its notoriety allows, respectively demands that I at least attempt to connect these considerations with some interesting case studies.

1. The Genuineness of the Artifact

When the Brazilian artist Eduardo Kac discovered holography in the eighties, he began with some basic questions:

> It is very important to emphasize that not all texts recorded on holographic film are holopoems. It is technically possible, for example, to record a symbolist sonnet on a hologram. Such a sonnet does not become a holopoem simply because it is displayed on holographic film. What defines a holopoem is not the fact that a given text is recorded on holographic film. What matters is the creation of a new syntax, exploring mobility, non-linearity, interactivity, fluidity, discontinuity and dynamic behavior only possible in holographic space-time. ("Biopoetry" 184-5)

Kac's aim is the differentiation of genuine holopoetry from one that only brings old artifacts into the new medium. He sees the necessity of this differentiation in the disparity between technical skills and artistic concepts while using new technologies—a general problem at that time (cf. Coyle 65-68). The holography-artist Margaret Beyon for example differentiates between the mere use of a technology and its utilization for aesthetic purposes: "The credit for exhibitions arranged by artists of laser beams as purely physical phenomena, for instance, should go to the inventor and manufacturers of the laser, rather than to the 'artist'" (Coyle 73). In a similar sense the art critic Peter Fuller is questioning the share of the artist in holographic objects:

> [T]he very *process* of making a hologram does not allow for the admission of a human imaginative or physical expressive element at any point. The representation is not worked, it is posed and processed . . . hence the hologram remains a particularly dead medium compared with painting. (Coyle 74)

It is just this disesteem of holography that is answered by Kac's demand for holopoetry emerging from the medium itself by using its technical potential for aesthetic and conceptual purposes.

I do not want to discuss the characteristics that holography as technology has in store for specific artistic statements. It was important to show that the problems surfacing with new media are not new ones. But we have to discuss the characteristics that digital technologies keep ready for specific artistic statements. In earlier texts (2002) I have identified these specific characteristics as interactivity, intermediality, and performance:

Interactivity aims at motivating the recipient to co-construct the work, including several possibilities:

- reacting to characteristics of the work (programmed interactivity: human–software). This includes first of all (but certainly not any more exclusively or even mainly) multilinearity in hypertexts requesting the readers to assemble navigational decisions on their own.
- reacting to activities of other recipients (network-bound interactivity: human–human via software). This includes cooperative writing projects requesting all readers entering the website to become authors of the project introduced.

Intermediality as a further characteristic of digital literature marks the (conceptual-integrative) connection between the traditional media of expression, language, image, and music. The academic discussion differentiates performative nuances that are reflected in terminological variants. In his analysis of intermediality Jürgen Müller for example differentiates between medial elements that perform alongside each other and those that conceptually cooperate. The first variant he calls multimedial, the latter intermedial (cf. Müller 83). It remains questionable, however, whether this differentiation is sufficiently discriminating. In scholarly discussions the term intermedia is used quite variably and in some instances only indicates a change of media or a reciprocal fertilization among the different media, like literature in film, literature in painting etc. regarding the contents (cf. Eicher and Bleckmann; Zima). Müller's demand regarding a conceptual fusion also lacks conceptual definitions. When do side-by-side events merge with cooperation? How can we, for example, answer this question in a simple TV documentary connecting image and sound commenting and deconstructing the statement by a sophisticated camera? The decision whether a TV documentary or a photo-novel—to add a popular genre of net literature—is intermedial or multimedial will depend on the decision of the viewer.

Performance, my third characteristic, designates the programming of an intrinsic performance or of one that is dependent on reception. One could inscribe aspects of the performance into the invisible textual level of the digital work so that words and images are allowed on stage. Such an invisible textual level can be found for example in animated images into which the activity of its individual images was inscribed. The change of the image in most cases was supplied by a text editor with a text that can be revealed again within such an editor by showing the succession of images, the number of loops, or the background color. Another textual level, much more easily accessible, is the HTML-Source with its executive commands and the JavaScripts that, while not appearing as text, continue their activity unseen in the background. The prompt of this performance can either come from the program or from the recipient. In the latter case, however, we are referred back to interactivity as the earlier characteristic of digital literature.

I don't have to underline specifically that in this concrete case we see the aforementioned characteristics acting in different degrees and encounter them in the most different combinations. But I do want to underline that in this list of characteristics I am referring to phenomena of the monitor and not to the underlying arithmetic operations per se. The very fact of encoding for me is not a sufficient element since in a computer every phenomenon by necessity is encoded, even the classic linear text. The mere existence of such a text in the computer makes it as little into digital literature as the recording of a poem onto holographic material makes this poem into holopoetry. The encoding for me is interesting only insofar as it takes presence on the monitor as an aesthetic element of expression.

By specifying characteristics I have actually determined after the event what I mean by digital literature. But why am I operating with this word at all and not with one of the competing terms? The answer is: for pragmatic reasons. The term digital literature seems to offer the least occasion for misunderstandings. It does not refer to concrete individual characteristics of digital literature like interactivity, networking or non-sequentiality as do the terms interactive literature, net literature or hypertext. It rather designates a certain medium, which I am, by the way, describing as digital and not as electronic in order to ensure the differentiation from other electronic media like cinema, radio or television. The linkage of my subject matter to this medium implies that it depends on that medium for reasons that refer to the form of this literature and not, for instance, its distribution. This dependence I call the genuineness of the medium. The relevance for this definition is not the fact but the *necessity* of its existence not only in the net but more comprehensively in digital media.[2]

However, *media relevance* has to precede the genuineness of the medium. This term *digital* in digital literature relates to the medium of its production and not to the semiotics of its material. Since language consists of discrete signs one could say that literature is always the result of digital encoding, contrary to images that are based on non-discrete signs. This is where semiotitians' objections against the prevailing expansion of the term language to non-linguistic signs reside.[3] Since digital literature by definition is different from traditional literature it also by definition has to surpass semiotic digitality. This is achieved by connecting to visual signs and performative elements that all can be described as non-discrete signs. The characterization of the term "digital literature" therefore points to the *technological* notion of the medium and not to its *semiotic* one.

With this specification we can also answer the objection asking what non-digital literature might mean contrary to digital literature.[4]

The consequence of this characterization of our object of analysis is a shift from linguistic hermeneutics to a hermeneutics of intermedial, interactive, and performative signs. It is not alone the meaning of a word that is at stake, but the meaning of the performance of this word on the monitor, connected to visual signs and as a reaction to reader activities. This hermeneutics cannot be based on a system of defined meanings, as in the case of linguistic signs. In taking up Umberto Eco's thoughts relating to cinema in his *Theory of Semiotics:* The semiology of digital literature is a semiology of *parole* that has no *langue* at its basis.

Matters are still getting more complicated: What legitimates us to talk of genuineness of media? Espen Aarseth reminds us in his book *Cybertext* that the digital medium is not even needed for hypertext or combinatorial poetry. Nor do the animated texts of the Young-Hae Chang Heavy Industries really need the digital medium nor does David Knoebel's "Click Poetry" that is set to sound and that indeed does not demand more interactivity of the user than one single click. They could indeed run on the movie screen as did some literary texts that were made into movies in the tradition of text-films since the end of the sixties (e.g. Michael Snow's *So It Is* in 1982). And what about a work like *The Child* by Alex Gopher and Antoine Bardou-Jacquet, which not only dynamizes the text and parallelizes it with sound but also uses it simultaneously as a linguistic and visual sign in the tradition of concrete poetry? Should we attest medial genuineness to such a work?[5]

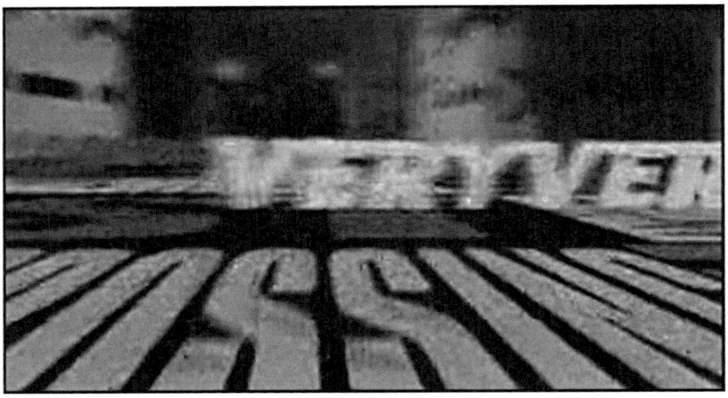

Fig. 1. Alex Gopher and Antoine Bardou-Jacquet: *The Child.*

The Child shows a woman in labor an d her partner speeding through Manhattan in a taxi to get to the hospital. This story, in itself quite unspectacular and also told with quite conservative camera work visualizes all characters and objects by their respective letters. For example in the beginning the skyscrapers of Manhattan are named by vertically readable names; the window through which the viewer is led into one of the apartments is formed by the word WINDOW formed by two lines of letters placed in a square. Inside the apartment the body of the woman is outlined in the vertical placement of the words "BROWNHAIR / PRETTYFACE / WOMAN / PREGNANT / REDDRESS / SNEAKERS," while the man standing next to her is shown with the words "BLACKHAIR / PLEASANTFACE / BIGGLASSES / HUSBAND / LITTLEMAN / DARKSUIT" that change into "BLACKHAIR / BIGGLASSES / *ANXIOUSFACE* / HUSBAND" after the woman cries out "The baby, its coming" with *ANXIOUSFACE* appearing in luminescent colors and fluttering handwriting. The third and main textual level is the script in addition to the words of the song and the sparing dialogue of the main actors (i.e. the exclamation of the pregnant woman, the anxious "OK" of the husband, the shouting for a taxi, the instructions to the driver, and the short dialogue with the police following the racing taxi.

The work is interesting not only regarding its visualization, but also regarding its concept. It is the digitalizing of the world in a semiological way and reminds us of Jeffrey Shaw's *Legible City*. But it works in a different way since it does not tell an invisible story behind walls, but reduces the visible objects to their momentary meaning for an unspecified narrator. A Cadillac stretch-limousine holding up traffic is only endlessly long if you are trying to reach your maternity ward. Is it beautiful nevertheless? Does one of the three people in the waiting taxi think that this Cadillac is beautiful? What would this mean for the character? If the other cars are just perceived as "cars," this already seems to fit with the situation described. In such moments one does not have a sense for aesthetic details. Nevertheless: the taxi-driver is a "DREADLOCKS RASTAMAN CABDRIVER." Why is it important to concretize this? And why aren't we told sex, ethnicity, age or build? Is it the hair that speaks most for such a chase, which, by the way, includes a serious accident with what looks like several casualties? Which cultural stereotypes were used here? Had we *seen* the "DREADLOCKS RASTAMAN CABDRIVER" we might have had the same associations, but probably without realizing it. But already the designation of this one characteristic points to the mechanism of typifying. The textualization of visual signs forces us to choose and is unavoidably direct by its discreteness. The work, by casting everything into language, speaks about the different methods of encoding visual and linguistic languages. It speaks, to use its own metaphor, about the birth of meaning.[6]

This is a very nice example of digital literature insofar as it is characterized by meta-reflexivity (by comparing linguistic and visual levels of meaning), intermediality (through the conceptual coexistence of medial elements) and performance (by staging the letters like a running film). It is a work that is most interesting because of its tension between ambitious and conventional aesthetics: The narrative level and the camera work are located in mainstream-aesthetics and seems to present the (necessary) opposite pole to the provocation of the viewers by presenting the narrated reality in such an unusual way. Nevertheless it is a video-clip that was shown in the movie theatres as Clip Cult Vol. 1—Exploding Cinema by the Cologne distributor Rapid Eye Movies in 1999. There is a lack of the terminologically relevant genuineness of the medium in it since the fact that it is using a digital technology for its production hardly is sufficient since nowadays every cut, every musical arrangement and every textual formatting is done with the use of the computer.

Maybe one should find pragmatic reasons assigning this work—just like the text-films by David Knoebel and Young-Hae Chang Heavy Industries—to the realm of digital literature in order to have a further example for showing how interesting the phenomena are that are brought forth by this field of investigation. We could counter the objection that these are merely filmed texts by pointing out their media-*typicality*: It is correct that there are also multilinear texts in the print-medium as well as dynamized writing in the medium film but in both cases these are atypical features in the medium crossing experimentally the borders of another, in which this exception then becomes a rule specific for the medium. Only digital technology provides a medial home for multilinearity and dynamization of texts.[7]

Does this mean that the terminological and taxonomic problems have been solved to a certain extent? By no means. We still have not talked about the term *literature* in digital literature. What is the literary element in digital literature?

2. The Disappearance of Genre-Boundaries

When Richard Ziegfeld in his *Interactive Fiction* acquainted his readers with the interactive, audio-visual widening of literature he knew that this widening also would make the term literature into a problematic one. He therefore is directly asking, "Is interactive fiction a literary or a visual art form?" arguing in his answer with proportions:

> Interactive fiction addresses language concerns—with visible rather than oral words—that readers can contemplate for so long as they

> wish and to which they may return for further study. While interactive fiction offers potent possibilities in the visual realm, it presents a proportion of word in relation to graphic device that sharply distinguishes it from the visual electronic media. Thus, interactive fiction is the first *literary* electronic form. (370)

For the time frame we are dealing with the borderline to the other visual forms—film, TV, video, software adventure games—is clear. In 1989, still some years before the arrival of the WWW, the interface of digital media was dominated by words. It was the time in which one could still exclaim, as did Michael Joyce in 1988: "Hypertext is the word's revenge on TV" (47). Even ten years later the argument of proportionality might have been plausible in order to characterize a work of digital media, or of net literature as Christian Köllerer calls it, as a "complexly structured, *mostly* linguistic-aesthetic artifact" (53) in the face of multimedia. But even if the criterion of quantity seems to be quite convincing at first sight it is hardly practicable since we have to ask ourselves soon: How can we measure the primacy of the word—by the space it takes on the monitor or in the memory? After the attention it captivates or the amount of information in comparison to visual elements? And what does it mean if through staging a word is emphasized mainly in its materiality as a typographic sign and therefore is not effective as a text but as an image? *The Child* already has shown this possibility even though it is aiming at both visual and linguistic functions, much like the tradition of concrete poetry. In a piece like Camille Utterback's and Romy Achituv's *Text Rain* the situation is already more difficult.

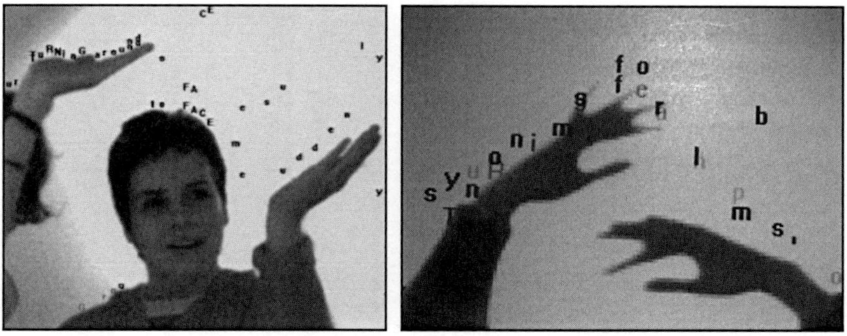

Fig. 2. Camille Utterback and Romy Achituv: *Text Rain.*

In the interactive installation *Text Rain* of 1999 the viewers stand or move in front of a big monitor in which they see themselves as black and white projections and on which letters fall from the top edge:

> Like rain or snow, the text appears to land on participants' heads and arms. The text responds to the participants' motions and can be caught, lifted, and then let fall again. The falling text will land on anything darker than a certain threshold, and "fall" whenever that obstacle is removed. (Utterback and Achituv)

Therefore, *Text Rain* is more about playing with the text than about narration. Nevertheless this work is indebted to literature and doubly inspired by it. On the one hand it picks up on Apollinaire's concrete poem "Il Pleut" (1918) distributing the letters in five slanted verticals across the page like falling raindrops. On the other it is using the poem "Talk, You" from Evan Zimroth's book *Dead, Dinner, Or Naked* (1993) that can become a reading object in spite of the game:

> If a participant accumulates enough letters along their outstretched arms, or along the silhouette of any dark object, they can sometimes catch an entire word, or even a phrase. The falling letters are not random, but form lines of a poem about bodies and language. "Reading" the phrases in the *Text Rain* installation becomes a physical as well as a cerebral endeavor. (Utterback and Achituv)

The lines of the poem create the following text:

> I like talking with you,
> simply that: *conversing*,
> a turning-with or -around,
> as in your turning around
> to face me suddenly …
> At your turning, each part
> of my body turns to verb.
> We are the opposite
> of *tongue-tied*, if there
> were such an antonym;
> We are synonyms
> for limbs' loosening
> of syntax,
> and yet turn to nothing:
> *It's just talk.*

Clearly, *Text Rain* is working with the statement of the poem by manipulating the way in which it can be read: The conversation between two people, wordless and aimless, which lets the movement of the bodies change body parts into words, becomes a playful conversation between the bodies of the viewers and the letters of the installation. The subject of the poem, namely the relationship of body language and verbal language is also the subject of the interactive installation. Therefore, the choice of the text is not accidental and therefore the reading process is important for understanding this work.

The documentation of *Text Rain*, however, shows that the lines can only be deciphered after a tenacious collection of all letters and even then only with difficulty. The viewers also don't really address the reading process but mainly test the interface. Utterback's description then deservedly puts "Reading" in quotation marks and at the same time divulges the fact that even the author does not expect a real reading from the viewers. The cerebral effort that was announced probably consists more in the "reading" of *Text Rain,* i.e. in the interpretation of the physical effort than that of Zimroth's poem. The work illustrates its entire attraction on the physical level; the fascinating elements of the installation are the movements that it creates in front of the monitor among the viewers in the interaction with the falling letters. Virtually as an inversion, this interaction brings forth what the poem describes: Letters change the bodies of the viewers while they are moving. The irony is that even still the non-reading of the text (or rather only its non-reading) actually realizes it, since the change of its function from the linguistic to the visual level while becoming an interactive partner corresponds to the aimless conversation that is celebrated in the end: ". . . turn to nothing: / *It's just talk*." As Evan Zimroth's poem reflects on the communication of two people going beyond the pettiness of mere concrete information, the installation frees the text of its representative function letting it become pure self-sufficient presence.

It should be mentioned that it is rather Utterback's work with Zimroth's poem that elevates it above the concrete message. The original is longer by seven lines and contains different line-breaks than Utterback's video-documentation of *Text Rain* and it does contain information exchange.[8] This interference shows most clearly that Utterback was aiming at the relationship between bodily and verbal language or rather at the playful interaction between letter and body. Poetry like Apollinaire's rain poem or some other text about rain or snow might have been more obvious but would have given away the deeper meaning by drowning *Text Rain* in the banality of a double rain metaphor.

Insofar as this installation does not aim at the reading of the text it acts rather as digital art than as literature. Therefore my suggestion of differentiation is not so much based on proportionality but on the role played by the text as a whole. If the text continues to be important as a linguistic phenomenon and can be perceived as such, we can talk of digital literature. If the text becomes a visual object of interaction, we are dealing with digital art, which may be specified as digital installation or performance art, depending on the taxonomic diligence.

With this differentiation we are not trying to define literature with regard to the letter (lat.: littera) but with regard to its characteristic of teaming up with other letters forming meaningful words. This seems to be a quite conservative method in the light of futurist and dadaist avant-garde efforts to free the letter from its stronghold of meaning in sentences or words, as did for example Marinetti in his *Parole in libertà* or shortly after him Hugo Ball, Richard Huelsenbeck and Kurt Schwitters in their sound poetry. Regarding a meaningful text however, and in light of recent decentralizing movements, this conservatism offers a practicable basis for distinguishing literature from other genres like painting, film, or music. But is that necessary? Some scholars, like for example Katherine Hayles, working in the areas of digital aesthetics question concerns of categorization and definition in order to remain open for the specific characteristics of the new inter- and transmedial phenomena (45).[9] We might add that this is done from the beginning in order to avoid battles of authority with literary scholarship that is both trying to expand its objects of research and, in an old pattern, to distinguish itself as a discipline from scholarship in neighboring aesthetic studies. In the interest of openness and underlining once more the earlier mentioned preference for interpretive work over a terminological one, I can only agree with this rejection and avoidance. I cannot do this, however, if the rejection of terminological questions leads to a loss in interpretation. As the discussion of examples so far has shown, very often only the question whether a work acts as literature, or film or installation sharpens the eye for all its levels of meaning.

Regarding *Text Rain* the classification remains more difficult and more exciting than it might look after the clear differentiation between the linguistic and the non-linguistic role of the text. *Text Rain* on the one hand only realizes the underlying poem when it is not perceived as a linguistic phenomenon; but we can only come to this conclusion after we have read the poem.[10] This installation first has to be treated as literature before we can realize that in reality it wants to be art. Only those who submit the interaction with the letters to the goal-orientation of deciphering realize that those who refrain from it do with the text what it is suggesting. This is where the deeper wit of *Text Rain* lies; it possibly wasn't even intended and in most cases probably only becomes visible by presenting a readable version of the poem in the documentation of *Text Rain*, which at the same time simultaneously invalidates it.[11]

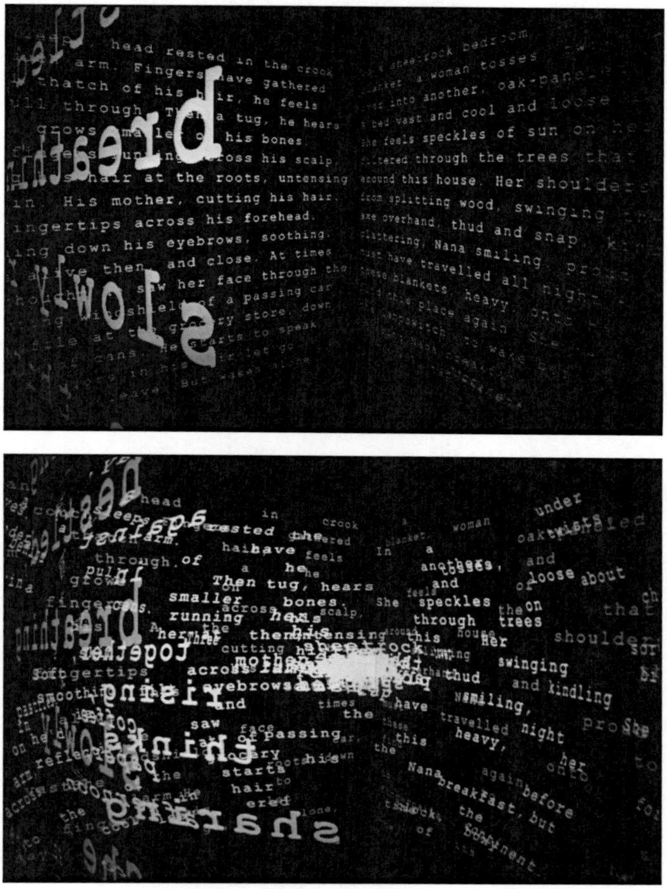

Fig. 3. Noah Wardrip-Fruin et al.: *Screen.*

In Noah Wardrip-Fruin's *Screen* we have a comparatively clear case (Wardrip-Fruin et al.). This work can be viewed in the "Cave," a small room onto the three sides and the floor of which information can be projected; the viewer wears spectacles allowing a three-dimensional perception of the information. S/He is also provided with a glove that allows calculating the position of his hand in space. The work starts by playing a recorded text about the loss of memory. Subsequently three walls are covered with the text of short episodes. Finally, single words from these texts drop from the wall approaching the viewer who now has the option to push them back with his gloved hand into their old position. The frequency and speed of this dropping increases in the course of the episode; in the end, all those words are falling off the wall that were not pushed back.

In Screen the text then first voices the general subject matter—the loss of memory through the loss of words. The interaction then builds on this in an illustrative manner by showing the dropping of the words from the storage medium "wall" and the reader's option to counter the dropping. The irony of this interaction lies in the fact that memorization becomes a sport or game in which winning is more important than the perception of the words threatened by loss. This playful interaction with memory is also expressed in the invitation for a new round expressed in the end: "If memories define us what defines us when they are gone? An unbearable prospect. We retrieve what we can and try again" (Wardrip-Fruin et al.). Learned in media theory we could say with Plato's Phaidros-dialogue: The renewal of that which is threatened by forgetting in mental memory is replaced by the questionable attempt to fixate it in the external medium wall.[12]

Noah Wardrip-Fruin calls *Screen* an "instrumental text" which he also sees as a sub-genre of "playable media" including computer games among others (cf. "From Instrumental Texts"). Without doubt we could also assign *Text Rain* to this sub-genre. What is said in the documentation of *Screen* applies to both: "It creates new experiences of text in relation to the reader's body" (Wardrip-Fruin et al.).

Contrary to *Text Rain* of course the relationship in digital literature of the literary to the physical or digital, i.e. those elements about which the digital text was expanded, is showing in *Screen* with the desired clarity. We might therefore call it an ideal case in the attempt at characterizing literaricity within the term digital literature.

3. Authorship in Digital Literature

Talking about the specifics of digital literature we finally also have to ask about the role of the author. In the following, however, I will not talk about the so-called death of the author in hypertext nor of collaborative authorship in co-writing projects.[13] But I will talk of the dismissal of the author from text production through narrative machines and bacteria.

Narrative machines are the contemporary form of experiments in automated writing that go back to the dadaists and surrealists of the 1920s and that later were taken up again by William Burroughs. The aim of these experiments in an aleatoric, non-intentional writing was not dismissing the author but surpassing the limits of his creativity, overcoming his personal perspectives. This could be achieved by narcotizing[14] or multiplying the author.[15] Automated writing in the reality of actual narrative machines is based on a previous schematization of characters, potential conflicts and their solutions. This, however, does not really lead to a widening of expressive forms as we can see in the story generator *Makebelieve* by Hugo Liu and Push Singh, Scott Turner's *Minstrel* or *TALE-SPIN* by Meehan.[16] In these cases the automatic text-production means in effect the opening out of alienation—as a social phenomenon of modern society—to include also the poetic process that in its turn produces the alienation on the reception side. Namely, if the reader is not confronted any more with the specific self-awareness and world-perception of an author what then is the point in the reading process? What message can a text still have without a real sender?

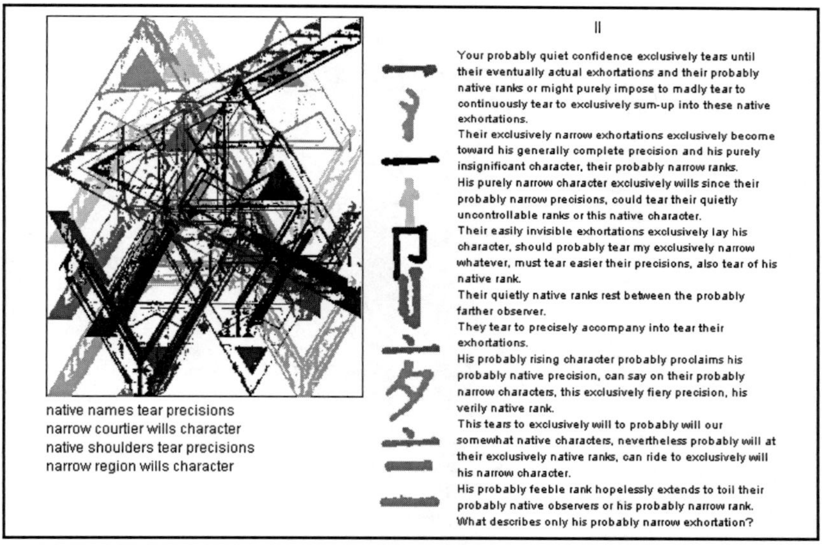

Fig. 4. Simon Biggs: *Great Wall of China.*

I do not want to get myself absorbed by this discussion.[17] I do, however, want to point out an example in which the producer of the software and the writer share the authorship. It is Simon Biggs' *Great Wall of China* generating syntactically correct but semantically meaningless sentences from the vocabulary of the English translation of Kafka's fragment *Beim Bau der Chinesischen Mauer.* Biggs' text generator is especially interesting because of its transmedial offer for the reading process. The reader has to read Kafka's text in order to understand the text of the text generator as a sensual permutation of the primary text. Like the human beings there are waiting in vain for the arrival of a messenger, here the reader is waiting in vain for the arrival of a message. In the same way as Hermes never arrives in the primary text, the hermeneutic activity in *Great Wall of China* is grasping at blanks. Biggs' authorship is superimposed on Kafka's authorship, as is the narrative machine on the written text. *Great Wall of China* does not tell a story; it is the commentary of a story nevertheless. It is the translation of a text into immediate perceptibility and its linguistic share is the fact that the linguistic material does not make any sense.

The model of a double reception—of the performance of the text generator and that of the text it is based on—reminds us of *Text Rain* and *Screen.* Comparable to those cases here as well we can speak of digital literature, which, however—if the reading of the text is not carried out—will be perceived rather as digital art. Returning to Eduardo Kac in the end, the next example is more complicated: He seems to make matters easy for us with his own terminology, but in reality he makes it rather more difficult.

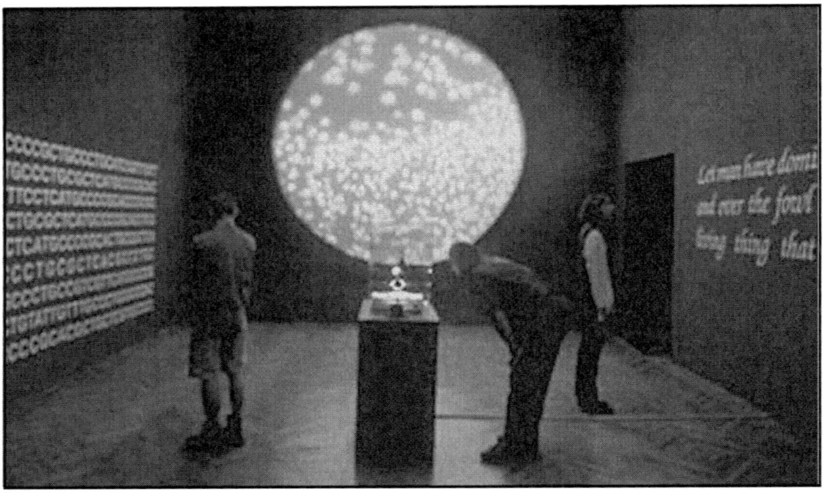

Fig. 5. Eduardo Kac: *Genesis.*

Kac not only addressed holopoetry, he also claimed biopoetry in which not narrative machines but living organisms take over the co-authorship of a text: "Now, in a world of clones, chimeras, and transgenetic organisms, it is time to consider new directions for poetry *in vivo.* Below I propose the use of biotechnology and living organisms as a new realm of verbal creation" ("Biopoetry" 184). The fourth of the following points reads as follows:

> Transgenic poetry: synthesize DNA according to invented codes to write words and sentences using combinations of amino acids. Incorporate these DNA words and sentences into the genome of living organisms, which then pass them on to their offspring, combining with words of other organisms. Through mutation, natural loss and

> exchange of DNA material new words and sentences will emerge. Read the transpoem back via DNA sequencing. ("Biopoetry" 184)

Here Kac is describing what he has experimented with in his project *Genesis*, begun in 1998, in which bacteria rewrite a sentence from the bible in cooperation with the visitors of a website: "Let man have dominion over the fish of the sea and over the fowl of the air and over every living thing that moves upon the earth." Kac transferred the sentence into Morse and replaced its signs with the letters of a genetic code: the periods were replaced by C (Cytosin), the lines by T (Thymine), gaps between words by A (Adenine) and spaces between letters by G (Guanine). From this results an AGCT-Chain representing a gene nonexistent in nature. Kac connected it to fluorescent protein and injected it into E. Coli bacteria. These bacteria are stored in a Petri dish on a counter of an exhibition hall under an ultraviolet light that can be switched on by the visitors of the exhibition or the visitors of the corresponding website. The light affects the DNA-sequence of the bacteria and speeds up their mutation. After a certain time the changed genetic code is translated back according to the same pattern. The result of the cooperative writing activity between visitors and bacteria shows only a small change, which however already disguises the subject of the originally promised human dominance: "Let aan have dominion over the fish of the sea and over the fowl of the air and over every living thing that ioves ua eon the earth" (Kac, "Genesis").

The message of this project is easily recognizable. By acquiring authority over all other living beings, humans destroy the scriptural foundation that had originally authorized them. The judgment passed by the rewriting of the sentence at the same time reminds us of the confusion of tongues inflicted by god as a reaction to the tower of Babel, another famous symbol for human hubris. The more hidden message of the project is talking about a conflict between the aesthetic experience and individual responsibility, since the click of the mouse speeding up the mutation of the bacteria offers up a fascinating play of colors, since the bacteria are glowing in cyan and yellow under the ultraviolet light. The curiosity and joy of seeing this play of colors suppresses the question of the actual results of the activity.

Hence, the future body in the present constellation is not written in a goal-oriented manner, but blindly, which therefore hardly corresponds to the logic of any responsible intervention in creation. Since this involvement lacks every concept, we cannot even say: The word is becoming flesh. Rather, here it is the subconscious that is becoming flesh, namely the temptation to use our own curiosity inconsiderately. The point is that the flesh again is becoming word, sentencing by rewriting the original sentence: The originally promised dominance is taken back as a direct result of the executed dominance.

The fact that the perpetrators are becoming victims is part of the aesthetic experience of the public. Jessica Wyman speaks of a trauma into which we are thrown: "What surprises is the degree of trauma into which we are psychically thrown by the creation of this gene and the action (sometimes read as violence) that we do to it as viewers by the activation of an ultraviolet light which speeds its mutation" ("On Eduardo Kac's Genesis"). And Steve Tomasula stresses the associative connection with the object of ones own activity:

> Standing in the box formed by the walls of *Genesis*, it's easy for viewers to reverse the scale and think of themselves in the position of the bacteria with ultraviolet light streaming down (possibly through a hole in the ozone layer?). We're invited to contemplate consequences of interfering with evolution when Kac translates, at the end of the exhibit, the DNA code of his original message back into English. . . . In Kac's *Genesis*, though, we see an icon for our new-found ability to rewrite ourselves—instantly, and in ways whose ramifications might not become apparent for generations. (255f.)

Genesis therefore is not only asking the question of authorship, it is itself an artifact on authorship. Beyond that it also asks the question whether this is poetry, as suggested by Kac's terminology, or rather an installation. Where could we note the difference between poetry and installation? The spontaneous answer again could lead us to the linguistic aspect: If the words produced by the bacteria in their singularity were meaningful the result of their work could possibly be called poetry, to be more exact: aleatoric poetry. But in *Genesis* the words generated by the bacteria are not meaningful in themselves. Once we have made this observation, we can continue with our argument in a twofold way. On the one hand we can refer to sound poetry as a playful form of literature and therefore, from the perspective of the text, assign *Genesis* to digital literature. On the other hand we can see the linguistic meaning of the words generated by the bacteria in the very fact that they become semantically fragile, since the halfway coherent words function like the exclamation or even the call for help of a degenerate author. In this case *Genesis* could be assigned to digital literature from the perspective of the installation.

However we are going to decide this question, it has become clear that the phenomena of digital aesthetics confront us with many borderline cases. It is advisable to verify what role is played by the text in each concrete instance. In general we should use categories sparingly and only provisionally. For the rest we will have to reckon with at least as many "taxonomic scandals" (Nies 327) in the area in which intermediality is the main characteristic, as they exist in literary studies. This at least is a continuity on which we can build.

Translated by Brigitte Pichon and Dorian Rudnytsky

Notes

1. I admit to have contributed to this terminological debate in my book *Interfictions*. In all probability instead of clearing up some of the questions I have produced new ones. Coming to my defense I can say, however, that the chapter "Begriffe, Merkmale, Typologie" (Terms, Characteristics, Typology) takes up only 5% of the book and that its greater part consists of extensive case studies.
2. The term "net literature" remains problematic even if it is disengaged from the net and widened into feedback-processes in the computer or even in social systems. With this widening of the expression "net," the term net literature looses its differentiating power regarding traditional phenomena in literature that are as well bound to social systems and into the feedback-processes of the computer, as the case may be.
3. Roland Barthes points out this problem at the beginning of his essay on the "Rhetoric of the Image" in: *Image, Music, Text.*
4. Cf. Schäfer. See in this context also Hans H. Hiebel's differentiation between primary digitality (discrete-distinctive signs) and secondary digitality (digitalization of signs as consequence of the computer) (8). Digital literature then can be related to secondary digitality. It has been pointed out that digital media strictly speaking should be called binary media (cf. Pflüger), so that the term digitality, relating to the computer, proves to be a questionable criterion of definition. But since the term "digital media" has generally been accepted—and has to be the chosen one compared to the even more inaccurate one of "new media"—it would be quite unserviceable to speak of "binary" instead of digital literature.
5. I am talking of Alex Gopher's arrangement *The Child* (1999) using a sample by the jazz-legend Billie Holiday. The details of the technical production of this Clip, camera work etc. are described in detail by Reither.
6. Saskia Reither underlines, however, that this textualization also leads to a de-stereotyped perspective since the architectural highlights of the city (MoMA, Guggenheim, Chrysler Building etc.) do not appear in their generally known form but as texts that are put alongside the other buildings (cf. Reither).

7. Since *The Child* despite its textuality belongs to the genre video-clip and is successful as such, the argument of a typical feature of the medium, however, stands on shakier ground than in the case of the text-film by David Knoebel and Young-Hae Chang Heavy Industries. A prime example for a piece of art bound to its medium is John Cayley's *Overboard* which is produced algorithmically in real time and could truly not be shown as a film.
8. In the original the poem reads as follows:
   "I like talking with you, simply that:
   conversing, a turning-with or -around, as in
   your turning around to face me
   suddenly, saying *Come*, and I turn
   with you, for a sometime
   hand under my under-
   things, and you telling me
   what you would do, where,
   on what part of my body
   you might talk to me differently.
   At your turning,
   each part of my body turns to verb.
   We are the opposite of
   tongue-tied, if there were such an
   antonym; we are synonyms
   for limbs' loosening of syntax,
   and yet turn to nothing:
   It's just talk."
9. For the differentiation of transmediality and intermediality cf. Simanowski, "Transmediale Kunst."
10. This consecutiveness of reception first as literature and then as visual art follows the principle of alternating reception that we also experience in concrete poetry (see for example Eugen Gomringer's *Schweigen* (1954) in which we first have to read the word "Schweigen" ('silence') and understand its meaning before we can meaningfully refer it to the gap between the nine words "Schweigen").
11. In the moment the underlying poem is identified in the accompanying text of the installation, thereby becoming accessible, the repressive power of the text over the interpretation of the image (in this case the installation) sets in, which Roland Barthes has underlined in his essay *Rhetoric of the Image.* Insofar as the viewers can only read the poem after their encounter with the installation, a "belated repression" sets in, or rather a correction that transfers the completion of the interpretation outside of the representation of the installation.

12. For a detailed consideration of *Screen* see my interview with Noah Wardrip-Fruin.
13. Cf. my articles "Autorschaften in digitalen Medien" and "Tod des Autors? Tod des Lesers!"
14. A specific form of "narcotizing"—apart from alcohol or drugs as the usual stimulants—is the diversion of the author in the process of writing by simultaneously reading texts to him, as in the experiments by Gertrude Stein and Leon Mendez Salomon in 1896. Cf. Schulze 42.
15. Cf. the *Cadavre éxquis* invented by the surrealists as the most popular form of collective automatism, in which two players each write one word one after the other without knowing the previous one.
16. *Makebelieve* works on the basis of 9,000 pre-structured sentences that are catalogued on a trans-frame regarding verb, object, and adjective and that are provided with the attributes *before* (cause) and *after* (effect). The automatic generation of the text is carried out in such a way that for a *cause* in a sentence entered by the user fitting *effect*-sentences are searched for in the catalogue of sentences. The program makes sure that verb, object, and adjective of the sentences connected with each other have a semantic relationship. In this way the program automatically produces a text of five to twenty lines in reaction to a sentence typed in by the user, like in the following example: "John became very lazy at work. John lost his job. John decided to get drunk. He started to commit crimes. John went to prison. He experienced bruises. John cried. He looked at himself differently."
17. For an extensive discussion see the chapter on narrative machines in my book *Ästhetik des Spektakels*, which will be published in 2007.

Works Cited

Aarseth, Espen. *Cybertext: Perspectives on Ergodic Literature.* Baltimore: Johns Hopkins UP, 1997.

Barthes, Roland. *Image, Music, Text.* Ed. and trans. Stephen Heath. New York: Hill, 1977.

Biggs, Simon. *Great Wall of China.* 1996. 25 Apr. 2006 <http://hosted.simonbiggs.easynet.co.uk/wall/greatwall1.htm>.

Cayley, John. *Overboard.* 2004. 25 Apr. 2006 <http://www.shadoof.net/in/overboard.html>.

Coyle, Rebecca. "Holography—Art in the Space of Technology: Margaret Benyon, Paula Dawson and the development of Holographic Arts Practice." *Culture, Technology and Creativity.* Ed. Philip Hayward. London: John Libbey, 1990. 65-88.

Eicher, Thomas, and Ulf Bleckmann, eds. *Intermedialität: Vom Bild zum Text.* Bielefeld: Aisthesis, 1994.

Gopher, Alex. *The Child.* 1999. 25 Apr. 2006 <http://www.youtube.com/watch?v=ZGWkdNWFZiI>.

Hayles, N. Katherine. *Writing Machines.* Cambridge, MA: MIT P, 2002.

Hiebel, Hans. *Kleine Medienchronik: Von den ersten Schriftzeichen zum Mikrochip.* Munich: Beck, 1997.

Joyce, Michael. "Siren Shapes: Exploratory and Constructive Hypertexts." *Of Two Minds.* Ann Arbor: U of Michigan P, 1995. 39-59. First in: *Academic Computing* 3.4 (1988): 10-14, 37-42.

Kac, Eduardo. "Biopoetry." *Cybertext Yearbook 2002-03.* Ed. Markku Eskelinen and Raine Koskimaa. Jyväskylä: Research Center for Contemporary Culture, 2003. 184-185. 25 Apr. 2006 <www.ekac.org/biopoetry.html>.

———. "Genesis." 1999. 25 Apr. 2006 <http://www.ekac.org/translated.html>.

Köllerer, Christian. "Die feinen Unterschiede: Über das Verhältnis von Literatur und Netzliteratur." *Literatur und Kritik* 339/340 (1999): 50-54.

Looy, Jan van, and Jan Baetens, eds. *Close Reading New Media: Analyzing Electronic Literature.* Leuven: Leuven UP, 2003.

Müller, Jürgen E. *Intermedialität: Formen moderner kultureller Kommunikation.* Münster: Nodus, 1996.

Nies, Fritz. "Für die stärkere Ausdifferenzierung eines pragmatisch konzipierten Gattungssystems." *Zur Terminologie der Literaturwissenschaft: Akten des IX. Germanistischen Symposiums der Deutschen Forschungsgesellschaft Würzburg 1986.* Ed. Christian Wagenknecht. Stuttgart: Metzler, 1988. 326-336.

Pflüger, Jörg. "Wo die Quantität in Qualität umschlägt." *HyperKult II: Zur Ortsbestimmung analoger und digitaler Medien.* Ed. Martin Warnke, Wolfgang Coy, and Georg Christoph Tholen. Bielefeld: Transcript, 2005. 27-94.

Reither, Saskia. "46 Cuts in 3 Minuten: Rhythmisierte Typografie im Musikvideo 'The Child'." *Geteilte Zeit: Zur Kritik des Rhythmus in den Künsten.* Ed. Patrick Primavesi and Simone Mahrenholz. Schliengen: Edition Argus, 2004. 126-137.

Schäfer, Jörgen. "Sprachzeichenprozesse: Überlegungen zur Codierung von Literatur in 'alten' und 'neuen' Medien." *Analog/Digital—Opposition oder Kontinuum? Beiträge zur Theorie und Geschichte einer Unterscheidung.* Ed. Jens Schröter and Alexander Böhnke. Bielefeld: Transcript, 2004. 143-168.

Schulze, Holger. *Das aleatorische Spiel: Erkundung und Anwendung der nichtintentionalen Werkgenese im 20. Jahrhundert.* Munich: Fink, 2000.

Shaw, Jeffrey. *The Legible City.* 1989-91. 25 Apr. 2006 <http://www.jeffrey-shaw.net/html_main/show_work.php3?record_id=83>.

Simanowski, Roberto. *Interfictions: Vom Schreiben im Netz.* Frankfurt am Main: Suhrkamp, 2002.

———. "Autorschaften in digitalen Medien: Eine Einleitung." *Digitale Literatur.* Ed. Heinz L. Arnold and Roberto Simanowski. Munich: Text + Kritik, 2001. 3-21.

———. "Tod des Autors? Tod des Lesers!" *p0es1s: Ästhetik digitaler Literatur / The Aesthetics of Digital Literature.* Ed. Friedrich W. Block, Christiane Heibach, and Karin Wenz. Ostfildern: Cantz, 2004. 79-91.

———. "Transmediale Kunst als Phänomen aktueller Ästhetisierungsprozesse." *Transmedialität: Zur Ästhetik paraliterarischer Verfahren.* Ed. Urs Meyer, Roberto Simanowski, and Christoph Zeller. Göttingen: Wallstein, 2006. 39-81.

Tomasula, Steve. "Gene(sis)." *Eduardo Kac: Telepresence, Biotelematics, and Transgenic Art.* Ed. Peter T. Dobrila and Aleksandra Kostic. Maribor: Kibla, 2000. 85-96. Rpt. in *Data Made Flesh: Embodying Information.* Ed. Phillip Thurtle and Robert Mitchell. London: Routledge, 2003. 249-257.

Utterback, Camille, and Romy Achituv. "Text Rain." 1999. 25 Apr. 2006 <http://www.camilleutterback.com/textrain.html>.

Wardrip-Fruin, Noah. "From Instrumental Texts to Textual Instruments." 2003. 25 Apr. 2006 <http://hypertext.rmit.edu.au/dac/papers/Wardrip-Fruin.pdf>.

———. *Digital Literature.* Interview with Roberto Simanowski. 2004. 25 Apr. 2006 <http://www.dichtung-digital.org/2004/2-Wardrip-Fruin.htm>.

———, et al. *Screen.* Quicktime documentation. 2004. 25 Apr. 2006 <http://www.dichtung-digital.com/2004/2/Wardrip-Fruin/Screen.mov>.

Wyman, Jessica, "On Eduardo Kac's Genesis." *YYZ Artists' Outlet.* 2003. 25 Apr. 2006 <http://www.ekac.org/jessica.yyz.html>.

Ziegfeld, Richard. "Interactive Fiction: A New Literary Genre?" *New Literary History: A Journal of Theory and Interpretation* 20.2 (1989): 341-372.

Zima, Peter, ed. *Literatur intermedial: Musik, Malerei, Photographie, Film.* Darmstadt: Wissenschaftliche Buchgesellschaft, 1995.

Frank Furtwängler

Human Practice

How the Problem of Ergodicity Demands a Reactivation of Anthropological Perspectives in Game Studies[1]

Abstract:
This article presents a critical review (not a rejection) of the concept of "ergodic literature" when applied to computer and video games. Therefore it goes back to some of the sources Espen Aarseth triggered when he appropriated the term from physics in 1997 for the subject of cybertext and explains the necessary consequences of the term "ergodicity" for literature and games when it is not merely used metaphorically. A more cautious use of terms and concepts from other disciplines is suggested, especially as the term "ergodic" in physics has a different but relevant meaning in the context of these games. The article tries to mediate between some of the general anthropological claims of cybertext theory/game studies and the understanding of "ergodic systems" in thermodynamics and statistical physics. The problems that result from this mediation can be seen as symptomatic for the challenges of game studies in the more general mediation between different perspectives on games.

I. Introduction

As the conference abstract that motivated this text argued, certain aspects of digitalization, of coproduction and cooperation, and new forms of literary interactivity between human beings and machines have been discussed as dominant features in a field defined by terms like *digital literature*, *internet literature*, *cyber-* and *hyperfiction*, *e-poetry* or *interfictions.* I could well situate my own interests in this field, although there has to be a liberal notion of the category *text*, which however should be unproblematic. Some years ago I had the opportunity to explore the very odd relationship between *interactivity* and *narrativity* in computer games. A goal was to sketch a historical and systematic framework for further explorations at a time when appropriate heuristics for the academic understanding of computer and video games were basically non-existent. Still, now that there actually has been a lot of work done in the new academic field called *game studies* many problems in coming to grips with the new medium and especially in the way methods were conceived to conceptualize this task have not ceased to puzzle me. Some researchers appear to be optimistic concerning the possibilities of general theories of computer and video games, possibilities I don't really trust. I also don't trust conceptions of terminology in many parts of the field, which is indeed the main topic of this paper. As a result, my own basic demands regarding computer game research were to *avoid terminological debates*, particularly concerning such terms and notions that are derived from traditional theories only to get projected on new phenomena straightaway so that accompanying new qualities get obscured. This is a kind of credo in game studies, although opinions about the consequences differ widely. This is not merely an allusion to the often-criticized application of narratological models, but also to the supposedly "new" approaches that force existing traditional theories about *game* and *play* onto the subject. We have seen some applications in this respect that are virtually meaningless for an understanding of the way computer and video games work as a medium, e.g. the typological classifications using Caillois' game categories as an approachable and straightforward example of an application that turned out as a self fulfilling prophecy. Nevertheless, discussions reflect a growing sense of respect for the diverse forms of these "games" and for the diverse (methodological) approaches derived from different disciplines since Espen Aarseth announced his "Year One of Computer Game Studies" in 2001, even though many of the approaches remain questionable and are not *common sense*.

When it comes to "terms" there is quite often no coming to terms. It seems difficult to build or even detect a network of communication based on key terms everyone agrees on. Helpful discussions are complicated by fundamental communication problems. The term "interactive," to take only one random example, may count as well established on the one hand, on the other hand its descriptive value is doubted. Eskelinen and Tronstad refer to this term as an "unspecified banality" (199), despite the efforts some spent in establishing it as a critical term. Skipping other comparable problematic implications, I can only indicate that there are some important questions the community of game researchers has to face: Do they work in the same field? Do they even work on the same subject or notion of the subject that allows for a sense of community and identity? Or are subgroups of a virtual community fighting each other for the sake of the big "theoretical shot": a great unified theory of computer and video games? What role do terms like *digital literature*, *internet literature*, *cyber-* and *hyperfiction*, *e-poetry* or *interfictions* mentioned above play? In which strategic function are terms and neologisms justified: as means of communication about the phenomena or to highlight one's own research (to put it in diplomatic terms)? Further: Will they ever win with the definition of terms and is it possible to use terms on a common basis that helps to overcome the hermetic borders of disciplines and research positions?

My impression is that communication by means of terms is becoming increasingly difficult. Authors introduce terms into discourse but they often become an end in itself, so that even their authors lose interest. This seems to be the case with the term "ergodic literature" introduced by Espen Aarseth as a synonym to "cybertext," a term he actually introduced on the first pages in his groundbreaking book *Cybertext: Perspectives on Ergodic Literature* (1997). Some authors accept such terms introduced by others for their own work and try to activate their potential value, basically doing the job the originators should have done before they lost interest. Others prefer to use established terms as some negative foil against which they try to establish their own inventions. In fact they are introducing yet some more synonyms providing simply new facets of diverse polemic discussions that dominated *game studies* in an unfortunate fight between disciplines claiming the subject as theirs from the beginning. Regarding this topic *game studies* have already spent too much energy and time for pointless (though interesting) discussions. Tragically or prophetically Aarseth foresaw this development when he summarized the problems I just sketched:

> But the prevailing attempts to rejuvenate and relocate existing theories by insisting on their relevance for the new media and their largely unsuspecting users, is a 'colonialist' strategy that is always a demonstration of (unnecessary) power and often a misreading of the theory being used. ("Aporia and Epiphany" 32)

Considering the damage done, it may prove fruitful to go back to some of the sources of either the problem or the discipline (which sometimes are the same). An appropriate beginning could be seen in a reconsideration of the triad "author"—"work"—"reader," the definition of which—following the abstract mentioned at the beginning—has to be extended considerably for further explorations in our field. To determine the state research is in, let us look at the interrelation of its individual parts again. There may indeed be quite some distance ahead before an extension of the triad is within grasp. And apparently there is still a lesson to be learned from the demands of traditional models, e.g. from the study of literature and the anthropological turn following reader-response theories. These lessons will be sketched as an outlook at the end. I emphatically imply this as lessons to be learned from the *experiences* of traditional disciplines, especially from their past failures and impasses, and I do not mean from their specific models or the application of their models. My intention is not to suggest another uncritical application of traditional tools, derived e.g. from narratology or the study of literature in general. I am obviously in agreement with Aarseth's basic plan, when he said almost ten years ago:

> Especially, I wish to challenge the recurrent practice of applying the theories of literary criticism to a new empirical field, seemingly without any reassessment of the terms and concepts involved. This lack of self-reflection places the research in direct danger of turning the vocabulary of literary theory into a set of unfocused metaphors, rendered useless by a translation that is not perceived as such by its very translators. (*Cybertext* 14)

But adding to this perspective, the uncritical, metaphorical use of terms (the vocabulary) from other disciplines should also be excluded, particularly when such terms were established to describe phenomena in the original context that might eventually appear as relevant for the discussion about digital texts or computer games. My starting point is the observation that way too often we are discussing the "work" alone, isolating the middle part of the triad from its crucial parts "author" and "reader" or the newly born hybrid form of "riter"/"wreader" who has some "production value" but doesn't kill the "author" on the computer and video game production side.

II. The Problem of "Ergodicity"—For a Change

Let me continue illustrating the observation and its consequences with a discussion of a crucial term that has been mentioned here before, well aware of the fact that this may be taken as a contradiction of my own plan not to do theory by a discussion of terms, established or not. The term in question is "ergodicity" and has been introduced in hypertext theory over a decade ago influencing game studies to a great extent. In advance I would like to emphasize my intention of delivering a decidedly constructive critical approach, and I do not intend to contribute to a devaluation of the concept of cybertext as such, although some people in the "community" would appreciate any new fuel for keeping things burning. The problems discussed here would have never occurred by sticking to the term "cybertext" alone. But the fact that the terms "ergodic literature" and "ergodicity" have been established rather well in cybertext theory and game studies perfectly illustrates the conceptual deficiencies in the very basic foundations of a greater part of our theoretical approaches to the subject and according to my arguments may be viewed as symbolic examples of these deficiencies.

The earliest definition of "ergodic" in this context is by Aarseth, found in his book *Cybertext: Perspectives on Ergodic Literature* from 1997. This definition is as often quoted, as it is relevant. Therefore I will present it here at full length:

> The concept of cybertext focuses on the mechanical organization of the text, by positing the intricacies of the medium as an integral part of the *literary exchange*. However, it also *centers attention on the consumer*, or user, of the text, as a *more integrated figure* than even reader-response theorists would claim. The performance of their reader takes place all in his head, while the user of cybertext also performs in an extranoematic sense. During the cybertextual process, the user will have effectuated a semiotic sequence, and this selective movement is a work of physical construction that the various concepts of "reading" do not account for. This phenomenon I call ergodic, using *a term appropriated from physics* that derives from the Greek words *ergon* and *hodos*, meaning "work" and "path." In ergodic literature, nontrivial effort is required to allow the reader to traverse the text. If ergodic literature is to make sense as a concept, there must also be nonergodic literature, where the effort to traverse the text is trivial, with no extranoematic responsibilities placed on the reader except (for example) eye movement and the periodic or arbitrary turning of pages. (1-2; accentuation by Furtwängler)

The concept of "ergodic literature" claimed to consider the user as a highly integrated figure, supposedly radicalizing the program of reader-response theory—integrating him or her on the very basic level of the materiality of the medium. This radical claim promised to be the ideal means of conceptually integrating the triad "author"—"work"—"reader." But unfortunately—or others may perhaps say necessarily as the first steps—the decision of user integration on the basic level of the medium's materiality actually led to a concept of the user who gets defined by parameters that reduce his or her levels of experience to trivial dimensions. It is quite possible to say that the user has become a more integrated figure than he ever asked or wished for, paying the price of identity loss.

So the emancipation from reader to user/player is paid with a high price in the model of "ergodic literature." While he[2] must "traverse the text" with non-trivial effort, his experiences may be quite trivial for the sake of "user control" in the model itself, which makes it necessary to dissolve the user in the (materiality of the) medium introducing the most radical rhetoric of technological immersion. That all this appears to be a critical approach to Aarseth's model lies in the merits of his concept that made some of the most vital contributions to the establishment of game studies. But the vast expansion and generalizations of this "text-model" into the realm of the phenomenon of electronic games has developed into a problem. This extension is decidedly not the work of Aarseth alone, who apparently lost interest in the cybertext model when he switched to the study of games, perhaps for good reasons.

In the last statements of *Cybertext*, Aarseth emphasizes his focus on "broad, highly visible issues" (183) and mentions e.g. the conflicts between "desires of users" and "ambitions of creators," which, for my money, should be among the more vital issues in the study of these new forms of media. So it is rather sad that Aarseth places this emphasis at the end of his book without really focusing on the desires of the user. To do that would have fulfilled two tasks: first, a relativization of the radical emphasis of the materiality of the medium as given in his approach; and second, a real revision or radicalization of reader-response theory, which was in fact a veiled theory of production. Nonetheless, while Aarseth mentions the problem of technological determinism as a crucial point in his theoretical approach right at the beginning of *Cybertext*, he sticks to the "material technology of the medium" (14) resulting in his research plan:

> In the context of this study, the question of the text becomes a question of verbal media and their functional differences (what role does a medium play?), and only subsequently a question of semantics, influence, otherness, mental events, intentionality, and so forth. These philosophical problems have not left us, but they belong to a different level of textuality. In order to deal with these issues responsibly, we must first construct a map of the new area in which we want to study them, a *textonomy* (the study of textual media) to provide the playing ground of *textology* (the study of textual meaning). (15)

I have doubts about the construction of an adequately useful map of the new area that leaves out textual meaning (which is the medium's most important link to the user), especially in computer games.[3] All of this is an expression of disrespect concerning user dimensions. In recent texts (Aarseth, "Playing Research"), Aarseth alluded to the concept of "textology" as an enterprise worth aiming at, although not in the framework of cybertext theory or theory of "ergodic literature," which only seems to be a logical consequence of the model itself and its deficiencies that don't allow for that kind of an anthropological perspective. As a result and with the growing relevance of computer and video games in the field of interactive media, the model Aarseth introduced in *Cybertext* and pushed towards the direction of *game studies* in his summarizing article *Aporia and Epiphany in Doom and The Speaking Clock*, became a static reference point in game studies. Meanwhile all have apparently agreed that computer games are describable if not as "ergodic literature" then at least as "ergodic art" and are in fact "ergodic." I will argue that the latter statement is either a misconception of the term or a straightforward decision.

Actually the impression might arise that the interest in recipients or consumers in this case (and several others) has only rhetorical functions. No one in game studies can justify research without attention to the role of the player or user. Then, there are two possible strategies: The first would be to tackle the problem and the second to integrate the user in the way just described. There is also an in-between strategy, and that is leaving the next step to the imagination, especially that of others. Gonzalo Frasca put it this way, calling the research tradition by its name in a laudable manner:

> However, formalism is not the flavour of the month in these posteverything times. Certainly, formal approaches are limited—and ludologists should always keep that in mind—but they are probably the easiest way to uncover the structural differences between stories and games. I personally see this structural approach as a first, necessary step in video game studies, which we will definitely outgrow once it helps us to better grasp the basic characteristics of video games. (222)

But will it ever be possible to consider some further steps within the framework of a theory that treats users or players in such a manner?

The status of the user in cybertext theory as described—especially if it is extended to the subject of computer and video games—may be well buttressed by the term "ergodic" in its original meaning as it was conceptualized in statistical physics first in the works of Boltzmann. But I think Aarseth did not have the consequences of this support in mind. They certainly are no reason for celebration either as they wipe out the supposedly radical claims in the theory of "ergodic literature" as a model that "centers attention on the consumer, or user, of the text, as a more integrated figure than even reader-response theorists would claim" (Aarseth, *Cybertext* 15). I even think that this backing disqualifies the model for further studies of the subject seeing that the notion of "ergodicity" only highlights the far more radical claim I suspected earlier in this paper: It advances a more or less "mechanical organization of the user" and does not support an anthropological perspective on user integration merely emphasizing the text.

The enterprise of going back to the original meaning of the term is legitimate here as Aarseth himself explicitly marks the appropriation of the term from physics in *Cybertext.* There was no need to do so, but it gave Aarseth's model a touch of "exact science" from the start. The "appropriation of the term from physics" is a piece of information that is perpetually mentioned when someone refers to the definition of "ergodic literature."[4] In this respect it is not derived from its etymological source *ergon* and *hodos*.[5] My first assumption here is that cybertext theory is *not* violating its very basic claims in appropriating the term "ergodic" as an unfocused metaphor, "rendered useless by a translation that is not perceived as such by its very translators," as Aarseth put it (*Cybertext* 14).

III. The "Ergodic Problem" in Physics

For a basic understanding of the concept of "ergodicity" in the so-called "ergodic hypothesis" and "ergodic theory" (which are separate things), no mathematical equipment is necessary. This makes the fact that Aarseth does not at all specify his "appropriation from physics" even more puzzling—I could find no reference to any source, not even a small hint in a footnote. Why then this appropriation?

A good starting point for a brief investigation is Boltzmann's (and presumably Maxwell's) belief in the so-called "ergodic hypothesis," saying:

> A mechanical system, if left undisturbed, will pass through every point of the phase space lying on a certain energy surface. In other words, the positions and velocities of all the mass points (representing the atoms) will eventually take every possible value consistent with the given total energy of the system. (Brush 363)

The system will pass through all kinds of possible states, leaving the system in *free* evolution and waiting for a (very) long time. Though working in statistical mechanics, this was in fact a hypothesis that proved to be too strong and in 1913 both Plancherel and Rosenthal showed independently that a trajectory passing through any "phase space" couldn't be a mechanical one. Their finding led to the weakened "quasi-ergodic hypothesis," that said that a trajectory can go by any point in phase space "as near as is wanted." We may only note that this weakened hypothesis was of not much use as it was the equality of "time average" with "phase average" that made "ergodic hypothesis" so convenient, which in fact was the reason why the hypothesis was put forward by Boltzmann after all. This equality made the method of statistical ensembles possible, a very basic foundation in statistical mechanics. In the early 1930s Birkhoff, Koopman and von Neumann tried to prove the equality without the assumptions of the "ergodic hypothesis," which resulted in "ergodic theory," being a more mathematical theory about general dynamics.[6]

You may well question the relevance of all this for the subject of hypertexts or computer games. I will try to show a possible relevance by breaking it down to something that at least looks like a game, because it uses its devices. Accordingly, a very simple and basic ergodic system may then be exemplified by a set of dice: There is no difference between the throwing of one dice one thousand times and throwing one thousand dice only once—the more dice one throws or the more throws one makes the better it is.[7] That is because the "ergodic hypothesis" says that time average and average over the statistical ensemble are the same—that is what the statistical example of the throwing of the dice means. And this is equivalent to the statement that all accessible microstates are equally probable over long periods of time.

Now, even in a most rough metaphorical treatment of all this: Who is interested in microstates in a macroscopic world in which the games we are interested in take place? There is no need to make guesses about relevant states of "phase space" in computer games, although it would be possible to construct such an abstract "space" for computer games on the basis of observable characteristics. But the treatment of the potential possibilities of different states by means of statistical methods is not evident at all. Even if such states were in fact possible, they would be highly irrelevant for players, as they have no function for semantic structures in games. Take for example a huge room in a 3D computer game with a door as an entrance and another as an exit. There are infinite trajectories possible to get from the entrance to the exit. Thus there is an infinite number of irrelevant states of the "mechanical" system, but only one semantically relevant route, that is the direct one. Research questions assuming homogeneous space not organized hierarchically in game environments are only possible if the user or player and his needs and intentions are kept in a blind spot. To assume that computer game design in the planning of game structures leaves these needs and intentions out of account is a fallacy. Especially the production side (author) has to anticipate what the user thinks to provide any sense and dramaturgy etc.

Without exploiting the metaphorical exaggeration of the analogy to physical systems any further, I think that the statement about the same probability of all states was in fact the main reason for treating cybertexts including computer games as "ergodic." And this has to be the source of many vital questions about our research practices and the different levels or layers of observation we are interested in. It is typical for ergodic systems not to be predictable; chaotic systems and real random processes are ergodic. It would be rather absurd to even compare such systems with computer games where contingency and randomness get permanently controlled via user's attention that is controlled via things the player recognizes and accepts as stereotypes etc.—computer games in particular rely on strict mechanics of anticipation. Aarseth commented on ergodicity in his understanding of the term after the appropriation: "Ergodic phenomena are produced by some kind of cybernetic system, i.e., a machine (or a human) that operates as an information feed back loop, which will generate a different semiotic sequence each time it is engaged." ("Aporia and Epiphany" 32) But information feedback loops in this sense are not part of ergodic systems, except perhaps only if information is considered to be deprived of meaning, which may just be the case if you separate "textonomy (the study of textual media)" from "textology (the study of textual meaning)" ("Aporia and Epiphany" 15).

What is dominant in ergodic systems is a strong tendency away from cognitive authorities of decision. So you may say that biological systems are in fact non-ergodic. Let me explain this with another simple example that may just illustrate the border between physical and biological systems, when we project the example of the throwing of the dice onto a swarm of mosquitoes: If you are interested you can measure the average (scalar) speed of this swarm. First, you follow a random mosquito for quite some time and measure its speed at certain times—which stands for an observable or a property—and then calculate the average. That is the time-average. Second, you measure the speed of quite a lot of mosquitoes at one time and calculate the average of the single speed measurements. That may be the phase average. If there is no statistical variation in the behavior of the mosquitoes then both methods are bound to result in the same statistical average. In this scenario, no mosquito leaves the swarm, has its own will, joins the swarm, dies, or gets eaten. The system is closed and isolated, without any feedback particularly from the outside. Ergodic systems are no systems for individualists or achievers: In such systems it is therefore advantageous to submerge both independent thoughts and emotions. The best strategy would therefore be to avoid giving any feedback to the system. Now we take the example to the next level:

Take one thousand players and let them play the same game (not together but separately) and measure their quality or quantity of experience (if not possible then measure their amount of sweat) and calculate the average. Then let one single player play this game one thousand times. Not the same? The player's motivation will probably sink endlessly or he will learn or discover and explore states of the game no one had in mind, not even the player. Loss of motivation or losing a game will result in the player's leaving the system/game thus changing the character of the system—resulting in loss of ergodicity.

How player experience affects the system "computer game" can be observed in myriads of different examples, e.g. and to name just two, training effects obvious in the popular "speedruns" through games like *Quake* or even *Super Mario Bros.*, or the player's experience relevant for effective navigation through virtual environments, which was shaped outside the game and gets into the game by a transformation of real-world experience.[8] Aarseth is quite aware of these effects and describes them in a more recent article "Playing Research: Methodological Approaches to Game Analysis" as "part of the game" and as significant for diverse types of players whom he calls—in reference to Bartle—*socializers, killers, achievers* and *explorers*.[9] Thus Aarseth concludes: "Like ergodic works in general, there are variations in the realization of the games which means that a collective pool of experience will always bring new aspects forward" (6).

But is there a way to integrate human dimensions of experience in the concept of "ergodic works"? I guess not and that is why I think it is time to be honest about the limits of the cybertext concept sketched by Aarseth and others. These questions about experience are not in line with the early assumptions of "ergodic literature" and simply in order to preserve some of the original meaning of the term ergodic—as it is understood in physics and also literary studies—it is necessary to mark these limits of scope. Talking about games as "ergodic" and taking user experience seriously at the same time is not really an option and as a consequence there has to be a change of attitude. In the following I will further explain why I think this is necessary.

IV. Consequences

Part III was by no means a strict and incontestable examination of the meaning the term "ergodic" has in mathematics and physics, but instead a quite liberal one that allowed a transfer of the properties of microscopic systems to our macroscopic world of experience. To assign ergodicity to systems that may represent the system "user—machine" in computer games is making a very strong assumption. To assign ergodicity to any kind of system in fact always makes a strong assumption. The application of the term only makes some sense when the integration of the user is radical in a sense that both "text" (in the original sense of the theory) and "user" are organized mechanically, which is another strong assumption. In this scenario games may be ergodic systems just in the sense that all possible states are accessible and playable, so that statistical conclusions about a certain player's state are actually possible. The probability of this state would then represent the frequency of that very state in the ensemble. Reduced to statistical quantities the loss of user perspective is evident in this statement. Let me repeat Brush's description of systems that are characterized by the "ergodic hypothesis": They are mechanical systems that will pass through every point of the phase space lying on a certain energy surface—*if left undisturbed*. A treatment of computer games as ergodic systems is unable to take diverse kinds of user action into account, thus reducing interaction to trivial dimensions so that the player is no source of disturbance of the system. The scope of such a model is then considerably limited in its anthropological perspective.

To sum it up, there are two views of looking at the problem of "ergodicity" in this context, both of them leading to desperate prospects for the conceptual foundations of cybertext theory as an appropriate model for computer and video games, not to mention a complete model. First, Aarseth literally integrated the term "ergodic" in the sense of its original meaning, as "a term appropriated from physics" as he says (*Cybertext* 1), which means that there is no place for an integration of the user in a non-trivial sense. This implies that the mechanical organization of the text has to be also the mechanical organization of the user. Or second, Aarseth was in fact practicing what he dislikes most, namely that "lack of self-reflection places the research in direct danger of turning the vocabulary of literary theory into a set of unfocused metaphors, rendered useless by a translation that is not perceived as such by its very translators" (*Cybertext* 14). Only this time—"the theoretical perspective of 'fill in your favorite theory/theoritician here' is clearly really a prediction/description of 'fill in your favorite digital medium here'" (Aarseth, "Aporia and Epiphany" 31)—"literary theory" is replaced by "statistical physics/mechanics."

I have already commented on the first view several times here. Let me add only this: I'm aware of the fact, as I have already mentioned here, that no one has ever seriously claimed completeness of the "formalistic approach" and this should be something worth considering for those who attack this model. But if we agree on the first view then it seems important to explicitly mark the first view as part of the research program in the line of cybertext theory, as the problematic status of the user is not self-evident for great parts of game studies and it is also not evident that the decisions of the cybertext model and its concept of "ergodic literature" in extensive use in game studies is a "first necessary step" (Frasca 222). Many misunderstandings would actually have never occurred by naming this limited scope more explicitly for others from the start.

The second view definitely needs some more comments: Eskelinen and Tronstad tell us in their article "Video Games And Configurative Performances": "Now, there's no doubt that games and computer games are ergodic if not art, because in games we have to produce, encounter, and respond to variable sequences of action" (197). While we all agree on the second part of this observation, which is partly a result of cybertext theory's success in focusing us on the importance of the materiality of the medium, there is no real evidence to assume ergodicity here from a physicist's point of view, who just might be a bit puzzled in the apparently improbable case that he or she stumbles across it. An easy way to reply to objections would be to stick to the etymology of the term Aarseth executed when he introduced it for the purpose of his cybertext model, despite the fact that the etymology in physics is not certain (cf. Gavalotti 36). This does not change the fact that Aarseth appropriated the term "ergodic" from a highly specific field of statistical physics for no evident reason. Such an appropriation doesn't have to be tragic at all at first sight, but it is not at all a laudable decision when it comes to a possible dialogue between the disciplines of the natural sciences and the humanities. Considering the relevance of information science and the subject of dynamical, complex systems etc., such a dialogue is not mistaken at all. The consideration of complex systems has supported an exchange between disciplines like physics, chemistry, biology, economy, ecology and sociology e.g. in "synergetics,"[10] as these systems are built by many interacting parts and are relevant for all disciplines that try to predict the behavior of systems and therefore understand the way they work. Their relevance is by far not purely metaphorical and a term like "ergodic" might just be a relevant interface in the discussion between disciplines that have a different sense for the role strict meanings of terms play in their theoretical models.

In addition to that, "ergodicity" in a more literal appropriation from physics has already been introduced as a topic in literary studies, e.g. as the "Ergodenproblem" of society in Christian Kassung's and Albert Kümmel's studies on Robert Musil's *Mann ohne Eigenschaften*, where "ergodicity" is explicitly linked to the ever puzzling discoveries of thermodynamics and statistical physics in their relevance for reversible and irreversible time order; a discussion resulting basically from the discovery of entropy, since reversibility, i.e. symmetry or time invariance, is with few exceptions a general property of physical laws.[11] This topic has considerable relevance for the studies of computer games, where experience of reversibility—the manipulation of the arrow of time—in "macroscopic" processes is one of the most central forms of non-trivial experience that affects causal order of events in the medium of the game. It even is a popular playfield.[12] I find it still striking that with all the focus on non-linearity in the study of new media, not much has been done to explore the question of the reversibility in causal order of events experienced in computer and videogames. Martin E. Rosenberg showed this relevance early in hypertext theory, saying:

> In this sense, what hypertext theorists mean by "nonlinearity" is more properly analogous to what physicists call "symmetry," or "reversibility." Since the focus in hypertext theory is on the functional direction of signification, the correspondence with symmetry and reversibility works more precisely, because these terms refer to the direction of the flow of duration or time. (Rosenberg 275)

Still, no one ever seemed to have paid attention to the semantic dimension of this process, meaning that the "functional direction of signification" is not reducible to the materiality of the medium. If you allow me a short digression about this other case of problematic disrespect against user dimensions: These omissions are obvious and significant in Rosenberg's article when he continues "Words must be read one after another, and cannot be read backwards; we might say that the reading process is linear in the sense that its trajectory remains irreversible" (275). Thus again, obvious and easily obtainable observations are blocked by the celebration of "mechanical" processes, most evident in the metaphorical use of "trajectory." One of these strikingly apparent observations is by far most notably obvious in the computer and video games of recent times. In these games it is often inevitably necessary that the order of events has to be repeatable; it has to be possible to reproduce every act or sequence of the game to correspond with the strategic development of user decisions in these sequences, where failure ("load a saved game") and success ("save the game") provide for continuous informational feedback loops in both respects: player and game/medium. This highly important issue for the interrelations of technological and anthropological questions has certainly to be on the list of reasonable future game studies. The substantial role of the user in this process prohibits making statistical assumptions about the equality of states or succumbing to the urge to understand the path the user takes through the "work" (his or her trajectory) as an approximately random one predominantly determined by the materiality of the medium. Like David Z. Albert puts it: ". . . ergodicity (if you think about it) is clearly also the sort of thing one has in the back of one's mind when one speaks of trajectories wandering aimlessly and randomly and all over the place and in no particular direction" (Albert 59). There is a distinct lack of the user's freedom to take such trajectories, which is determined predominantly by the "semantics" of or in the medium that give a sense to the users' actions. This may only come into view by a substantial integration of anthropological perspectives. To further illustrate my point, Rosenberg sees or says that one might see the series of "lexias," units of text and information, in a hypertext system as a means to enable "users to leap forward or backward, lexia by lexia," and that "each lexia remains separate, virtually autonomous; the wreader's experience of the sequence of lexias remains discontinuous, except as the author or reader establishes linkages among them" (280). My task here is not to decide about the right or wrong of this in hypertext models. In the transition to computer games, however, "sequences of lexias" most likely became denser, so that on the same level of observation discontinuity is not a dominant structural feature of the audiovisual worlds anymore. Of course, there are still comparable discrete structures of events, but if we leave out the even more complicated questions about corresponding levels of observation for the moment, I think that the user plays an integral role in building dense sequences of lexias or semiotic sequences (Aarseth) by filling the gaps, bridging the space between lexias.

To cut a long story (particularly about Rosenberg's article) short and to come back to my more specific exploration of Aarseth's appropriation from physics, I would like to stress here that my interest is not some repetition of arguments in the line of Rosenberg who said that the term "nonlinear" is primarily a trope from physics (and chaos theory). Aarseth legitimately finds this statement to be "somewhat overstated" ("Aporia and Epiphany" 43) as it is significant for many statements of Rosenberg's impressive article. The problem is also not located in the definition of "ergodic literature," which is sufficiently clean and sharp. The problem lies in an obvious gap of communication between certain disciplines that may be bridged by a proper use of terms by not using unfocused metaphors, rendered useless by a translation, to come back to Aarseth's admonition once again. The relevance of "ergodicity" for this is, I think, an important one in the sense I have just tried to sketch as an unfortunate historical occurrence that seems to have come to pass: the original meaning of an appropriated term challenges its metaphorical use in the same field.

But if we agree on loosening the strict meaning and definitions of the term "ergodic" in physics and allow a certain kind of metaphorical appropriation, the systems described as "ergodic" in cybertext theory are only "ergodic" in the scope of this theory and only on its formalistic level of reflection, examination and observation. What is at stake then is the anthropological dimension of game studies which we might lose by the radical integration of users, resulting in their fixation. The project of "textual anthropology" is easily in danger, using an expression Aarseth put at the end of an earlier article he wrote before he worked on "perspectives on ergodic literature" ("Nonlinearity and Literary Theory" 82) and which is published in the same volume as Rosenberg's. Aarseth described the object of study in the anthropological approach as "a process (the changing text) rather than a project (the static text)" in 1994, assigning the source of the change in text to the user. But I suspect that the study of "textonomy" instead of "textology" is equal to the decision which part is attributed dynamic properties: the text or the user. In this respect the appropriation of the term "ergodic" is highly significant and a symbol for this decision. If we can't alter the path cybertext theory would prescribe for game studies then we must ask for complementary theoretical approaches that are able to save the user in his dynamic significance by formulating a comparable radical rejuvenation of anthropological perspectives. Such an agreement is, I think, possible.

In my introduction I already mentioned a lesson to be learned from the demands of traditional models, and I meant it in that very respect e.g. from the study of literature and the anthropological turn *after* reader-response theory.[13] Wolfgang Iser has coined part of this turn with statements like "Literature is not self-sufficient, so it could hardly bear its own origin within itself.

What it is is the result of its function" (Iser 210) only partly in connection to the reader-response model Aarseth wanted to radicalize. Iser was aware of the fact that reader-response theory and its famous concept of the "gap" ('Leerstelle') was not sufficient for an anthropological turn within the discipline and in many respects reader-response theory reminds us of some of the properties and shortcomings that I described here for cybertext theory. Perhaps as a logical development of this, the anthropological turn was distinctly oriented at the question why we have a medium. This is in line with an obvious triviality, i.e. that media are human constructions usable for human constructions. Media result from their functions and taking media as a starting point does not have to imply a strict focus on material sign production—it can actually point to a place of consensus. This brings us once again back to the fallacy that "the question of the text becomes a question of verbal media and their functional differences (what role does a medium play?)" without considering the "question of semantics, influence, otherness, mental events, intentionality, and so forth" (Aarseth, *Cybertext* 15). The question "what role does a medium play?" is not separable from semantics and should provoke an altered concept of media. The (also) radical claim of literary anthropology even was to have developed an explorative instrument for finding out about the "human equipment" in reverse direction. Iser once responded to a question about the connections between reader-response theory and his concept of literary anthropology that "reading as text processing also means—and this was an implication which may not have come sufficiently to the fore—finding out something about the human makeup: namely, the way in which the letters we perceive translate into a stream of imagery in our minds" (Iser and van Oort). According to Iser this orientation can only foster our understanding about the phenomenon that literature really is. Furthermore this concept frees us from the tiresome questions about what's literary in literature and poetic in poetry and hopefully also from the burden of working on genre typologies. As in other forms of the anthropological turn in several disciplines ontological determinations fade out of the discussion when the reader, recipient, user or player is subject in his relation to the "work." In the case of Iser this was a direct consequence of certain impasses he diagnosed in literary theory. According to this view, literary theory was occupied mainly with building frameworks for methods of interpretation before the 1990s and texts existed for confirmation of their model's claims (Iser 209), which is a highly problematic issue—in general and particularly in game studies (as mentioned here before)—because theoretical claims are not meant to construct the subject but the other way around.

Eskelinen und Tronstad rightly say: ". . . there's a limit to formalizations, as they don't say much about the player's experience and deeper motivations" (214).

To agree on this limitation means to accept the combination of complementary models that support each other to compensate their deficiencies. I don't want to answer to the question whether it makes sense to follow one specific path alone without considering others from the very beginning or accepting these paths as intertwining. An anthropological perspective may help to wake the mechanically organized user from its potential, virtual death in cybertext theory by a new focus on user's experience and needs, while the insights into the structural makeup of games provided by cybertext theory may balance possible generalizations of the anthropological view.

Iser's observation that "literature as a medium is also indicative of the needs to which it responds" still seems to be an important step in the right direction. As he continues: "In addressing itself to this issue, literary theory is bound to change direction. Instead of providing a matrix for model-building, it has to explore the sign-function of the medium, thus turning the text into a reflection of the needs in question" (Iser 209), it becomes apparent once more that *sign-production* (cybernetic or not) is inseparable from *sign-function* on the way to a deeper understanding of a medium. In interactive media of course we are challenged to quite another degree as we have a system of user and game or machine that consists of two interacting subsystems that are dynamic in themselves. So the decision of cybertext theory to fixate the dynamics of the user to some degree or radically for the moment is just an understandable way of reducing the complexity of the system as a whole. Literary anthropology may easily accept the recipient, the reader, as the dynamic part of the system, which is obvious as the texts that for example Iser had in mind are static especially from the view of cybertext theory. So my account and the close reading of the term "ergodic" as something Aarseth should have done in the past should definitely not be taken as an offence but as an honest explanation why things are the way they are in game studies. To take close-reading into account is necessary because the supposedly radical anthropological claims of cybertext theory and its concept of "ergodic literature" are in fact a pack of lies, most evident in close-reading the appropriation from physics that was in fact not a lie but just misleading. To reconsider the term "ergodic" and reflect the way in which it now is appropriated from Aarseth's cybertext model in game studies is important for a growing respect towards the terminology necessary in the future course of our studies and a possible subsequent and desirable respect game studies still have to earn from other disciplines and even from within the own field to seriously establish a new discipline.

Notes

1. This article is a revision and more detailed version with some changes in argumentation of Furtwängler, "Menschliche Praxis."
2. The male personal pronoun is of course only used by convention.
3. As I have argued several times elsewhere, e.g.: Furtwängler, "Mensch-Maschine Computerspiel."
4. I experienced this practice in several texts in the field of *game studies*, also in semester works by students.
5. The strategy of shielding concepts with appropriations from the "hard" sciences against objections is well known and especially understandable considering unfortunate objections against decidedly "soft" anthropological concepts that often get mocked. Regarding Aarseth's "vulnerable" explorations of his own gaming experiences he describes in "Playing Research" I have not only come across silly mockery in internal discussions of the forum *spielkultur.net*, but I myself also have been subjected to this.
6. For an excellent, brief introduction to the subject cf. Badino.
7. Time usually plays an ambivalent role in ergodic systems, because on the one hand time gets "transformed away," on the other hand time has to be a long time. As Manfred Eigen and Ruthild Winkler show in their book *Das Spiel: Naturgesetze steuern den Zufall* the observed times quickly surpass the life-span of the universe: "Beim Nachrechnen der Zeiten aber, die zur Reproduktion einer bis ins kleinste fixierten Verteilungskonstellation nötig sind—der Physiker nennt diese die (Poincaréschen) Wiederkehrzeiten—, kommt heraus, dass diese im allgemeinen größer als das Lebensalter des Universums sind, das ist größer als zehn Milliarden Jahre. Boltzmann berechnete, dass man für eine Reproduktion der Lagekoordinaten aller Atome innerhalb von zehn Angstrom sowie für eine Reproduktion der Geschwindigkeiten innerhalb von 0.2 Prozent ihres mittleren Wertes in einem Kubikzentimeter eines verdünnten Gases (ca. 1/30 des Atmosphärendrucks) bereits mehr als $(10^{10})^{19}$ Jahre benötigen würde." (Eigen and Winkler 196)
8. For an extensive account on that topic illustrated with examples, cf. Furtwängler "Mensch-Maschine Computerspiel." Unfortunately the article I present in this volume is deprived of specific examples due to its focus on concepts, which is usually a thing I very much despise in game studies practice.
9. "The four types are *socializers* (the players who play to enjoy the company of other players), *killers* (players who enjoy preying on and harassing other players), *achievers* (players who like to win and triumph) and *explorers* (players who enjoy discovering the game's secrets and hidden mechanics, including discovering and exploiting programming errors)" (Aarseth, "Playing Research" 3).

10. For an introduction to the aims of this field in interdisciplinary perspectives cf. Haken.
11. In discussions with the authors it took some time to explain the assumption of "ergodicity" in the cybertext model due to compatibility problems in the conceptions. But I want to thank Kassung and Kümmel for very fruitful discussions. Furthermore I want to thank all the colleagues particularly from physics who were a great help in the mediation between disciplines over the last years, especially Wolfgang Dieterich, head of the department "Statistische Physik" at the University of Constance and Hans Diebner, head of the *Institute for Basic Research* at the *Zentrum für Kunst und Medientechnologie ZKM*, Karlsruhe at that time.
12. This is by the way also a perpetually recurring theme in popular media. To balance the sad absence of examples from computer and video games in my text, I want to name only some of the games that are actually linked to that subject and are central for my main research on that topic. While most games have some kind of save/load possibilities, the following are examples of games that reflect time reversibility as part of their mechanics: The new *Prince of Persia* series (Ubisoft Montreal Studios, Ubisoft 2003-2005), *Viewtiful Joe 1* and *2* (Capcom Production Studio 4, Capcom Co., Ltd., 2003-2004) and the forthcoming *Timeshift* (Saber Interactive, Atari, 2006), where manipulation of time dominates game mechanics and dynamics. This may also be seen in several examples of temporal slowdown, cf. *Max Payne* (3D Realms Entertainment, Remedy Entertainment Ltd., Gathering, 2001) or *F.E.A.R.: First Encounter Assault Recon* (Monolith Productions, Inc., Vivendi Universal Games, Inc., 2005), important as a preliminary stage of specific time reversal. Furthermore examples affecting "global" organization of reversibility in thematic and narrative content affecting game mechanics indirectly, cf. *TimeSplitters* series (Free Radical Design, Eidos/Electronic Arts, 2001-2005); *The Legend of Zelda: Ocarina of Time* (Nintendo EAD, Nintendo Co., Ltd., 1998), *The Legend of Zelda: Majora's Mask* (Nintendo EAD, Nintendo Co., Ltd., 2000), *Chrono Trigger* (Square Co., Ltd., 1995), *Chrono Cross* (Square Co., Ltd., 1999), *Dark Chronicle* (Level-5 Inc., Sony Computer Entertainment Incorporated, 2003). These are all more recent examples, hinting at a logical historical developments of computer game structures, first apparent in the 90s, cf. *Timequest* (Legend Entertainment Company, 1991), *Day of the Tentacle* (Lucasarts, 1993), *Lost in Time* (Coktel Vision, Sierra, 1993).
13. Perhaps a hint to the irony as hypertext or cybertext theory is in fact part of literary studies.

Works Cited

Aarseth, Espen J. "Nonlinearity and Literary Theory." *Hyper/Text/Theory*. Ed. George P. Landow. Baltimore: Johns Hopkins UP, 1994. 51-86.

———. *Cybertext: Perspectives on Ergodic Literature*. Baltimore: Johns Hopkins UP, 1997.

———. "Aporia and Epiphany in Doom and the Speaking Clock: The Temporality of Ergodic Art." *Cyberspace Textuality: Computer Technology and Literary Theory*. Ed. Marie-Laure Ryan. Bloomington: Indiana UP, 1999.31-41.

———. "Computer Game Studies, Year One." *Game Studies* 1.1 (2001). 30 May 2006 <http://www.gamestudies.org/0101/editorial.html>.

———. "Playing Research: Methodological approaches to game analysis." *Game Approaches / Spil-veje. Papers from spilforskning.dk Conference*, August 28-29, 2003. 30 May 2006 <http://spilforskning.dk/gameapproaches/GameApproaches2.pdf>.

Albert, David Z. *Time and Chance*. Cambridge, MA: Harvard UP, 2000.

Badino, Massimiliano. "The Foundational Role of Ergodic Theory." Preprint 292 of the Max Planck Institute for the History of Science. 2005. 30 May 2006 <http://www.mpiwg-berlin.mpg.de/en/forschung/preprints.html>.

Bartle, Richard. "Hearts, Clubs, Diamonds, Spades: Players who suit Muds." 1996. 30 May 2006 <http://www.mud.co.uk/richard/hcds.htm>.

Brush, Stephen G. *The Kind of Motion We Call Heat. Book 2: Statistical Physics and Irreversible Processes*. Amsterdam: North Holland Publishing Company, 1976.

Eigen, Manfred, and Ruthild Winkler. *Das Spiel: Naturgesetze steuern den Zufall*. Munich: Piper, 1976.

Eskelinen, Markku, and Ragnhild Tronstad. "Video Games and Configurative Performances." *The Video Game Theory Reader*. Ed. Mark J. Wolf and Bernard Perron. New York: Routledge, 2003.195-220.

Frasca, Gonzalo. "Simulation versus Narrative." *The Video Game Theory Reader.* Ed. Mark J. Wolf and Bernard Perron. New York: Routledge, 2003. 221-235.

Furtwängler, Frank. "Menschliche Praxis: Wie das Ergodenproblem die Re-Animation anthropologischer Perspektiven in den game studies herausfordert." *Dichtung Digital* 7.34 (2005), 30 May 2006 <http://www.brown.edu/Research/dichtung-digital/2005/1/Furtwaengler/>.

———. "Mensch-Maschine Computerspiel: Über eine notwendige Paartherapie in der Medienrealität." *Jahrbuch für Computerphilologie 7.* Paderborn: mentis, 2006. 111-130. 30 May 2006 <http://computerphilologie.uni-muenchen.de/jg05/furtwae.html>.

Galavotti, Giovanni. *Statistical Mechanics: A Short Treatise.* Berlin: Springer, 1999.

Haken, Hermann. "Synergetics as a Strategy to Cope with Complex Systems." *Interdisciplinary Approaches to Nonlinear Complex Systems.* Ed. Alexander S. Mikhailov and Hermann Haken. Berlin: Springer, 1993. 5-11.

Iser, Wolfgang. "Towards a Literary Anthropology." *The Future of Literary Theory.* Ed. Ralph Cohen. New York: Routledge, 1989. 208-228.

Iser, Wolfgang, and Richard van Oort. "The Use of Fiction in Literary and Generative Anthropology: An Interview with Wolfgang Iser." *Anthropoetics III.2* (Fall 1997/Winter 1998) <http://www.anthropoetics.ucla.edu/ap0302/Iser_int.htm>.

Kassung, Christian. *EntropieGeschichten: Robert Musils "Der Mann ohne Eigenschaften" im Diskurs der modernen Physik.* Munich: Fink, 2001.

Kümmel, Albert. *Das MoE-Programm: Eine Studie über geistige Organisation.* Munich: Fink, 2001.

Rosenberg, Martin E. "Physics and Hypertext: Liberation and Complicity in Art and Pedagogy." *Hyper/Text/Theory.* Ed. George P. Landow. Baltimore: Johns Hopkins UP, 1994. 269-298.

Philippe Bootz

The Problem of Form

Transitoire Observable, a Laboratory for Emergent Programmed Art[1]

Abstract:

I will present some conceptions of programmed art focused on the problem of form. I will not explain here the different approaches but only open the question in the perspective of the procedural model. I will start from the basic common point of view of the collective *Transitoire Observable* and, after an overview of some aspects of the procedural model, I will pose the question of form as a specific management in the programming of arbitrary aesthetic constraints that are posed by the author in his management of the situation of communication created by the work whatever the surface aesthetics is on screen. In this sense, we will speak of "programmed forms" as forms in programming and not as forms of the programmed multimedia event.

1. *Transitoire Observable* and the Question of a Specificity of Programmed Forms

1.1. Programming as a Matter

E-poetry first focused on a programming approach. But since the end of the 1990s, many e-poetry works have focused on video art aspects of multimedia events on screen or have explored "traditional" usage of the link in hypermedia. Many others use the electronic medium only as a simple medium and as a support for a classical visual poetic approach or as classical visually illustrated texts. These orientations are encouraged by firms that produce software in which the specifics of the numerical medium are masked by video metaphors. I do not deny that these kinds of poetic projects change the traditional understanding of the text. They also produce different and non-classical manners of expression. But it seems to me, and also to several other poets and artists, that this popular approach does not meet the specific needs of this medium because it does not engage programming. Thus these approaches are unable to propose a situation of communication that is truly new.

This is why, in 2003—on the suggestion of Alexandre Gherban—Alexandre, Tibor Papp and I decided to break with these approaches and to create a new collective that relocates programming at the core of electronic art. We thought that certainly there existed specific forms close to programming; forms we did not know and had to explore. We wrote a manifesto in which we explained that we consider programming (and not the program) the material of this art; that the multimedia event that appears on screen is only a transient observable state (a "transitoire observable" in French) that occurs while a computer is operating. We think that forms exist that engage the complexity of the situation of communication created by programmed works. We also think that these forms are close to programming and highly independent of the nature of the *transitoire observable*. Programming is a new kind of material that artists can sculpt and model. It is a matter of a formal approach to the duality of algorithms/processes. What is pointed out in this approach is the relationship between the algorithmic nature of the code and the pragmatics of reading.

So, at this level, e-poetry certainly does not exist but one can indicate a general art of programming. This is why several poets and artists quickly joined the collective: Jim Andrews, Wilton Azevedo, Jean-Pierre Balpe, Bluescreen, Patrick Burgaud, Philippe Castellin, Xavier Leton, Loss Glazier, Antoine Schmitt, Eric Serandour, and Reiner Strasser.

1.2. Major Directions of Exploration

These artists employ numerous different surface aesthetics, generally issued from previously programmed works, often published in *alire*. These works can be interactive or not, but they always put the reader in a particular position where he is responsible for an experiment of reading, and is not the focus of the work: The reader is a partner. He has a special role to play in the work. His reading no longer remains outside of the work. Through its dark moments, its failures, its shortcuts, its questionings, and its great place of non-meeting, the reading activity fully participates in the work, it is a process of the work, it is an intrinsic component of the work. Reading is truly a sign, as is the operational process that it confronts, even in non-interactive works. The work is not designed to satisfy a reader nor is it to be read in the same manner as a book or a video. It is designed to operate because physical running is its raison d'être. In *alire12*, interactive works by Burgaud, Strasser, Schmitt, Gherban and Andrews satisfy such an assertion on different levels.

Interactive or non-interactive works are conceived to confront the intentionality of the author and that of the reader. The intentionality of the reader passes through interactivity and expressive non-interactive behavior. The intentionality of the author is expressed on screen and in the program, not only in lines of code that generate visible materials or that govern expressive visible behaviors, but in more specific meta-stylistic rules applied as arbitrary constraints on the procedures that manage the observable multimedia transient event.

It appears in *Transitoire Observable* that meta-stylistic rules can be similar in e-poetry and in more general programmed art; it is the basis of the collective. This convergence of points of view asks whether specific programmed forms exist, independent of the aesthetical surface level on screen. It also opens, perhaps, the possibility of a-media multimedia in which meta-rules would be managed on a very high abstract level, without directly taking into account the specificity of media.

2. Main Theoretical Features Appearing in *alire*

To explain how these authors arrived at these conceptions, it is useful to know the most important concepts that emerged from works published in *alire*. These conceptions are relative to the problem of meaning and to the relations that exist between signs and author, signs and reader, author and reader.

These works have broken new ground on three levels. The second and third levels constitute what I will call the "procedural model."

2.1. The Surface Rhetorical Level

First, *alire* has added some new specific rhetorical possibilities, namely syntactical animation. It has also published other new specific rhetorical possibilities: Poetic hypertext, kinetic poetry, and algorithmic poetry (combinatory or automatically generated poetry).

In each case, generally, we only need traditional semiotic theories to explain this surface rhetorical level. For example, all features in syntactical animations can be explained by using a traditional linguistic definition of text and a traditional approach to syntax. We only need to consider the difference between the temporal and the spatial construction of signs inside the same set of words and letters (spatial reading and temporal reading).

Unfortunately, this level fails to explain most of the significant features in interactive works. It also fails to explain the changing of meaning with time or the difference between combinatory and automatic generation. So, we need to consider the totality of the situation of communication between the author and the reader that constitutes a system. This communication passes through the work. Writing and reading are internal functions of this system, which constitutes the second level of signs.

2.2. The Communicative Level

a. Cognitive and Technical Understanding

At this level, we cannot neglect the technical behavior of the system in the construction of meaning. Communication features, ergodic activity of reading, semiotic features, cognitive features and technical features cannot be disjunctive. Notably, *the reader cannot construct a meaning of the work without constructing at the same time a meaning of the totality of the situation of communication.* From a systemic point of view, the work has no reality. It can only be understood as a subsystem of the situation of communication that cannot be extracted from this situation. Two important, non-classical rhetorical features (namely double reading and aesthetics of frustration) use this relation between the construction of the meaning of the work and the construction of the meaning of the communicative situation.

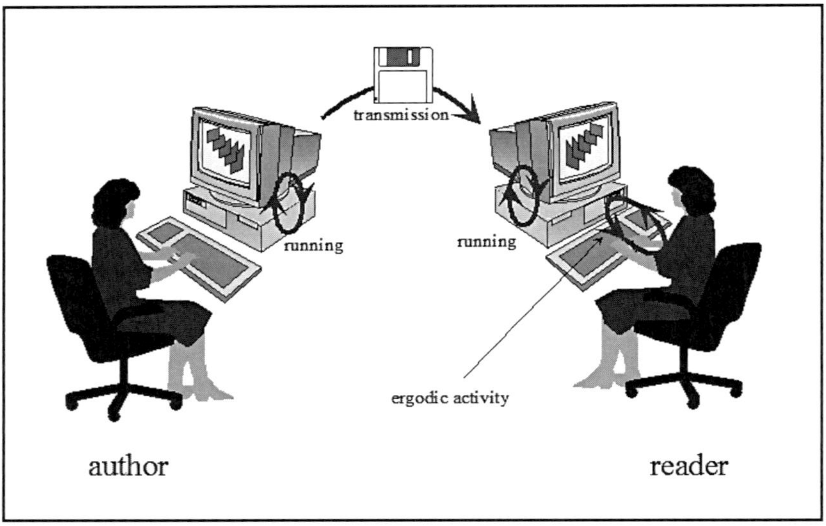

Fig. 1. The system's structural scheme.

It is useful to consider the cognitive level because the archetypal mind representations that are used by a reader are not stable; they are based on experiments of reading or writing in his particular medium and do not come from a deep cultural tradition. Thus they can differ greatly from one person to another. This mind representation has been called the *profondeur de dispositif* ("system depth") in the procedural model. A *profondeur de dispositif* is applied as a filter by the cognitive strategy of reading or writing in order to decide what is the significance of the signs in the transient observable event that is produced during operation. We can find many relations between different works in *alire* that show that a number of these pieces, independently of the nature of the work, have constructed the actual mind representation (which is described by the procedural model) by using in fact exactly the same strategy to manage reading.

At a technical level, author and reader are only users of the computer. Notably, the author in the working process does not manage the totality of the rules that are used by the computer while it is running. Hence the computer no longer is a Turing machine. Another model must be used. I have constructed the model of the "machines de monstration" ('showing machines'). The program written by the author does not totally manage the physical process during the computing operations. That is, the algorithmic level of the program is not completely responsible for the functioning of the *transitoire observable*. We can say that the author is the author of the program and data, but only the co-author of the physical process that appears to the reader while the machine is running. Using more traditional literary language, we can say that the author's program contains a large level of the "non-said." But this non-said does not play the same role as the non-said of the classical printed text: This non-said will be interpreted by the machine and not by the reader to produce the observable sign. Thus it is necessary to consider that while the computer is running, program (algorithmic level) and physical process are two different and complementary parts of the work. Adaptive generation uses this unsaid as a constraint for programming.

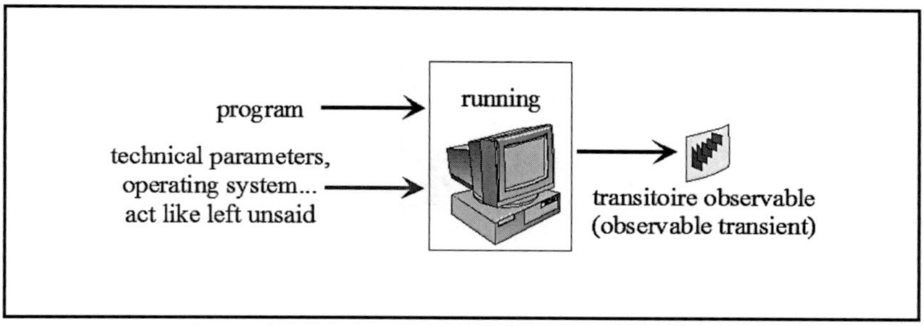

Fig. 2. Technical left unsaid.

b. Semiotic and Communicative Understanding

From a semiotic point of view, we can separate the classical and general semiotic notion of text (the text is the object of interpretation) into three different parts that do not act in the same space. Program and data (in high level forms) constitute the *texte-auteur* ("author-text" or "text-of-inscription").[2] This is a sign that is only accessible by the author. It is in the domain of the author. The second sign is constituted by what will be considered as "the text" by the reader. It is the *texte-à-voir* ("text-to-be-seen" or "text-of-visualization"). It is a part of the observable transient event that can differ from a reader to another because readers will not apply the same *profondeur de dispositif* on the *transitoire observable.* The physical process itself is a function. From a semiotic point of view, it transforms the *texte-auteur* into the *texte-à-voir.* Because it generates the *transitoire observable*, it is called "generation" in the model.

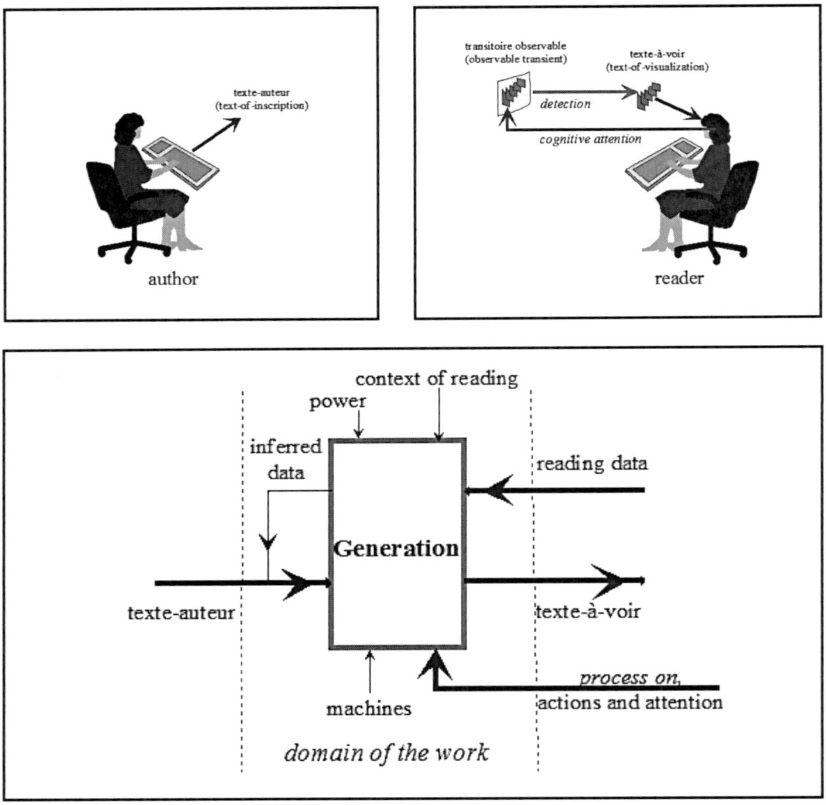

Fig. 3. The different structural and functional components of the work.

The totality of this subsystem constitutes the "domain of the work." It plays the role of an interface between the domain of the author and the domain of the reader. Thus it is the part of the model that is most similar to the classical notion of work. A work can no longer be regarded as an object, but as a complex set of components of different natures (object for the *texte-auteur*, state for the *texte-à-voir* and process for the generation) that can only exist while the program is running.

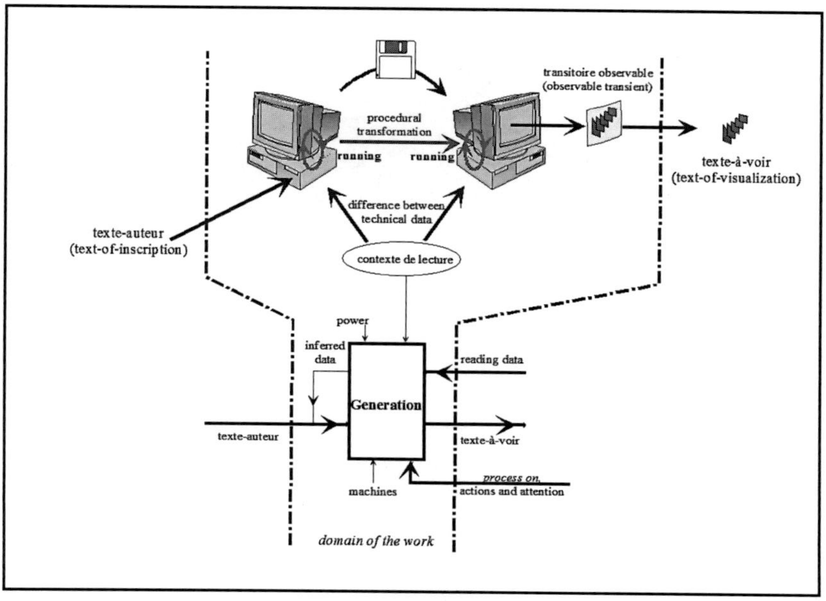

Fig. 4. Structural and functional schemes of the domain of the work.

This set of signs must be completed by two mind representations that are the classical meanings of the work and the totality of the situation of communication. These two mind representations are the *texte-écrit* (text-as-written) and the *texte-lu* (text-as-read). The *texte-écrit* is used by the author to construct the *texte-auteur* and, notably, to elaborate strategies of temporal, space, and media behaviors on screen, to delimitate what he will consider as interface and what he will consider as *texte-à-voir* in the transient event, to manage interactivity.

As he is inside the system of communication, the reader can give meaning to his own ergodic activity while he is interacting. This special meaning is called "double reading." Generally, it arrives during reading to reorient the construction of the work's meaning in another way. Double reading uses the activity of reading as a sign intended for the reader. In double reading, interactivity acts as a reflexive sign: The reader interacts with himself. Double reading has been used since the beginning of the 1990s.

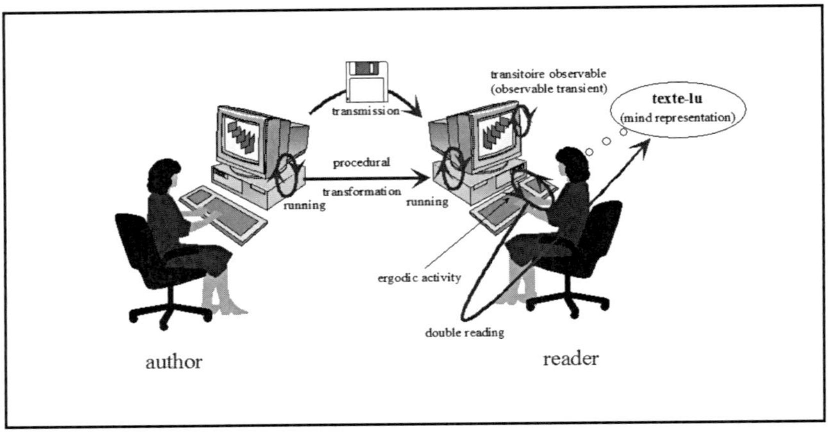

Fig. 5. Double reading.

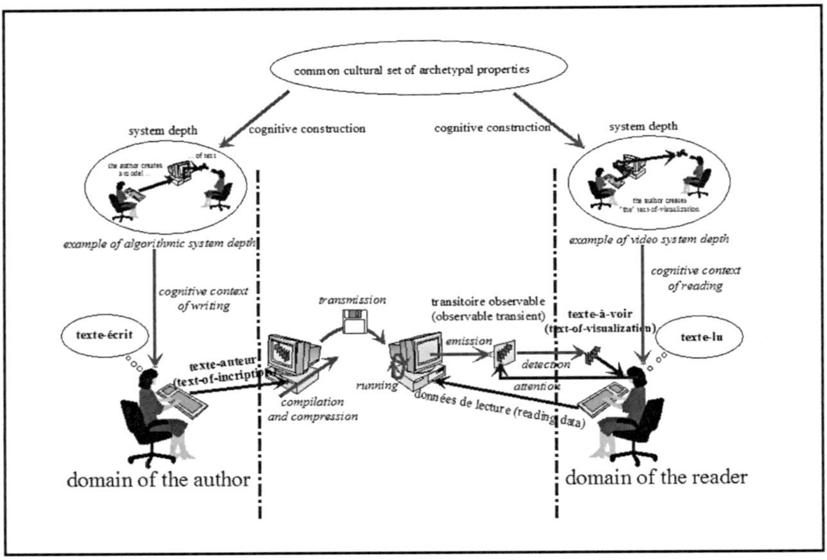

Fig. 6. Complete description of the situation of communication in an example where the author has an algorithmic *profondeur de dispositif* and the reader a video *profondeur de dispositif*.

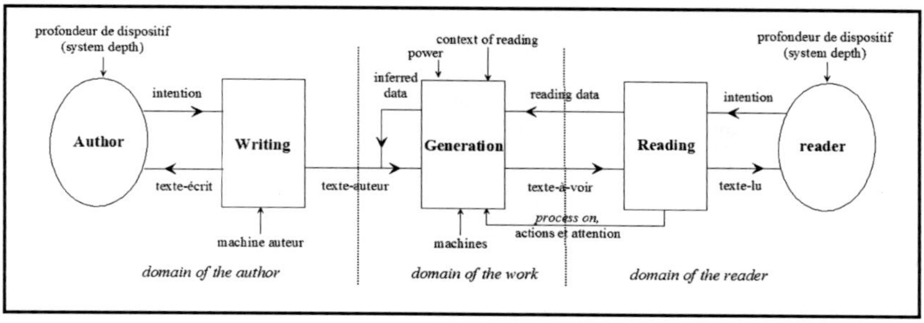

Fig. 7. Functional scheme of the situation of communication.

2.3. The Meta-Level

The theory of double reading has shown that reading can act as a sign. When it does, for example in the aesthetics of frustration, this sign is a part of the situation of communication and not of the domain of the work. Thus the totality of the situation can become a sign, a meta-work. It is an iconic sign of a conceptual reality of the work that does not reach the surface rhetorical level. The referent of this sign is usually a situation or a process in life. The meta-work is iconic in the situation it creates or in its functioning. This meta-work is intended for a person watching the reader while he is reading. This person is not reading himself, but watching somebody else reading. He is a meta-reader. Reading and meta-reading do not apply to the work by the same canal. They are two different functions applied to the work, even if they can be successively applied by the same person. The meta-reader can understand the reactions and actions of the reader in a certain manner that differs from the reader's double reading. But he can only do this if he knows the meaning the author gave to the reading activity. This meaning is expressed in papers or other paratexts.[3] It is not expressed at all in the *texte-auteur*. Some features in the program or other data of the *texte-auteur* can only play the role of an index for this meta-level of the author's intentionality. Meta-reading is not an interpretation of the situation but a recognition of it. The meta-reader must also perceive the specific influence of the program non-said while operating. Thus meta-reading can been seen as "a most intellectual understanding" of what happens in the situation of communication whereas reading can be regarded as "a most affective understanding" of what happens in the work. For double reading, the activity of reading is a significant level of the communication between the author and the reader. For meta-reading, the activity of reading is an expressive level of the communication between the author and the meta-reader.[4]

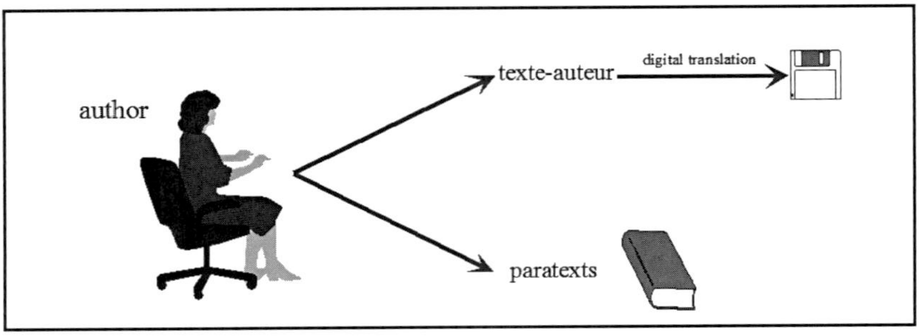

Fig. 8. Total activity of the author.

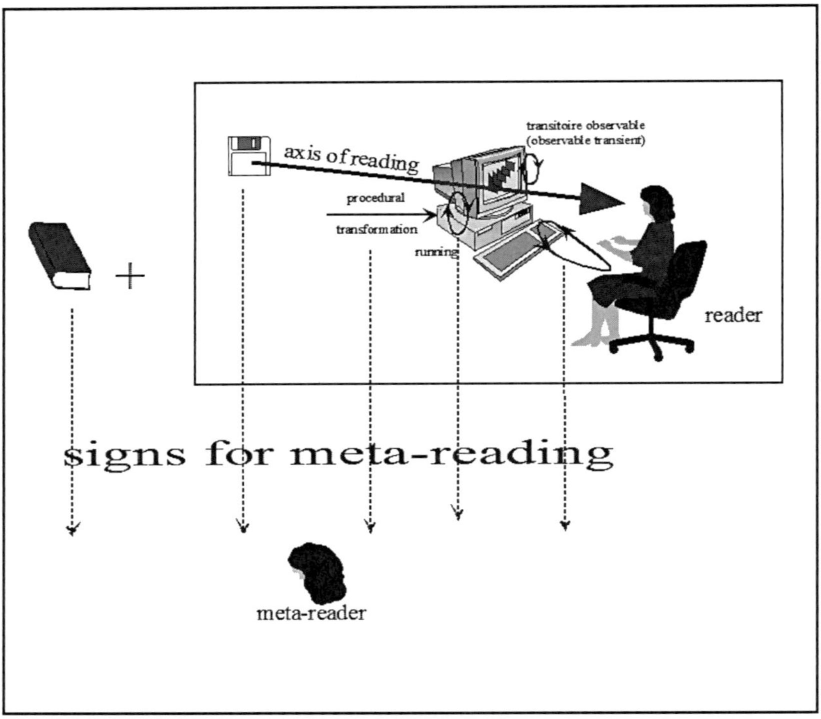

Fig. 9. Difference between reading and meta-reading.

The aesthetics of frustration gives a specific role to reading and considers that the reader is not the only person addressed by the work. The reader plays a specific, unexplained role in the reading process because reading is the scene of this role. Expressive reactions, or actions, are signs. They are in an iconic relation to real life activities. This aesthetics considers that the traditional failure of reading or rereading is a specific modality of reading; it has a specific meaning at the scene of activity. Thus the work can be very frustrating for the reader but, from the point of view of meta-reading, a failure of reading cannot exist. Even a failure produces meaning but this meaning doesn't pass through a reader's understanding of the work when reading fails.

When using meta-reading, e-poetry is fundamentally concerned with communication by natural language and by the real nature of reading and writing in a society of communication. Many works argue that the cultural basis of societies of information and communication is a myth. These works are based on miscommunication. They are using the most mythic device of communication, the computer, because it is fundamentally a machine of language, a machine of programmed language, whatever the nature of signs on the surface of the screen may be.

3. The Question of Form in Programmed E-Poetry

3.1. Different Levels of Constraint

The question of form in this system is relative to the role that the author gives to each part of the domain of the work. It is fundamentally relative to the *profondeur de dispositif* of the author. It depends only on the author, not on the reader. I define a form as a creation and a management of an arbitrary set of constraints. It is an efficient point of view because we can find different levels of constraints by taking into account the relationship between the different components of the domain of the work.

3.2. Simulation and Intersemiotics

A first set of constraints is only relative to the surface rhetorical level. There are two possibilities nowadays: The *transitoire observable* simulates a traditional form (from traditional or visual poetry) or it mixes different semiotic behaviors in an interpoetic (or intersemiotic) physical behavior. Intersemiotic works can confront different kinds of observable signs (for example in video poetry) or with time can be a transformation of one sign into another from the same semiotics (a manner explored by Jacopo Baboni Schilingi in music).

When the author only uses these kinds of constraints, the computer is used as a tool, as an amplification of complexity, and the only question of form that remains to be solved by the author is the transformation of the *transitoire observable* to a *texte-à-voir*. The other components of the domain of the work do not play any role at all. These kinds of systems enter in the class of "computer-assisted literature" and not in the class of *Transitoire Observable*'s "programmed literature." In programmed literature, the aesthetic problem of this pure observable level is the same as in computer-assisted literature; it concerns the rhetorical level, but the general problem of form does not restrict itself to this level.

For instance, temporal programming uses the transformation of signs of the same semiotic system. In this approach, the beginning and the end of the transformation can be recognized as a certain sign (or set of signs) by the reader, but the transient part is only interpreted by a cognitive interpolation with the beginning or the end. It seems that this part uses a specific temporal semiotics that preserves the global coherence of reading. This temporal semiotics is not interpreted in the same semiotic area. Thus, this transformation has special properties for the reader: It occurs as a continuous physical process but also as a semiotic discontinuity, a step between two non-simultaneous interpretations, and, while the physical process is linear and limited, the cognitive interpretation is looping between the two interpretations: They finally coexist in the mind. This constraint can be described by two rules: Beginning and end of the sequence must be recognizable signs in a certain semiotics and the transformation time must be a physically continuous temporal event. This event, probably, is a unique UST.[5]

3.3. Using Meta-Stylistic Rules

The communicative and the meta-levels present types of constraints that are more specific to programmed art because they use real time running while reading. The aesthetics of frustration or the strategies of managing the reading process I have presented above concern one type of constraint: How to manage the interactivity. In contrast to hypertext philosophy, we see that the reader's intentionality is not the focus of the work. In *Transitoire Observable*, the old adage introduced in *alire* in the 1980s, always applies: Reading forbids to read.

Another set of constraints is based on the relationship between the *texte-auteur* and the *texte-à-voir* and concerns the generative level of the work. In this relation, the *texte-auteur* is a model of the *texte-à-voir*. Jean-Pierre Balpe discussed this point in his theory of the meta-author.[6] Historically, the first kind of programmed texts using this conception was automatic generation. But adaptive animation made by L.A.I.R.E. also abides by this conception, even if the observable behavior is very different. The difference is due to the nature of the algorithms that are used. Automatic generation uses algorithms of creation of linguistic material although programmed animation uses algorithms to manage the expressive aesthetic behavior of the observable material. But this difference only concerns the transformation of the *texte-auteur* into the *texte-à-voir*; it neither concerns programming itself nor the conception of model. In each case, the author has to write two different kinds of rules: Rules of the semiotic level of the *texte-à-voir*[7] and rules of the management of these rules. The reader cannot see this second set of lines of code. Adaptive[8] generation shows that today it becomes a model of models. Thus these works manifest two aesthetic levels. The first is classically showing the reader the *transitoire observable* through observable media, processes, and behaviors. The second is only indexed in the *transitoire observable.* To perceive it, it is necessary to re-run the work, sometimes on another computer.[9] This second level is the manifestation of meta-stylistic rules. Because they do not directly appear to the reader, we can be sure that—through the operation of reading—the reader no longer is the person addressed in the work. In fact, the person really addressed in the work is a social and cultural meta-reader. A given person, as he is an instance of this cultural meta-reader, can play the two roles but they cannot be assumed at the same time: Meta-reading does not occur through the reading of the work. There exists a real rhetorical unity between interactive works that use the aesthetics of frustration and those that are not interactive but use meta-stylistic rules.

Generally, the meta-stylistic rules take the form of parameters, or a logical set of rules for dialogue between parts of the program, or special non-visible behaviors like measures. These meta-stylistic rules are purely arbitrary rules of creation. Technical features do not justify them. Thus, they are really artistic constraints. When meta-stylistic rules have been chosen, the problem to manage them consists of constructing a suitable tool to describe (on the conception level) and to program them. For complex projects, the conception level of the constraint is the most difficult to describe. Authors in *Transitoire Observable* use different models of conception: Jean-Pierre Balpe uses the system of generation he has created; Tibor Papp uses a grammar in which paradigms are individual behaviors and syntax as a set of rules of assembling; Alexandre Gherban also uses a set of behaviors (temporal behaviors), but these behaviors are chosen and used as parameters by using cellular automata; Antoine Schmitt uses physical mechanical laws and a predictive calculus on temporal specific objectives (aesthetic of causality); Xavier Leton and I use measures and rules of adaptation, I use generative editing (the horizontal editing) to construct observable behaviors with elementary components; Bluescreen uses arbitrary laws of interaction between programmed objects to construct a collective behavior. . . .

To cite examples from *alire12*, the visible processes can be managed by algorithms of artificial life (Gherban, Bluescreen), by physical models (Schmitt), or by temporal programming (Bootz). But, in each case, *transitoire observable* is not understood as a simulation. If life exists, it does not occur in the result but in the principle of the program's functioning. It is not biological life that would obey the laws of duplication and competition, but an aesthetic life in which these algorithmic rules are constrained by meta-aesthetic rules that do not directly appear in the transient observable event. Aesthetic forms on the surface of the screen are instances of forms produced by these meta-aesthetic rules. To do this, the program can measure while running (Bootz). The program uses the results of these measures in order to modify its own strategy. In this way it is an adaptive generator. For example, the result of a calculation by an algorithm can be changed by the result of a measure to obey a certain aesthetic idea, or the measure can forbid certain parts of the program to function. More generally, the program is constructed as a set of independent rules that have a relationship while running in order to construct an observable transient state constrained by aesthetic meta-rules. This is the case, even if the program does not use a specific algorithmic model (Papp). The author does not exactly know what will happen, but the author knows the classes of the main aesthetic features that will be made manifest in the observable multimedia transient event when the technical unsaid does not trouble the programmed process.

Conclusion

Thus, even if each author explores different modalities of meta-rules, with sometimes very different objectives, I think that the existence of these meta-rules is sufficient to speak of a common aesthetics. The common direction for the author is to search for the best form of conception and realization of the class of meta-rules he explores. This unity is not manifested on the screen, but in the similitude of the management of the reader's and the author's role: It is not an aesthetics of text; it is an aesthetics of creating.

Notes

1. This article was written with the linguistic assistance of Loss Pequeño Glazier.
2. The *texte-auteur* is not the digital form that is stored in the machine but the form it takes in which a human being can understand this *texte-auteur* as signs. It is, for example, the source program written in a high-level language of programming and not the binary code in which it appears for the machine after compilation. In the case of visual data the *texte-à-voir* is the picture that the eye of a human being can see and not the file. The *texte-auteur* must be translated into digital files because the machine cannot use it as is.

3. The role of paratext in the interpretation of an electronic artistic work was studied using several examples in the paper "Approche du rôle de la mémoire dans la conception et la réception de quelques œuvres en littérature numérique."
4. Cf. the paper "Approche sémiotique d'un certain art programmé: œuvre signe et méta-lecture."
5. A "Temporal Semiotic Unit." The theory of UST was created in the laboratory of computer music of Marseilles (MIM) in order to analyze music cf. <http://www.labo-mim.org>. Its extrapolation to visual and multimedia art is being explored today.
6. Cf. Balpe, "Un roman inachevé—dispositifs" and "Méta-auteur."
7. In this case, it is the *texte-à-voir* for the author, not for the reader. The author manages what happens on the screen of his computer, the semiotic behavior of what he considers on this screen as the *texte-à-voir.*
8. An adaptive generator tries to manage the technical left unsaid. It changes the logical set of rules used to construct the observable transient by interpreting physical measures made by the program during its operating time.
9. Adaptive generation is totally transparent on a computer because the measures it makes always produces the same results. Thus, the change in the logic of operation is the same from one re-reading to another.

Works Cited

Balpe, Jean-Pierre. "Un roman inachevé—dispositifs." *Littérature* 96 (1994): 37-53.

———. "Méta-auteur." *alire 10* (1997): 95-99.

Bootz, Philippe. "Approche du rôle de la mémoire dans la conception et la réception de quelques œuvres en littérature numérique." 2003. *@rchivesic.* 6 June 2006 <http://archivesic.ccsd.cnrs.fr/sic_00078598>.

———. "Approche sémiotique d'un certain art programmé: œuvre signe et méta-lecture." 2004. *@rchivesic.* 6 June 2006 <http://archivesic.ccsd.cnrs.fr/sic_00078604>.

Typologies, Histories, and Case Studies

Mela Kocher

The Ludoliterary Circle

Analysis and Typology of Digital Games

Abstract:
How has interactive storytelling been developed in the past 30 years? Where do problems and chances in game narration lie? The ludoliterary circle is a typology that provides tools for analyzing different ludic forms of storytelling with regards to these questions. It focuses on the perspective and the interactive engagement of the user and the narrative mode of the "text." While showing how these parameters correlate with each other, it will also be demonstrated which and why certain combinations are more successful than others.

Intro

Whichever innovative forms of digital storytelling we are demanding and hoping for in the field of computer games, it is observable that most games follow certain worn-out patterns even though they might have been changing over the last decades and different tendencies can be distinguished. Traditional narrative game genres like adventure games melt more and more into other genres (like *Fahrenheit*, a mix of adventure game and interactive movie, the *Max Payne* series or *Hello Kitty: Roller Rescue*, unusual narrative action games), following the general genre-mix-landscape nowadays. New storytelling forms are also emerging: less prestructured and static but more dynamic and user-input-friendly. This article aims to present a model that captures digital narration patterns, takes a snapshot at past and current digital storytelling and provides a tool to analyze and describe—understanding that it is a formal model, presenting abstract types, which can be discussed with cybertext genres such as computer games, hyperfiction, interactive movies and, hypothetically, VR environments.[1]

Talking about digital storytelling one has to admit that there obviously are significant limits of traditional narrative theories when encountering digital media. Especially the concepts of "interactivity" and "perspective of the user" can only be partly captured by narratology, which often focuses on the narrator—who does not play an important part in games—and handles static texts. These media-specific conditions are taken into account by the model to be presented here, which focuses on the parameters perspective, interactivity and narrative mode. Taking these parameters as a common base for cybertext analysis we can discern differences and similarities between different types of cybertexts that are linked insofar as they do in some way tell digital stories[2]—sharing family resemblances sensu Wittgenstein. Last but not least the model—including both narratological and digital media tools—tries to build a bridge between the narratological and the ludological approach towards computer games in game studies, a relatively new scientific field that indulges in heavy ideological debates on the legitimacy of the theories to be applied.[3] In the following paragraphs I will first explain the features of the ludoliterary[4] circle and then illustrate them with specific examples.

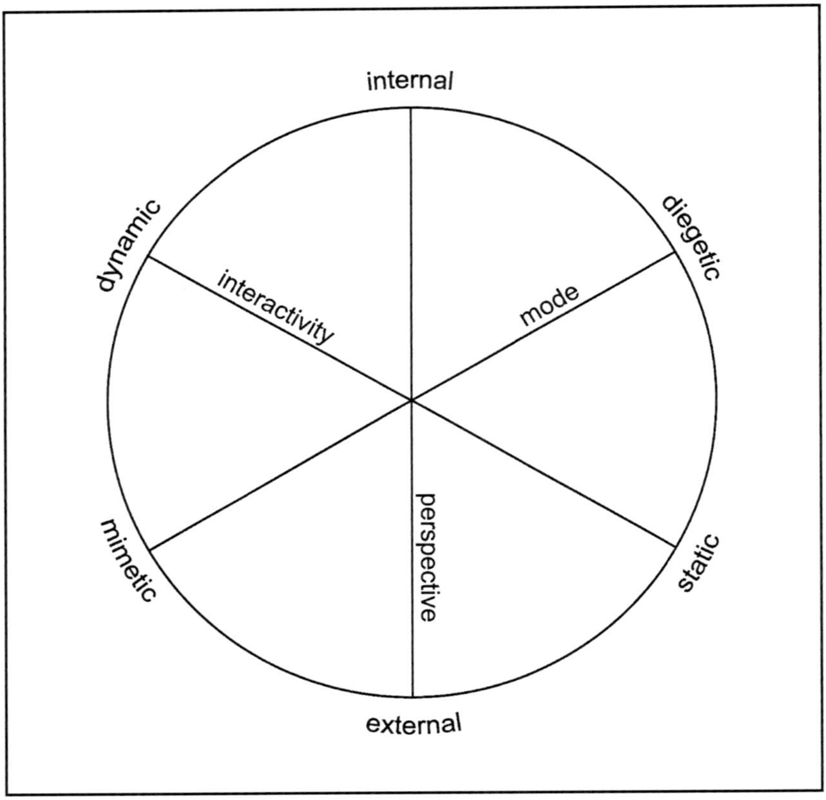

Fig. 1. Ludoliterary circle.

The Ludoliterary Circle

The ludoliterary model represents a circle that is equally cut by three axes meeting in the circular centre. Each axe represents a continuum of values, whereby the two opposite axial poles each mark one extreme. The binary code of perspective would therefore be internal/external, of narrative mode diegetic/mimetic, of interactivity dynamic/static. Positioning a game genre or a certain game on the circle line means that it can be characterized and discussed by means of the three axial poles that are closest on the circle line. Crucial is therefore the circle line, especially the spots where the line is crossed by the three axes. For example, a game positioned around the diegetic pole could be discussed by diegetic, static and internal parameters. On the other hand, a game lying on the opposite side of the narrative mode axe is referred to as mimetic, external and dynamic.

The idea of the model is to catch transitions in game features that cannot be captured by a—rather static—matrix. By allowing the positioning of cybertexts on any given spot on the circle line one can break binary classifications.[5] Therefore it is possible, for example, to characterize two cybertexts that are both static and diegetic but vary in their way to include the user's perspective. Besides showing and discussing different constellations it will be possible to see which constellations are more successful or intriguing than others and why, and also taking into account the history of digital storytelling.

The three parameters perspective, interactivity and narrative modes can be further differentiated. An internal perspective is given when the user has a "limited point of view" on the events and characters of the fictional world, when she shares the knowledge of, for example, the avatar of a third person adventure game.[6] Furthermore, the role the user slips into is a function inside the ontological time and space of the fiction (Genette's homodiegetic sphere), which can result in different ways of identification and empathy. Visually the internal perspective is usually represented by a first, second or third person perspective. Contrarily, taking an external perspective means the player is omniscient and also located outside of the time and space of the fictional world (Genette's heterodiegetic sphere). This is usually represented in games as the more distant bird's eye or the isometric view.

Where the interactive involvement of the user is concerned, the model focuses on media-specific extra-noematic—with Aarseth: ergodic—interactivity, leaving aside the noematic, "psychic" interactive process that takes place between any text and a user as it is described by reader-response theory (1). Borrowing Brenda Laurel's multidimensional categories, which she sees applied in all human computer interaction, it is possible to discern different shades of interactive involvement. Laurel considers the frequency, the range and the significance of the user's actions; frequency as "a measure of how often user input is enabled," range as "the range of choices available to users at a given moment in the interaction" and significance as "a measure of the impact of the users' choices and actions upon the whole" (78). At the dynamic pole of the model we set cybertexts that show high degrees of those three criteria and vice versa at the static pole.

Different ways of digital storytelling can be distinguished by the criteria of the narrative mode. The two extreme poles represent the classic categories of the diegetic and the mimetic—deriving from Aristotelian drama theory as much as from Goethe's and Schiller's interpretations. A game that is placed close to the diegetic pole shows therefore some or all of the following narrative features: The plot typically consists of a fixed set of events which can partly vary in their order.[7] A narrator is central to the way of presenting the story. The fictional world is experienced in an episodic way, meaning that events are linked thematically rather than causally. The epic extension of space is typical for many diegetic games where the user strolls through countries and islands in search for clues. Time is loosely fixed; playtime—in narrative terms: "Erzählzeit" ('narrating time')—and event time—"erzählte Zeit" ('narrated time')—are not related to each other. The time related focus can lie heavily on the past. All in all the narrative game experience can be captured with Laurel's expression of "extensification" (*Computers as Theatre* 94) where emphasis is put not so much in building up suspense but rather on a meditative experience.

Mimetic cybertexts in contrast ideally produce different events every time the game is started anew.[8] The narrator is much less present than in diegetic games, and story seems more to be enacted or performed than prestructured. Events are linked more causally than in diegetic games and can be described with Artistotle's term of unity of action. This is the same with space and time: Unity of space suggests that events take place in a limited area. Time is also condensed and play and event time are closely related with emphasis of things happening right now, in the present. All in all the term "intensification" (*Computers as Theatre* 94) can be used to describe this type of game experience.

Cybertext Genres

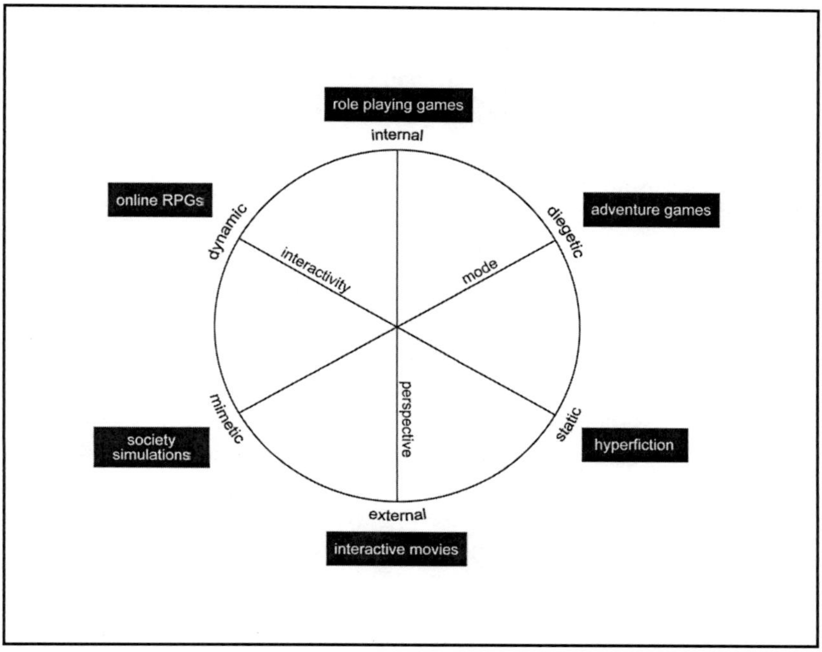

Fig. 2. Genre typology.

Originally constructed to describe computer game narration, the constellation of the ludolitary circle also allows the discussion of other cybertext genres like hyperfictions and interactive movies. In the following "quests" I will shortly play each genre respectively representative cybertexts with the corresponding parameters.

Quest 1: Adventure Games

Adventure games like *Monkey Island*, *Myst* or *Runaway: A Road Adventure* are the typical tokens when speaking about ludic storytelling, because they most closely represent the classic storytelling genre relatively similar to print literature. Intriguing the player by an internal perspective, she plays first or third person by taking up either stereotyped roles or figures that show some ironic and self-reflecting features. The knowledge of the player is equal to the knowledge of the avatar; according to that "limited point of view"-concept the main task of the game play is solving riddles, uncovering murders, saving princesses etc. The fictional world is often narrated by a narrator through diaries, letters, cutscenes etc. that are dispersed all over the game worlds. The player travels on her mission through many places and visits them repeatedly, whereas not only space, but also time seem to be extensive, since narration time stops when the player does not move on in her quest. Often the events do not seem to be intrinsically related to each other and make sense, if at all, only retrospectively. The plot's course—and its few endings—are determined; its nodes or riddles may only vary in their order. The interactive offer to the user corresponds well to this diegetic constellation: For the coherence of the story to be satisfactory, the player is not allowed to be too active—one of the crucial problems of interactive fiction. Adventures games are therefore very static: The player does not interact very often (disregarding the minimal navigational clicking), and she is not offered a wide range of options, either. It seems though, at first hand and especially when looking at older games like *Monkey Island* that are partly text based that there is a large set of objects to be combined with many commands. Mostly just a handful of combinations remain possible even though usually only one is successful, which makes that range only seemingly wide. The player's actions are not significant for the course of the story insofar as it could be influenced or changed. This gaming concept of internal perspective, diegetic mode and static interactivity works best when narration is of aesthetic quality and the characters are of innovative design, since the restrictions of problem solving and interactive influence can often be quite annoying. LucasArts' games like *Monkey Island*, *Grim Fandango* etc. are the ever-loved classic adventure games with great stories and hilarious characters, but also newer games like *Runaway* and *Fahrenheit* or genre mixes like the narrative action game *Max Payne* are satisfying from a story point of view.

Quest 2: Hyperfiction

A similar profile concerning interactivity and narrative mode is shown by many hyperfictions. The frequency of interaction with the interface is limited to the minimum and consists of clicking links—again the minimum needed for navigation. There is not really a wide range of options to choose from: Usually you find one to three possible links on one screen; some lead you further in the story, some loop you back. Since the text chunks are already programmed there is no significant impact of the user's action on the whole since her part consists of choosing the order in which they appear on the screen—a classical feature of the diegetic mode. As in the narrative mode of adventure games, there is also a strong narrator. Jumping from link to link the user is often taken to various places in different time zones; considering that a lot of combinations of those text chunks make little sense the experience is quite an extensive one regarding time, place and action. Other than in first or third person adventure games, narration is often carried out through a variety of characters. Changing perspectives and hopping from character to character puts the user into an external perspective, which makes it hard to identify with one character. On the one hand this perspective can be quite irritating since the position of the user is often reflected and she gets addressed as an outsider—which is the case in most of Susanne Berkenheger's work.[9] On the other hand the effect of clicking links, choosing the next text chunks and presumably the course of the story suggest a feeling of power and omniscience since the user seems to mutate to some sort of co-author.[10] The play with the author-concept is also part of the deconstructing and reflecting on literary processes—a great aesthetic potential of hyperfiction, a feature which sometimes wins over the lack of user activity and narrative coherence, and a potential that interactive movies do not quite reach.

Quest 3: Interactive Movies

Interactive movies[11] like *I'm Your Man* or *Swiss Love* are rather mimetic than diegetic and often show a clear dramatic structure of exposition, complication, climax, resolution and epilogue.[12] The dimension of time depends not anymore solely on the actions of the user, but rather on the text which controls play and event time, with some exceptions: A movie sequence runs in a linear way for minutes and stops only for the user to make her next choice which consists of clicking one link, then another movie sequence runs again. Since interactive movies play with the idea of different people running into each other (so the user can choose to follow one, then another), the unity of place is often given since characters meet at central spots. Also, typically mimetic, a narrator is not present and events are enacted and not told. For the course of the action this genre behaves in a more diegetic way, since the text consists of a certain set of frames that alter only in their order. This is consistent with the position on the ludoliterary model (which puts interactive movies closer to the mimetic pole but also in reach of the diegetic) and makes sense regarding the static interactivity: The user may choose from convictions of the characters or determine where the action will lead. For an interactive text it seems very little, since the interactive engagement of the user is reduced to a few clicks (neither worth mentioning considering frequency, range or significance). For the classical movie watching the interruptions are definitely too much since the player repeatedly falls out of immersion. The setting of the perspective is depicted accordingly: Changing often between the different characters makes it very difficult to immerse in one character, so the user ends up taking an external perspective, an objective distance outside of fictional time and space. This becomes very clear in the following scene of *I'm Your Man*, an interactive movie which plays with the concept of omniscience to a high degree: After he had been attacked with a pepper spray by a woman he was chasing, the villain turns to the camera and addresses the user. Since she was able to observe the action—"you haven't been maced"[13]—she is supposed to help him. "That puts us in position of mutual need. I need you to tell me where they went. You need me not to flip the switch, which will lock the doors to the theatre and begin a 24-hour marathon of menudo music. So tell me, where did they go, back to the party or up on the roof?" The metafictional subtext and the experience of different facets of a story from the perspective of various characters sound like exciting ideas, but this kind of constellation of external perspective, mimetic narration and static interactivity fails mostly because of similar reasons like in hyperfiction. The narrative coherence is only given when following one character but lacks much when using the variety of links. Also, while hyperfiction allow you to wander around and discover the story from various perspectives at your own pace, interactive movies are both too hectic and incoherent to be more than an experiment.

Quest 4: Society Simulations

Much luckier proves the combination of mimetic narration, external perspective and dynamic interactivity that can be observed in society simulations like *The Sims* series. The player focuses alternately on the different avatars, the members of a family, and coordinates their everyday life. This is external perspective at its most: The player is truly omniscient. She knows about the aspirations, wants, fears, needs and even memories of the Sims. As a goddess she is able to influence their way of life, their choice of jobs, husbands and wives, hobbies, skills, social interactions and more. The degree of interactive involvement is equally high concerning frequency, range and significance. Since the Sims constantly interact with their environment and produce needs and hopes to be fulfilled and fears to be prevented, the player is forced to act very frequently. Given any object or person to be manipulated there is always a variety of context sensitive options to choose from. The Sims you choose can dance, quarrel, flirt and make love with or talk, hug, tickle, entertain etc. another Sims, depending on how their relationship is developing at the moment. A fridge you can fill or empty or sell or move to another place, etc. There are no rights or wrongs in choosing from those options but just actions and their consequences. Therefore the player's decisions do greatly matter to the course of the action, the action itself and the behavior of the Sims.[14] Every time the game is played again it most likely becomes a different life scenario—unless you make exactly the same decisions again which is almost improbable regarding the pace that game moves in. The dynamics of interactivity correlate really well with the mimetic mode of the story. It is not being narrated[15] but acted out by the Sims. The stage is the house and the garden—and a few more places to go to like a pool or a shopping mall, but they are secondary—which forms a unity of place. Unity of time is also given since playtime and event time are strictly related to each other: One minute of Sims-time corresponds to one second of real time. Unless being manipulated by the user[16], system time just moves on in a steady pace. The events are coherent, action is followed by reaction, the processes are—due to capitalistic ideology—very logic: the better job, the more money, the more friends, the happier housewife or husband and greater life. And vice versa. An arrangement so flawless it is an ideal environment for those many subversive and creative attempts to let the Sims burn, die, rot, choke and just get as poor as can be. All in all, this constellation of perspective, interactivity and narrative mode appears to be very immersive and also very popular.

Quest 5: Online Role-Playing Games

The ludoliterary model also suggests a cybertext genre that is structured by means of mimetic mode, internal perspective and dynamic interactivity. This pattern is found in utopian cyberspace scenarios of the 1990s, in Star Trek's "Holodeck"[17] or in movies like Cronenberg's *eXistenZ*, where players get immersed in a digital world and caught up in intriguing events that are evolving ad hoc. The players in such "cyberdrama" (Murray 271) strongly identify with their role and influence the action to a great extent—the ultimate game experience with the utmost ludic immersion. Current developments in the online role-playing game (RPG) section—like the massively multiplayer RPG *World of Warcraft*—offer the user an immensely broad spectrum of actions to choose from while interacting with other characters and engaging in quests. Similarities to mimetic performative game forms like *The Sims2* are obvious. Though in online RPGs, other than in the external *The Sims2*, you don't switch between different characters during one game session since the main task lies in building up your avatar through an elaborate level and honor system. Therefore, internal perspective is maintained, which is, in combination with the large interactive and communicative potential, the main reason of the great success of the game.

Quest 6: Role-Playing Games

The cybertext genre that can be described and analyzed with the parameters internal perspective, dynamic interactivity and diegetic narration is that of the offline role-playing game (RPG) like *Final Fantasy X-2* or *Baldurs Gate II*. Taking an internal perspective, the player chooses, often creates an avatar with specific properties (gender, race, nationality, profession, looks, skills etc.). By this determination of the role, but also by the many dangers that awaits her avatar (and also by communicating with others in online RPGs), different kinds of identifying processes take place that are stronger than in the genre of adventure games which often present stereotyped and ready-made characters. The narrative mode consists both of diegetic and mimetic aspects: Mimetic is the enacting feature which is very central to online RPGs, which not seldom have different quests to be solved for every player and do not follow a specific set of events. The action in offline RPGs is mainly diegetic. Narrating and structuring elements like letters, diaries or cutscenes, that lead the player on her mission, are frequent. Dimensions of time and space are epic like in adventure games, where the traveling through and discovery of the fictional world take up a lot of playtime. The events of offline RPGs are usually set with a certain beginning and certain endings with variable middle parts, therefore not changing every time the game is played anew (although small parts, like the contents of chests and hiding places for keys can vary). Other than in adventure games the interactive potential is far bigger, even though the player's actions do not have much impact on the whole (since that is prestructured). Not only is the frequency of interactions much greater, considering the many fighting scenes one has to survive, but also the range of options is wider: Each avatar has different weapons or spells and other objects that can be chosen from. This correlates with the nonlinear structure of the story: While adventure games demand a much more correct order of solving riddles, RPGs are more tolerant and sometimes offer subquests that can be indulged in whenever wanted. Role-playing games sometimes have different endings, which at least suggests a bigger influence of the user on the events. The influence is not by far as big as in society simulations like *The Sims*, whose mimetic and more flexible narration is more suitable for dynamic interactivity.

Extro

The model presented here reflects historical processes as it shows different digital storytelling modes from the 1970s up to now. Having on the one side rather traditional forms dispersed around the diegetic pole, like adventure games, role-playing games and hyperfiction, it also captures new evolving aspects in digital narration. Those more mimetic forms are being observed in society simulations like *The Sims* and in online role-playing games like *World of Warcraft*—game forms that are drawing in more people than ever and are also very attractive for female players who search for different and innovative gaming concepts.

As a closed and two-dimensional model the ludoliterary circle clearly shows restrictions. The setting of the axes proposes that game narration is heavily determined by perspective, interactivity and narrative mode, although other features might be equally intriguing, like motives, more semantic values, iconic language and more. Of course there will also be narrative forms that cannot be described by the typology, which, as a closed circle, of necessity suggests some sort of coherence and continuity. It might be a further step to quantify and calculate the values along the axes to allow the whole circular space to be used in positioning game genres and specific cybertext tokens. For the moment, though, this circle aims at capturing and analyzing tendencies and self-imposed rules that digital storytelling is currently following—hoping that patterns will be broken and more experimental game play will surprise us.

Notes

1. For a more extended version on that subject cf. my German article "Der ludoliterarische Typenkreis." Also cf. Kocher, *Follow the Pixel Rabbit!*
2. For a computer game to be talked about as narrative, I consider at least following two of Celia Pearce's "narrative operators" to be decisive: meta-story and experiential. Meta-Story: "A specific narrative 'overlay' that creates a context or framework for the game conflict." Experiential: "The emergent narrative that develops out of the inherent 'conflict' of the game as it is played, as experienced by the players themselves." (145).
3. Cf. the "duel" between Julian Kücklich and Markku Eskelinen in their roles as "narratologist" vs. "ludologist" (Eskelinen and Kücklich).
4. Ludoliterary: a neologism of ludus—Latin for game—and literary.
5. This fact seems appealing, considering other circular models like Johann Wolfgang Goethe's circle, which is a typology of literary genres, Franz K. Stanzel's circle that captures narrative situations in novels, the "Spider" of

Hermann and Leuthold, a sociological model to evaluate political opinions and voting (Hermann and Leuthold; Fivaz, Schwarz, Hermann and Leuthold).

6. These criteria correspond with Stanzel's categories of perspective and person. It is legitimate to discuss them both together for texts where there is no narrator, nobody "who speaks" (Genette). Therefore the fictional role of the user is of bigger interest than of the narrator. Cf. also Marie-Laure Ryan's article presenting a matrix working with the criteria.
7. In Aarseth's terms: The textons do not change, but the scriptons do (62).
8. Therefore both scriptons and textons might be changed.
9. Cf. *Zeit für die Bombe*, *The Bubble Bath, Hilfe: Ein Hypertext aus vier Kehlen.*
10. This is very nicely illustrated in *Die Aaleskorte der Ölig* by Frank Klötgen and Dirk Günther. At the beginning you can choose on a chart which text/image-sequences you would like to see. Klötgen and Günther's artwork lies somewhere between the "hyperfiction" and the "interactive movies" poles of the model and shows a mixture of diegetic and mimetic narration: unity of action, seemingly different outcomes every run (1.6 millions, say the authors) but a strong narrator and structure and different play and event times.
11. The term "interactive movies" is a little misleading since the degree of interactivity is quite low in comparison to other digital texts. As a commonly accepted technical term it nevertheless makes sense to use it.
12. Cf. the chart depicting those stages according to the movie scenes in Ryan, *Narrative as Virtual Reality* 273.
13. A nice paronomasic pun: "Mace" is not only the brand of the pepper spray, but shows a phonetic resemblance to "maze," the expression for confusion, labyrinth, a place to lose orientation.
14. Even scriptons can be changed, considering the fact that you can create and import new objects and faces for this game.
15. Except for some occasional comments that are something between explanations of events and advices for game play (like: "Imre and René are being sent to military school. They didn't do their homework"). The instance of those comments is rather an implicit author than a narrator.
16. Dynamic interactivity in *The Sims* allows the user to fast forward time or pause it, the latter to sneakily gain some time to stash commands.
17. Cf. Murrays notions on daringly futuristic scenarios in *Hamlet on the Holodeck*.

Works Cited

Aarseth, Espen. *Cybertext: Perspectives on Ergodic Literature.* Baltimore: Johns Hopkins UP, 1997. 1.

Berkenheger, Susanne. *Zeit für die Bombe.* 1997. 13 June 2006 <http://www.wargla.de/index1.htm>.

———. *The Bubble Bath.* 2005. 13 June 2006 <http://www.thebubblebath.de>.

———. *Hilfe: Ein Hypertext aus vier Kehlen.* 1999. 13 June 2006 <http://www.wargla.de/hilfe>.

Eskelinen, Markku, and Julian Kücklich. "Ludology's Not Dead." 2004. 13 June 2006 <http://blog.cyberfiction.ch/C1855973910/E365298011/index.html>.

eXistenZ. Dir. David Cronenberg. Perf. Jennifer Jason Leigh, Jude Law, Ian Holm, and Willem Dafoe. Miramax, 1999.

Fivaz, Jan, Daniel Schwarz, Michael Hermann, and Heiri Leuthold. *Das neue Ratsinformationssystem (RIS): Perspektiven bezüglich Transparenz und Öffentlichkeit.* Berne, 2004.

Genette, Gérard. *Narrative Discourse: An Essay in Method.* Ithaca, NY: Cornell UP, 1980.

Hermann, Michael, and Heiri Leuthold. "Die 46. Legislatur im Profil." 13 June 2006 <http://sotomo.geo.unizh.ch/spider/>.

I'm Your Man. Dir. Bob Bejan. Perf. Mark Metcalf, A. Whitney Brown, Kevin M. Seal, and Colleen Quinn. ChoicePoint, 1998.

Klötgen, Frank, and Dirk Günther. *Die Aaleskorte der Ölig.* 1998. 13 June 2006 <http://www.internetkrimi.de/aaleskorte/>.

Kocher, Mela. "Der ludoliterarische Typenkreis: Analyse und Kategorisierung von Cybertexten." *Dichtung Digital* 7.1 (2005). 13 June 2006 <http://www.dichtung-digital.com/2005/1/Kocher/>.

Kocher, Mela. *Folge dem Pixelkaninchen! Ästhetik und Narrativität digitaler Spiele.* Zurich: Chronos, 2007.

Laurel, Brenda. "Interface as Mimesis." *User Centered System Design: New Perspectives on Human-Computer Interaction.* Ed. Donald Norman and Stephen Draper. Hillsdale, NJ: Lawrence Erlbaum, 1986. 67-85.

———. *Computers as Theatre.* Amsterdam: Addison-Wesley Publishing, 1991.

Murray, Janet H. *Hamlet on the Holodeck: The Future of Narrative in Cyberspace.* 3rd ed. Cambridge, MA: MIT P, 2000.

Pearce, Celia. "Towards a Game Theory of Game." *First Person: New Media as Story, Performance, and Game.* Ed. Noah Wardrip-Fruin and Pat Harrigan. Cambridge, MA: MIT P, 2004. 143-153.

Ryan, Marie-Laure. "Beyond Myth and Metaphor: The Case of Narrative in Digital Media." *Game Studies* 1.1 (2001). 13 June 2006 <http://www.gamestudies.org/0101/ryan/>.

———. *Narrative as Virtual Reality: Immersion and Interactivity in Literature and Electronic Media.* Baltimore: Johns Hopkins UP, 2001.

Swiss Love. Dir. Fulvio Bernasconi. Perf. Soraya Gomaa, Marino Campanaro, Noemi Steuer, and Hans-Joachim Frick. Maximage, 2002.

Wittgenstein, Ludwig. *Tractatus logico-philosophicus: Tagebücher 1914-1916. Philosophische Untersuchungen.* Frankfurt am Main: Suhrkamp, 1984.

Jörgen Schäfer

Gutenberg Galaxy Revis(it)ed

A Brief History of Combinatory, Hypertextual and Collaborative Literature from the Baroque Period to the Present[1]

Abstract:

Literature in computer-based media cannot be contemplated without a long literary tradition. This article aims at substantiating this assumption with numerous examples of combinatory, hypertextual and collaborative texts from German literary history since baroque times. Therewith it provides us with a historical basis in order to work out the common features and differences that with computers have entered literary texts.

Espen Aarseth and his followers have repeatedly pointed out and proved with many examples that their theories of "cybertext" and "ergodic literature" are focused on the mechanical organization of text and on the actions of users/readers on a broad and general level; they are not limited to texts in computer-based media. Although the computer, which strictly separates storage devices from interfaces may be the best-suited medium for "net literature," "ergodic literature," "digital literature"—regardless which term is to be preferred[2]—it may be rewarding to have a closer look at the long pre-history of "net literature" in non-computer-based media, especially in print media.

In modern literatures, it most notably was the international (though predominantly French) Oulipo group (the acronym is an abbreviation of *Ouvroir de la Littérature Potentielle,* in English: *Workshop of Potential Literature)* whose members tied in with the long tradition of combinatory literature. Raymond Queneau, the creator of the famous sonnet-machine *Cent mille millards de poèmes [One Hundred Million Million Poems]* (1961), explicitly regarded the constraints imposed by combinatory procedures as catalysts of writers' creativity. The Oulipians were dealing with combinatory procedures in a twofold way: On the one hand, they investigated and reinvigorated poetic possibilities from the past ("anoulipism"); they even ironically qualified their predecessors from the Ancient world and the Baroque era as "plagiarism by anticipation" (Motte 31). On the other hand, they aimed at developing additional possibilities which were unknown to their predecessors and which were to be based on latest scientific findings and technological novelties ("synoulipism").

As my focus in this article is on German literature, the paramount importance of Oulipo cannot adequately be considered here. But I will certainly come across many of their predecessors in German literature and thus keep up with Oulipo's analytic tendency. For this purpose, I will focus mainly on three tendencies for characterizing and classifying those many literary texts and procedures in which recursive processes can be identified.

First, current text generators can be traced back to previous forms of *combinatory literature*. In the German-speaking part of Europe numerous writers since the Baroque era were experimenting with literary forms that did not only consider a literary text a symbolic expression of a person's subjectivity but also considered a text as determined by the level of programming and processing of signs. On the one hand this is reflected in the tradition of word games such as anagrams, palindromes or proteus verses, on the other hand it is presented in mechanical text-generating machines.

Secondly, *hyperfictions,* too, are not necessarily dependent on computers: If the basic idea of hyperfictions is letting the reader determine how he traverses the text by choosing from different story threads, then this is possible in all storage media in which texts can be divided into segments which are connected to each other by hyperlinks. Readers have the choice between multiple links and thus need to make decisions during the reading process.

It goes without saying that man-man communication has always been possible prior to the installation of computer networks. It thus may be sufficient to point at two tendencies of *collaborative writing* rather sketchily. Collaborative texts have always been produced on site, e.g. in literary salons or writers' groups. However, it was not until the implementation of postal systems and of technological transmission media that long-distance collaborations were to become possible, ranging from varying writers' correspondences to Mail Art or Correspondence Art projects of the 1960s and 70s, from telephone and fax performances to simultaneous communication via computer networks.

Poetics and Literary Games: The History of Combinatory Literature

It is quite evident that literature has not only been confronted with the challenge of interactivity since modern computers were invented, but also that a "permanent mutability" of signifiers is a fundamental feature of *all* creative processes, which do not aim at permanent storage of strings. It is certainly true that mechanical text generators are not computers in a strict sense as they are neither able to convert analog signals into universal machine codes nor do they operate as fast as modern digital computers.

However, there have been some similarities between these machines as the algorithmic processing of "literary" signs has already been possible prior to the invention of digital computers. From the Kabbalah to Ramon Lull's *Ars magna generalis ultima* (1305-08), from Athanasius Kircher's famous *Ars magna sciendi* (1669) and Gottfried Wilhelm Leibniz's *Dissertatio de arte combinatoria* (1666) to Queneau's *Cent mille milliards de poèmes,* and finally to computer-based text generators there have been lots of examples, which can be traced back to three fundamental features. Firstly, there is a compact source code from which an abundance of texts can be generated. Secondly, this generation of texts requires that the processes of word and sentence formation be reproduced in a sort of micro-grammar. Finally, algorithmic procedures are used for processing linguistic signs (Cramer 245).

Any reconstruction of this tradition in German literature has to begin in the Baroque period when writers like Georg Philipp Harsdörffer, Philipp von Zesen and Justus Georg Schottelius worked out a poetic theory in the tradition of Ancient paradigms. They did not consider the successful work of art as an achievement of creative genius but insisted upon the teachability and learnability of poetic methods and procedures. Harsdörffer claims: "Ob nun wol etliche zu wolermeldter Kunst geboren / so ist doch die Kunst nicht mit ihnen geboren; sondern muß erlernet werden / wie alles / was wir Menschen wissen wollen" ('Even though quite a few persons are born to create respectable art, this art is not born with them; it has to be learned, as everything that humans want to know') (*Poetischer Trichter* 2).

These two aspects of combinatorics in the Baroque era had a strong impact on a variety of games in poetry such as anagrams or proteus verses and eventually led to the implementation of combinatorial procedures in mechanical text and poetry generators. According to the linguistic universalism represented by Leibniz and others, the whole world is considered a closed system, an order of things that can be algorithmically produced and varied from a preexisting and limited set of elements. This applies both to cosmic elements from which God was believed to have created the world and to the elements of language. According to this theory, any divine acting is regarded inaccessible while language cannot represent the given order of things as congruently. Thus both Baroque linguists and poets expected to draw conclusions about hidden organizational and creative principles either from investigations of rules of word and sentence formation or from implementation of mechanical principles into their poetic production. Of course, all these metaphysical premises were to get lost due to the technological execution of combinatorial procedures. The basic principles of theories of order, however, had not been abandoned. It is according to the logic of combining that so-called "blind" words can be generated from the lexical elements. These words have no meaning but are only legitimate because they have been formed according to word formation rules. Schottelius—and this is the missing link to literature—to a good part aims at "die Teutsche Sprache aus der Teutschen Sprache ferner zuerheben" ('lifting the German language far away from the German language') (Schottelius 98). Baroque literature, this is to be kept in mind, is very much based on playful linguistic practices. But this does not mean that at the core of such games lies the imagination and creativity of a writer. There rather are hidden mechanisms at work in mannerist experiments: "Ob nun wol der Poet bemühet ist neue Erfindungen an das Liecht zu bringe / so kann er doch nichts finden / dessen Gleichheit nicht zuvor gewesen / oder noch auf der Welt wäre" ('Even if the poet aims at bringing new inventions to light, he can find nothing the likes of which had not been already or still is in the world' (Harsdörffer, *Poetischer Trichter* 8).

Rules and Constraints: Anagrams, Proteus Verses and Other Literary Games

The extensive works of Georg Philipp Harsdörffer may serve as an example for demonstrating how countless examples of rule-governed text production derive from the reflection of language. Harsdörffer explicitly acted on the assumption that the subject matters of literature are not only to be found in the real world but also in language itself:

> Die Erfindung wird entweder herbeigeführet *von dem Wort* / oder *von dem Dinge* selbsten / darvon man handelt / oder von den Umständen desselben / oder von gehörigen *Gleichnissen*. Erstlich das *Wort* giebet eine Erfindung entweder in seinem angebornen Laut / unbekanter Deutung / oder mit versetzte *Buchstaben* / wann solche eine gantze Meinung schliessen / . . . . (*Poetischer Trichter* 10)

> The invention is either brought about *through the word,* or *by the thing* itself of which one is talking, or from the circumstances or adequate *similarities*. It is the *word* in the first place that gives to an invention its innate sound of unknown meaning or with mixed-up *letters* if they include a whole opinion. . . .

Many samples can be found in Harsdörffer's eight-volume *Frauenzimmer-Gesprächspiele* (1641-49), a collection of parlor games framed by a rather simple story: Six interlocutors, three women and three men, get together in a mansion to talk about social, scientific and poetic matters—and for just having a chat. Above all, however, they are setting each other riddles or other exercises and engage in literary games. These games are an integral part of the *inventio* prior to the real creative act. As exercises, their main function is to make the words available to the poet and thus to stimulate his imagination.

The *anagram* is certainly the best-known genre of such a literary game. It is a word game that is based on an interplay of two levels of text that, according to Aarseth, could be called "scriptons"—defined as strings as they appear to readers on some material surface—and "textons," which are "strings as they exist in the text" (Aarseth 62): A word or a complete phrase is made by transposing the letters of another word or phrase. At first, the connection between a signifier and a signified is dissolved, then the signifiers are rearranged—and only finally semantics comes into play because signifieds (and referents) need to be found for the newly established strings of signs. Hence the anagram is a comparatively simple example of the indeterminate transformation of cohesive structures into semantically coherent statements.

For centuries, the anagram has very often been discredited as baublery or magical practice. But there have also been periods every now and then in which it was quite popular: In German Baroque it was used as a creativity-stimulating parlor game at first, before anagrams were eventually incorporated into literary texts. For Harsdörffer, the anagram or "Letterkeer" ('lettertwist') was a technique of poetic invention because "so kan man die Buchstaben versetzen und eine andere Meinung heraus bringen" ('like this one can move letters and bring forth a different view') (*Poetischer Trichter* 17). The anagrammatic method is only due to the reordering of letters, without any references to the real world. Hence the Baroque treatises contain a variety of instructions, which make obvious the causal connection between permuting letters and mechanical devices. In his *Helikon,* Philipp von Zesen e.g. gives recommendations to facilitate writing anagrams by using cardboard letters (Zesen 174). Considerations on which means and tools are best suited for producing anagrams have not only been made in the Baroque era. Oskar Pastior—the only German member of Oulipo and certainly the best-known contemporary anagram writer—recommends various methods and exercises with subtle irony:

> Es gibt Hilfsmittel (Karton, Schere, Blockbuchstaben). Es gibt herrliche Vor- und Begleitübungen: Puzzles (besonders die schwedischen, handgesägten), Zeitunglesen und Fernsehen (Legasthenietraining), ei-

> nige gute Autoren, bewußtes Gehen mit beiden Füßen (hintereinander) auf dem Kies, Schüttelreime, komplizierte Stundenpläne oder gewisse Atem- und Abzähltechniken. Und es gibt die Strategie 'Steht der Tropfen, höhlt der Stein.' . . . Eine Menge Tüftelei also, der Autor konnte fast verschwinden. . . . (Pastior, *Anagrammgedichte* 82f.)

> There are aids (carton, scissors, block-letter). There are wonderful practices and dummy runs: puzzles (especially the Swedish ones that are hand-made), reading the paper and watching TV (training legasthenics), some good authors, consciously walking with both feet (one after the other) on gravel, spoonerism, difficult timetables or specific techniques of breathing or counting. And there is the strategy "Drip standing stone hollowing" [Pastior here invents a word game in German with the saying "steady dripping hollows stone"]. . . . Real brainteasers, the author was virtually to disappear. . . .

The author was virtually to disappear: This assertion, which encouraged Pastior to re-explore the possibilities of the anagram in 1985, is similar to statements from poets and poeticians of the Enlightenment—although their conclusions were diametrically opposed to Pastior's. As from the age of Enlightenment, the Baroque *ars combinatoria* was largely discredited. Instead, creativity and spontaneity became the paradigms of literary theory. The scriptural logic of letters was replaced by the phonocentrism of the Goethe era, and the anagram and other forms of combinatory poetry were discredited as baubleries ("Kindereyen") by poets and philosophers such as Johann Christoph Adelung (518).

The anagram was only rediscovered by various 20th century avant-garde movements. Whereas the popularity of the anagram in the Baroque era was due to the prospect that the secret order of things may be discovered by permuting signifiers, its modern renaissance is due to the expectation of getting access to the unconscious by the play of letters. The artist Hans Bellmer e.g. relates his breaking up and rearranging of the female body of his famous sculptural construction *The Doll* to the anagrammatic practice of language:

> Sie ist aus Division, Subtraktion und Multiplikation, aber auch aus jener Vertauschbarkeit geboren, die von den Mathematikern "Permutation," von den Philologen "Anagramm" genannt wird, und deren Bedeutung auf folgendes hinausliefe: Der Körper, er gleicht einem Satz—, der uns einzuladen scheint, ihn bis in seine Buchstaben zu zergliedern, damit sich in einer endlosen Reihe von Anagrammen aufs neue fügt, was er in Wahrheit enthält. (Bellmer 95)

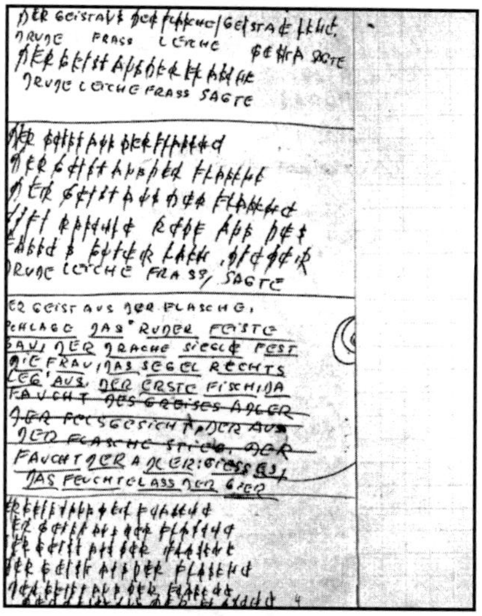

Fig. 1. Facsimile from Unica Zürn's notebook.

> She has been born from division, subtraction and multiplication but also from this permutability, which mathematicians call "permutation" and philologists call "anagram," and whose meaning adds up to the following: The body, it resembles a sentence inviting us to break it up into its constituent letters, so that these again connect with what they contain in an endless array of anagrams.

Unica Zürn, the longstanding partner and lover of Bellmer, is considered to be the most important anagram writer of German post-war literature. From 1953 to 1964, she wrote as many as 123 anagram poems without using cardboards or other devices; she only worked on paper, as can be seen from her "Der Geist aus der Flasche (IV) ('The Spirit from the Bottle [IV]')" (1960). First, she wrote down the first line from which she then crossed out the letters of the emerging words. She repeated this process until she succeeded in writing a new line (fig. 1):

> Schlage das Ruder, feiste
> Sau. Der Drache sieg'l fest
> Die Frau, das Segel rechts
> Leg' aus. Der erste Fisch! Da
> Faucht der Adler: giess' es,
> das Feuchte lass der Gier. (Zürn 86)

In the 1980s, German literature saw a sort of boom of the anagram initiated by writers such as Oskar Pastior (*Anagrammgedichte,* 1985) or Elfriede Czurda (*Fälschungen,* 1987). Following the Oulipo theories, Pastior considers his texts test arrangements in which he applies both self-invented rules and models from the history of literatures. He describes his basic principle as "Aufknacken von Wörtern und Wendungen in Bedeutungsklumpen von unbestimmter mittlerer Größe (sozusagen ein molekulares Cracking) und dann Zusammenfügen in irgendwo stupenden, aber exotisch einleuchtenden neuen semantischen Verbindungen" ('cracking of words and phrases into chunks of meaning of undefined size (a molecular cracking so to speak) and then joining them together in somehow startling but in their exotic way convincing new semantic connections') (*Das Unding an sich* 40).

He regards his anagrams as a "permutative und viel engmaschigere Buchstabenalchemie (also ein inframolekulares Cracking)" ('permutative and very strictly meshed alchemy of letters (a sort of inframolecular cracking)') (*Das Unding an sich* 40). Hence the anagram was well suited as a "super-metaphor" for the processing of literature, for "etwas, das nicht still steht und nicht weitergeht" ('something that neither stops nor continues') (*Anagrammgedichte* 9). In his *Anagrammgedichte* (1985), he took the titles of calendar stories from the table of contents of an edition of Johann Peter Hebel's complete works as first lines of his anagrams:

> Unverhofftes Wiedersehen
> Sehr oft wird Vene, ehe es fun-
> diert, frueh von Wehes Senfe
> in des Heftes Heu verworfen.
> Es fuehrt es Hefe-Vorwinden
> sehr diffuse vor. Wen Teehen-
> ne vorsieht, der fuenf Seh-
> westen oder Hufweisen fehver-
> weist. Von Fehes Duerfen her
> verseift Fernsehen, wo due-
> ster Huf es feindverwoehne.
> Huste, Revers—nen Wiedehoff! (45)

Proteus verses—named after the Greek sea-God who could change his shape at will—vary the elements of a verse in any combination. Only the positions of the first and last words of each line are unchangeable, so that a multitude of verses can be generated by permuting the elements either in horizontal or in vertical direction. In Germany, proteus verses were only known as from the

Baroque period when they were discussed in any major poetic book. In 1657, Stanislaus Mink von Weinsheun—a (slightly incorrect) anagrammatic pseudonym of Johann Justus Winkelmann—published an entire book under the title *Proteus. Das ist: Eine unglaubliche Lutznützliche Lehrart / in kurzer Zeit ohne Müh Deutsch= und Lateinische Vers zumachen / auch einen Französischen und Lateinischen Brief zu schreiben* ("Proteus. This is an Unbelievably Useful Way of Teaching How to Rhyme Verses in German and Latin in a Short Time and Without Troubles and Even Writing French and Latin Letters"), which was entirely dedicated to the proteus verse and to permutation of language.

In his *Poetischer Trichter*, Harsdörffer gives the following definition of the proteus verse or "Wechselsatz":

> Diese Reimart könte man einen Wechselsatz nennen: dann wann man die ersten Wort (auf folgt) und die letzten zwey (Fleiß und Preiß) unverändert auf solcher Stelle behält / können die andern Wörter 39916800 / das ist / neun und dreissig tausendmaltausend / neunhundert und sechzehentausend / und achthundert mal versetzet werden / zu welcher Veränderung der allerfertigste Schreiber / der täglich 1200 Zeile abschriebe / gantze 91 Jahre / und 49 Tage würde haben müssen: wolte man aber die Reimwort Fleiß / Preiß / auch versetzen / und Krieg und Sieg darfür gebrauchen / so kan man noch etlich tausendmal öffter wechseln. (*Poetischer Trichter* 51f.)
>
> One could call this way of creating poetry a jumble-sentence: if you keep the first words and the last two unchanged in the same place, the other words could be mixed up 39,916,800, that is thirty-nine thousand times one thousand nine hundred and sixteen thousand and eight hundred times changed, for which change the most learned writer who would write 1,200 lines daily would need 91 years and 49 days. But if one would also mix up the rhymes *Fleiß* ("diligence") and *Preis* ("price") and use *Krieg* ("war") and *Sieg* ("victory") instead one can mix up words several thousand times more.

For Quirinus Kuhlmann, the proteus verse has the function to illustrate the volatility of being. In his *Geschicht-Herold* (1672), he argues that all human abilities were "poured" from divine wisdom and that both heaven and earth are operating like a "changing wheel": "der Allmächtige Himmels- und Erdenschöpfer hat Himmel und Erden wi ein wechselrad eingerichtet / die Geschöpffe stat der wechselwörter genommen: Alle Weltdinge wechseln / alle liben / alle hassen" ("The almighty creator of heaven and earth has created heaven and earth like a changing-wheel and used beings instead of the change-words. All things on earth change; all love, all hate') (qtd. in Neubauer 32).

If it was the case that "di Natur anagrammatisiret und buchstabenwechselt" ('nature anagrammatizes'), this must be reflected in language, too. Kuhlmann wrote the best-known proteus of the Baroque period, "XLI. Libes-kuß: 'Der Wechsel Menschlicher Sachen' ("41st kiss of love: 'The change of human matters'") (1671):

> *Auf* Nacht / Dunst / Schlacht / Frost / Wind / See / Hitz / Süd / Ost / West/ Nord / Sonn / Feur / und *Plagen* /
> *Folgt* Tag / Glantz / Blutt / Schnee / Still / Land /Blitz / Wärmd / Hitz / Lust/ Kält / Licht / Brand / und *Noth:*
> *Auf* Leid / Pein / Schmach / Angst / Krig / Ach / Kreutz / Streit / Hohn / Schmertz / Qual / Tükk / Schimpf / als *Spott* /
> *Wil* Freud / Zir / Ehr / Trost / Sig / Rath / Nutz / Frid / Lohn / Schertz / Ruh / Glükk / Glimpf / stets *tagen.*
> *Der* Mond / Glunst / Rauch / Gems / Fisch / Gold / Perl / Baum / Flamm / Storch / Frosch / Lamm / Ochs und Magen
> *Libt* Schein / Stroh / Dampf / Berg / Flutt / Glutt / Schaum / Frucht / Asch/ Dach / Teich / feld / Wiß / und Brod:
> *Der* Schütz / mensch / Fleiß / Müh / Kunst / Spil / Schiff / Mund / Printz / Rach / Sorg / Geitz / Treu / und *GOtt* /
> *Suchts* Zil / Schlaff / Preiß / Lob / Gunst / Zank / Ort / Kuß / Thron / Mord/ Sarg / Geld / Hold / *Danksagen*
> *Was* Gutt / stark / schwer / recht / lang / groß / weiß / eins / ja / Lufft / Feur/ hoch / weit / *genennt* /
> *Pflegt* Böß / schwach / leicht / krum / breit / klein / schwartz / drei / Nein / Erd / Flutt / tiff / nah / *zumeiden* /
> *Auch* Mutt / lib / klug / Witz / Geist / Seel / Freund / Lust / Zir / Ruhm / Frid / Schertz / Lob *muß scheiden* /
> *Wo* Furcht / Haß / Trug / Wein / Fleisch / Leib / Feind / Weh / Schmach / Angst / Streit / Schmach / Angst / Streit / Schmertz / Hohn *schon rennt*
> *Alles wechselt; alles libet; alles scheinet was zu hassen:*
> *Wer nur disem nach wird-denken / muß di Menschen Weißheit fassen.* (*Himmlische Libes-küsse* 54f.)

The *sestina,* a poem composed of six stanzas of six lines each with a half-stanza of three lines at the end, was invented by troubadour poets in Italy and Provence. It has only six rhyming words at the end of the lines, which change according to a strict rhythm from stanza to stanza: 123456 / 715243 / 364125 / 532614 / 451362 / 246531 / 123.

In German literature, the sestina was not cultivated before the 17th century. There are some sestinas in the works of the usual "suspects" from the Baroque era such as Schottelius, Zesen, Harsdörffer, Kuhlmann, Martin Opitz or Georg Rudolf Weckherlin. Throughout the 18th century, it was rather disregarded. It was not until the Romanticist period that August Wilhelm Schlegel drew the attention to the sestina in his critical writings. This impulse was then taken up by poets such as August Wilhelm Iffland, Wilhelm von Schütz, Ludwig Uhland and Joseph von Eichendorff.

In contemporary literature, it was Oskar Pastior again who rediscovered the genre when he published a book of 34 sestinas under the title *Eine kleine Kunstmaschine* (1994). For the Oulipian Pastior, the sestina guarantees the "text-generative Potenz zusätzlicher Einschränkungen" ('text-generating power of additional constraints') (*Eine kleine Kunstmaschine* 81). This is what brings him to programmatically entitling his book a *Kunstmaschine* ("art machine"):

> Der Gedanke, wie er sich erwähnt, reimwortläufig, in dem vertrackten Rhythmus mit dem transitorischen Moment von 123456 zu 615243. Um so "in aufbrechender Umarmung" zu "denken," hat die Sestine permanent ihr genetisches Kürzel "im Kopf," eben diese Fähigkeit, wie ich nach einiger Erfahrung mit ihr denke, in ständig verschobenen Rückversicherungs- und prospektiven Falsifikationsschlaufen sich herzustellen. (Pastior, *Das Unding an sich* 81f.)

> The thought, as it appears in the course of rhymes, in the intricate rhythm with the transitory moment of 123456 to 615243. The sestina, to "think" like this in a "breaking embrace," it permanently is aware of its genetic grammalogue, its ability (as I think of it after some experience with it) to create itself in constantly slanting loops of reassurance and prospective falsifications.

Pastior even escalates and subverts the sestina paradigm by multiplying its constraints with those of other genres. In his "sestine mit diabetes" for instance, he combines the sestina with the anagram (*Eine kleine Kunstmaschine* 40f.), and in "heureka mit euter am pneu," he only uses two vowels per line (42f.).[3]

Dadaists as well as Surrealists referred back to aleatoric conceptions. But in general, they preferred "automatic" and improvising methods to combinatory ones. Tristan Tzara's famous recipe for composing a Dadaist poem does not define any specific collection of words but only describes a method for processing any data whatsoever (*Œuvres completes* 382). Hans Arp's creative approach is even less determined as can be seen from a pivotal statement on "automatic" writing in his essay "Wegweiser":

> Viele Gedichte aus der "Wolkenpumpe" sind automatischen Gedichten verwandt. Sie sind wie die surrealistischen automatischen Gedichte unmittelbar niedergeschrieben, ohne Überlegung oder Überarbeitung. Dialektbildung, altertümelnde Klänge, Jahrmarktslatein, verwirrende Onomatopoesien und Wortspasmen sind in diesen Gedichten besonders auffallend. Die "Wolkenpumpen" aber sind nicht nur automatische Gedichte, sondern schon Vorläufer meiner "papiers déchirés," meiner "Zerreissbilder," in denen die "Wirklichkeit" und der "Zufall" ungehemmt sich entwickeln können. Das Wesen von Leben und Vergehen ist durch das Zerreissen des Papieres oder der Zeichnung in das Bild einbezogen. (*Wortträume und schwarze Sterne* 7)

> Many poems in *die wolkenpumpe* ("the cloudpump") are related to automatic poems. They were written down, like the Surrealistic automatic poems, immediately without thought or revision. Dialectal constructions, outmoded sounds, Vulgar Latin, confusing onomatopoetic words and verbal spasms are particularly noticeable in these poems. The "cloud pumps" are, however, not only automatic poems, but already anticipate my "papiers déchirés," my "torn pictures," in which "reality" and "chance" can be developed uninhibitedly. The essence of life and decay is incorporated into the picture by tearing up the paper or drawing. (Qtd. in Döhl, "Hans Arp and Zurich Dada" 118; my revisions)

Arp addresses one of the core problems of any concept of automatic writing here. In all his texts and artworks, the "laws of chance" can only unfold within the frame defined by both the author's preconception and the specific peculiarities of the material. It is not surprising that he took many words and phrases from daily newspapers, particularly from the advertisement sections and "schlang und flocht leicht und improvisierend Wörter und Sätze um die aus der Zeitung gewählten Wörter und Sätze" (*Wortträume und schwarze Sterne* 46) ('interwove the words and sentences selected from the newspapers with freely improvised words and sentences of my own') (qtd. in Döhl, "Hans Arp

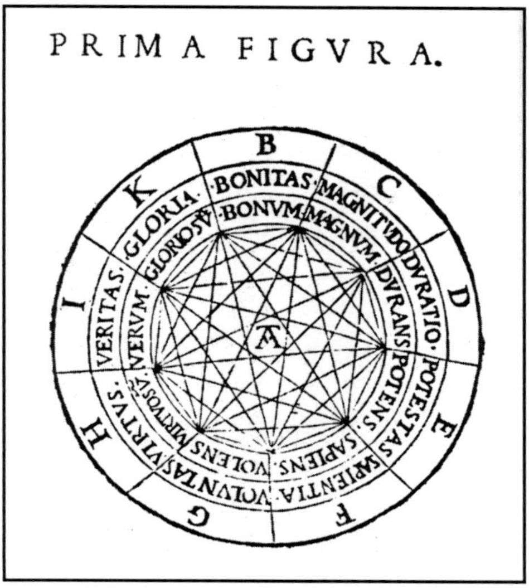

Fig. 2. Ramon Lull's *Ars magna generalis ultima.*

and Zurich Dada" 117). Unlike the Futurists, Cubists and Berlin Dadaists, Arp never pasted press clippings as pictorial representations into his pictures. For him, newspapers rather provided a habitual vocabulary whose transitory character fit well into his poetics of permanent change.

Combinatory Machines and Text Generators

The rather simple principle of generating an abundance of signs from a limited repertoire is the basic idea of various mechanical devices, which have been constructed in the course of the last centuries for the generation of texts. I already cited some very simple means of mechanizing the literary invention by using cardboard plates or other tools and media. The combination of these plates remains completely up to the user. However, there have been many more sophisticated mechanisms and machines for facilitating literary production by combining words or attributes selected from a number of lists. The prototype of such logic machines was the *Ars magna* devised by the Catalan monk Ramon Lull. Lull's machine consists of a stack of three concentric disks mounted on an axis where they are able to rotate independently (fig. 2). The disks were progressively larger from top to bottom. Nine fundamental terms, the so-called "principia absoluta" which comprise the main topics of scholastic philosophy, were related to the letters from A to K. By rotating the disks, a large number of random statements could be generated from the alignment of words. This mechanism generates "scriptons" which can be read as soon as the rotating disks with "textons" on them come to a complete standstill. At a random point in time, it results in an artifact that could not be predicted in advance. It is, of course, still up to the user—as Aarseth's triad of text/machine, collected words and user indicates (Aarseth 62)—to interpret the resulting strings of signifiers.

The *Ars magna sciendi,* Athanasius Kircher's adaptation and elaboration of Lull's *Ars magna,* illustrates a common tendency of the Baroque era: In addition to books, alternative *Aufschreibesysteme* ("systems of notation") (Kittler) appear that are storage media of traditional knowledge *and* generators of new knowledge at the same time. Neither the storing nor the production of knowledge, however, is ascribed to an author, and the knowledge is not written or printed in syntagmatic chains but machines are designed, which generate the knowledge from a compact source code by combinatorial procedures.

Therefore Harsdörffer implemented the methods of generating stem words and word formation rules in his *Fünffacher Denckring der Teutschen Sprache* (1651) (fig. 3), and Quirinus Kuhlmann, too, came up with the idea of a changing wheel (but, unfortunately, without giving any details, cf. Neubauer 33ff.).

These word-generating machines are the missing link between the *langue* of the German laguage and Lull's *ars combinatoria.*

Harsdörffer claims that his machine is able to mechanically reproduce all possibilities of German without having to compile voluminous dictionaries. The *Denckring* consists of five rotating disks, which the user must cut out from the book at first: "Dieses Blätlein muß heraus geschnidten / in fünff Ringe zertheilet / und auf fünff gleich-grosse Scheiben von Papyr / also aufeinander gehefftet werden / daß man jeden Ring absonderlich umbdrehen kan / wann solchs geschehen / muß man dises fünfffache Blat wider hinein pappen" ('This leaflet has to be cut out, parted into five disks and fastened onto five equal leafs of paper so that each disk can be turned around separately and when this has happened one has to glue that five-fold leaf back in') (qtd. in Hundt 283).

On the inner disk, there are 49 prefixes followed by 60 initial letters, 12 medial and 120 final letters as well as 24 suffixes. Aarseth would call these elements "textons" from which millions of different "scriptons" can be generated. It is a specific feature of Harsdörffer's machine that by setting the storage itself in motion it simultaneously turns out to be a generator. The linear reading process is transformed into a rotating mechanism. The user has to start the mechanism, and he can either stop the rotating disks or just wait until they come to a complete standstill. From the point of view of production aesthetics, the

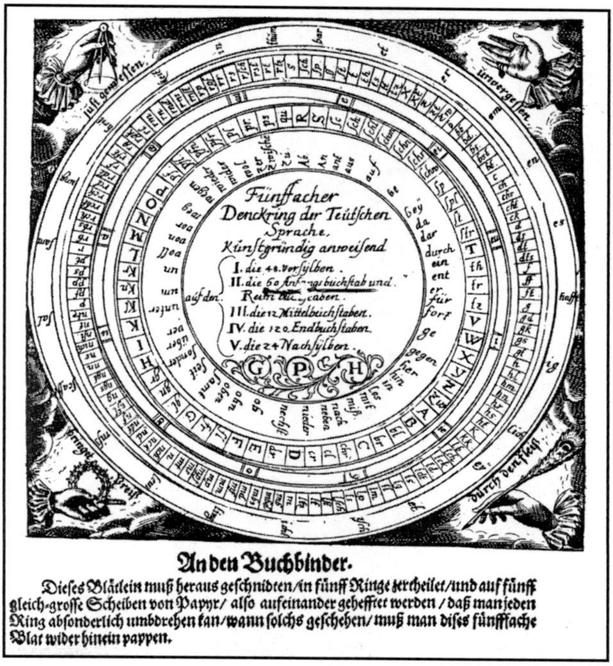

Fig. 3. Georg Philipp Harsdörffer's *Fünffacher Denckring der Teutschen Sprache* (1651).

outcome of such a machine-based combinatorics cannot substantiate any concept of artistic originality, and from the point of view of reader-response theories, we are confronted with a reader who can only interpret transitory strings of signifiers. Thus Harsdörffer did not only use his *Denckring* for representing all possibilities of word formation but he also considered it a useful tool of literary writing:

> Ist also dieses eine unfehlbare Richtigkeit / ein vollständiges Teutsches Wörterbuch zu verfassen / und beharren wir in der Meinung / daß alle solchen zusammen gesetzte Wörter / welche ihre Deutung würcken für gut Teutsch zulässig / sonderlich in den Gedichten / ob sie gleich sonsten nicht gebräuchlich / . . . . (*Deliciae mathematicae et physicae* 518)

> This word-generating procedure then is completely accurate in creating a complete German Dictionary and we retain our opinion that all these composite words should be allowed as good German, especially in poems, even though they might not be used otherwise.

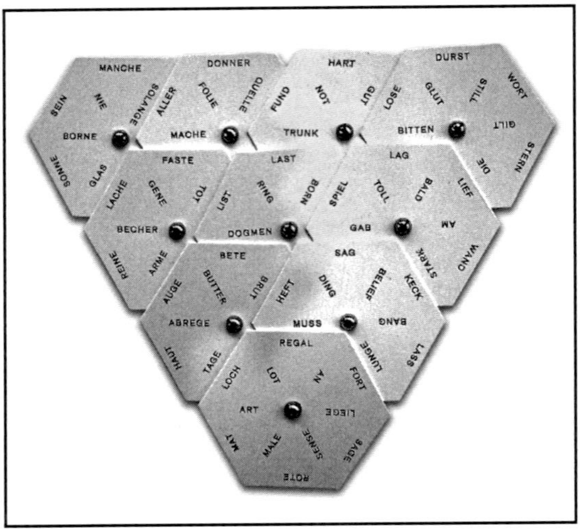

Fig. 4. André Thomkins: *dogma-mot* (1965).

This is Harsdörffer's solution of the problem of the non-semantic "blind" words. He thus resolves the problem that his machine inevitably generates words, which make no sense in the German language by declaring the outcome of his *Denckring* to be the proper language. This is the poetic potential of the machine; it becomes a generator of poetic invention or a mobile rhyming dictionary "Erfindung der Reimwörter / wann man die Reimsilben auf dem dritten und vierten Ring suchet / und die Reimbuchstaben auf dem zweyten Ring darzu drehet" ('for inventing rhymes by looking for the rhyming syllables on the third or fourth disk and then turning the second disk to add the rhyming letters') (*Deliciae mathematicae et physicae* 518).

Text generators based on rotating disks, however, are no peculiarity of the Baroque era but there also have been some interesting examples in recent times, e.g. in the Fluxus movement. André Thomkins produced his polyglot machine *dogma-mot* (1965), which allows the user to produce "mobile dogmas" (fig. 4). This machine consists of ten hexagonal cards, each with 12 words that are common in each of the three languages German, English and French (there are 48 German-French, 60 German-English and 12 German-French-English words). The hexagons pivot on a plane and form phrases in three directions on the hexagonal roof. They can even be arranged differently on the ten slots, so that the number of combinations can further increase. Thomkins gives an instructive German-English example on the blurb of his Fluxus Box:

WORT HAT BALD SENSE
ein engländer liest:
wurzel hut kahler sinn
(eierkopf, haargenau)

a german reads:
word has soon scythe
(words mutual short-cut) (*dogma-mot*)

Other writers and artists also worked with rotating disks or other moving media: Ferdinand Kriwet wrote many of his texts on disks, among them are three-disk-texts whose disks overlap and thus make combinations of elements possible. Dieter Roth produced so-called *Leserollen* ("reading rolls")—paper webs with abstract ornamental patterns—for the *Apparat zum Simultanlesen* ("Apparatus for simultaneous reading"). This machine was invented by his fellow-artists Daniel Spoerri and Jean Tinguely and consists of a rotating rod driven by an engine. On this rotating rod, paper webs can be fixed from which readers can simultaneously read varying texts.

Besides his rotating disks, Harsdörffer designed further mechanisms of word formation: He suggested inscribing the letters of the German alphabet in the dice. If the user throws two dice, he can create syllables. This method can be extended by using additional dice. If we apply a thematological approach to the history of literature, we also find at least one very prominent example of a mechanical apparatus for the production of texts which consists of revolvable wooden dice: the famous machine of the Grand Academy of Lagado in Jonathan Swift's *Gulliver's Travels* (1726).

Although none of the following examples can keep up with the fame of Swift's novel, there have been some writers who experimented with dice. Max Bense generated some of his so-called "Dünnschliffe" by rolling dice, e.g. for combining clippings from a newspaper and sections from a Franz Kafka novel (246). It is only a small step from such permutative procedures based on mechanical devices such as rotating disks or dice to the use of computers for producing literature. In Germany, this was done in the late 1950s and early 1960s by the "Stuttgarter Gruppe" ('Stuttgart Group'). In 1959, the mathematician and software engineer Theo Lutz produced so-called "stochastic texts" using the ZUSE Z 22 mainframe of the Stuttgart University's computer center. Lutz had generated a vocabulary from Kafka's *Das Schloss [The Castle]*, which he then permuted by using a program written in ALGOL. The computer scientist Rul Gunzenhäuser also published some computer-generated poems. These first pioneering experiments may not have been satisfying for traditionalists but for the

"Stuttgarter Gruppe" they appeared as the "incunables of 'artificial poetry,'" Reinhard Döhl recalls ("Vom Computertext zur Netzkunst"). This concept of "artificial poetry," however, which was explicitly based upon permutative and stochastic procedures was soon extended to other literary genres and accompanied by a variety of programmatic essays and manifestos such as Abraham Moles' "Erstes Manifest der permutationellen Kunst [First Manifesto of Permutational Art]" (1962) or "Zur Lage" (1964) by Bense and Döhl.

The interest in working with computers was not limited to the mathematicians of the group but also inspired writers and artists like Döhl and Klaus Burkhardt who jointly wrote the *poem structures through the looking glass* (1969). Bense and Döhl incorporated computer-generated passages into their so-called "Mischtexte" ("mixed texts") like Döhl's *fingerübungen* (1962), *Prosa zum Beispiel* (1965) or the radio play *Monolog der Terry Jo [Monologue: Terry Jo]* (1968) by Bense and Ludwig Harig:

> Der Monolog beginnt mit einem Computer-Text. Es sind neun synthetische Annäherungen an die Sprache des Mädchens. Die Tatsache, daß gewisse Analogien zwischen dem zu Anfang unbewußten Zustand des Mädchens und der Unbewußtheit eines Computers bestehen, ließ diese erste Verwendung eines mit einer programmgesteuerten Maschine hergestellten Textes in einem Hörspiel gerechtfertigt erscheinen. Diese Computertexte des Monologs werden in der Realisation übersetzt in eine durch ein kompliziertes Vocoder-Verfahren hergestellte synthetische Sprache, die im Verlauf des Monologs mehr und mehr abgebaut und von der natürlichen Stimme abgelöst wird. (Qtd. in Schöning 58)

> The monologue starts with a computer-text. There are nine synthetic approaches to the language of the girl. The fact that there are certain analogies between the initial subconscious state of mind of the girl and the subconscious of a computer might let us consider that the first use of a programmed, machine-generated text in a radio-play can be justified. These monologous computer-texts are realized by translating it into a synthetic language created by a complicated vocoder-system that is reduced more and more in the course of the monologue and thereby replaced by the natural voice.

The idea of generating entire texts on computers was soon taken up by some other computer centers during the 1960s: In 1966, Gerhard Stickel produced his *Monte-Carlo-Texte* on the IBM-7090 mainframe of the Deutsches Rechenzentrum in Darmstadt which he subsumed under the genre name "Auto-Poems." One year later, Manfred Krause and Götz F. Schaudt published their book *Computer-Lyrik: Poesie aus dem Elektronenrechner* ("Computer Poetry: Poems from the Electronic Computer"); the programs included both rhyme schemes and measures.

From File Boxes to Hypertext: Digressions and Permutative Novels

Alfred Döblin once argued: "Wenn ein Roman nicht wie ein Regenwurm in zehn Stücke geschnitten werden kann und jeder Teil bewegt sich selbst, dann taugt er nichts" (Döblin 21) ('If a novel cannot be cut up like a worm into ten pieces so that each bit moves independently, then it is no good') (qtd. in Murphy 21). Although every biologist would repudiate this argument, Döblin's *bon mot* refers to a fundamental precondition of all hyperfictions in both print and computer-based media. The producer of such a text has to compile a collection of text segments and to define relations between these fragments regardless of the specific medium he is writing for. In the history of literature, this ranged from fragments, paper slips or excerpts which were collected in file boxes and notebooks and then presented to the reader—either in books or in text objects like cases, boxes, sometimes even installations or environments such as Michael Badura's *Zettel-Werke.* It is then up to the reader to recombine the segments in the course of reading.

This development, too, was initiated in the Baroque era when various alternative storage devices and discourse networks challenged the fixed chains of signs of printed books:

> Wenn Datenplanspiele und Letternphantasien jenseits aller Metaphern im drucktechnischen Sinne real werden, wird der statische Raum der Bücher selbst zum zwischengeschalteten Medium, zu jenem beliebig (um)sortierbaren Karteikasten, in dem die Zettel potentieller Registermacher landen. (Rieger 100)
>
> If the data games and letterphantasma become established as printing technologies beyond all metaphors, the static space of books transforms into an (inter-)medium itself: transforms into this (re)sortable index box in which the memos of potential indexers end up.

But how did writers react to this challenge? For the purpose of this historical (re)construction, the rather pragmatic approach of Monika Schmitz-Emans may be helpful. She defines "book labyrinths" on a material level as follows:

> Als Buch-Labyrinthe verstanden seien darum im folgenden auf materiell-visueller Ebene nicht-linear konstruierte Bücher, die den Leser explizit hin und her "schicken," durch typographische Mittel, Leseanleitungen, nicht-lineare Numerierungen oder andere Signale, sowie Text-Baustein-Sammlungen. (179)
>
> As non-linear constructed books, which explicitly "send" the reader to and fro by means of typography, reading instructions, non-linear pagination or other signals, or as collections of text segments.

Regarding the arrangement of text segments by the author, two types can be distinguished: Segments can either be put together in succession in a book and highlighted only as loosely associated elements by cuts, numbers, chapter headings or various other paratextual elements or, alternatively, they can be printed on loose leaves such as untacked sheets of paper, cards or lots. In each case, familiar reading conventions have to be relearned, and thus most of these books contain paratextual features such as reading instructions, game rules, etc. But unlike the rules of football or board games, these instructions are not intended as a body of rules imposing sanctions on those who break them. In fact, these instructions are ironically broken, and the writers of such texts usually discuss and question the rules themselves; in many cases, they even call on the readers to challenge or to expand them by themselves, so that literature explicitly transcends the space of the rule-governed game in favor of a "meta-game" (Schmitz-Emans 193).

Most historical descriptions of "print hyperfictions" have focused on examples from Romanic and American literatures. Examples of novels and stories which link the text segments by hyperlinks, indices, annotations, etc. include Jorge Luis Borges' short stories, particularly *El jardin de senderos que se bifurcan [The Garden of Forking Paths]* (1941), as well as Georges Perec's novel *La vie mode d'emploi [Life A User's Manual]* (1978). According to Schmitz-Emans, Perec's novel on the one hand is a novel about a jigsaw puzzle, and on the other hand, the whole novel can be seen as an intricate jigsaw puzzle itself (Schmitz-Emans 180). Further dazzling literary experiments include Julio Cortázar's *Rayuela [Hopscotch]* (1963), which follows the model of the hopscotch game and its follow-up novel *62/modelo para armar [62: A Model Kit]* (1968), Vladimir Nabokov's *Pale Fire* (1962), Milorad Pavić's *Hazarski rečnik [The Dictionary of the Khazars]* (1984) and *Predeo slikan čajem [Landscape Painted with Tea]* (1988) and Italo Calvino's *Il castello dei destini incrociati [The Castle of Crossed Destinies]* (1973), which was inspired by tarot cards. Queneau's sonnet machine *Cent mille millards de poèmes,* though, is considered a model of innovative literary forms on loose leaves just as Marc

Saporta's card-game novel *Composition No. 1* (1962) or B.S. Johnson's "novel-in-a-box" *The Unfortunates* (1969). Both consist of unpaginated sheets on which fragments of a story are printed. The reader can shuffle the pages and read the resulting random order.

It is obvious that there seems to be a lack of prominent examples from German-language literature though it would be easy to contradict this impression by pointing to a long tradition of German print hyperfictions. The printing press and the distribution systems of the Renaissance had established accurate "address-systems," which for the first time allowed accessing any text passage. At the same time, German Baroque writers and linguists established indices or "Blatweiser" for making this access to data easier. In his *Mathematische und philosophische Erquickstunden* (1651), Harsdörffer developed an elaborate file-card system for keeping the imminent disorder of data under control:

> Wann man nun das Register machen will / so schreibet man den Inhalt / gehöriger Massen / auf ein Papyr / schneidet es in absonderliche Stücklein / und leget jedes in sein Buchstabfach: von dar nimmt man sie zuletzt wieder heraus / ordnet einen Buchstaben nach dem andern / und klebet entweder die Papyrlein ordentlich auf / oder schreibt sie noch einmal. (57)

> If you want to create an index system, you note the contents onto a piece of paper correctly, cut it into distinct pieces and place each one into its corresponding box for that letter. From there you finally take it, sort the letters into their correct order and then glue the pieces one after another or rewrite them correctly.

Hence Stefan Rieger described the Baroque knowledge management systems as predecessors of modern hypertext technologies:

> Für das Barock sind hybride Textformen die Folge, die hart an den Grenzen des Buches operieren. Im Zeichen von Ökonomie und Effizienz barocker Datenspeicherung erhält die lineare Organisation von Büchern eine mediale Konkurrenz durch alternative Aufschreibeformen, die Zugriffe durch Register organisieren und damit eine strikt lineare Abfolge des Buches durch andere Präsentationsweisen—etwa die Möglichkeit der Synopsen—überflüssig machen wollen. (88)

> In the Baroque period, hybrid text forms developed that were touching the borders of the printed book. Facing the economy and efficiency

> of Baroque data storage, the linear organization of books began to compete with alternative forms of notation. These organized access by indices and thus tried to replace the strict linear sequence of books by other ways of presenting knowledge, such as synopses.

Harsdörffer and his Baroque contemporaries implemented efficient data storage devices and data mining tools by compiling complex registers. This happened in order to get the overflowing knowledge of their times under control whereas poets emptied the file boxes in a rather playful way, as could be seen from Zesen's novel *Assenat* (1670). The plot of the novel—the life and love story of biblical Joseph—is narrated on 344 pages, which are supplemented by an appendix of another 200 pages with mythological and historic digressions, annotations, etc.

The works of Jean Paul have also been repeatedly characterized as hyperfictions *avant la lettre*. His excessive use of digressions, annotations, footnotes and periphrases that interrupt his narration again and again and lead him astray are an expression of the strong sense of individuality of early Romanticism. Hence the narrator of his novels establishes subjectively motivated relations between entirely unrelated elements:

> Es wäre daher die Frage, ob nicht eine Sammlung von Aufsätzen nützete und gefiele, worin Ideen aus allen Wissenschaften ohne bestimmtes gerades Ziel—weder künstlerisches noch wissenschaftliches—sich nicht wie Gifte, sondern wie Karten mischten und folglich, ähnlich dem Lessingschen geistigen Würfeln, dem etwas eintrügen, der durch *Spiele* zu *gewinnen* wüßte; was aber die Sammlung anlangt, so hab' ich sie und vermehre sie täglich, schon bloß deshalb, um den Kopf so frei zu machen, als das Herz sein soll. (Jean Paul, *Vorschule der Ästhetik* 202f.)

> One could ask whether it would not be helpful and pleasant to have a collection of essays in which ideas from all sciences without any definite direct (artistic or scientific) goal were mixed together not like poisons but like cards; like the mental dice of Lessing they would be profitable for anyone who knew how to *win* in *games*. I have this collection and add to it daily, if only to free the head as much as the heart. (Jean Paul, *Horn of Oberon* 144)

This method was only possible due to Jean Paul's excerpts, which he collected from borrowed books since he was an adolescent preparing for studying theology. Since 1782, these miscellaneous excerpts became the generative moment of his literary writing (Müller, *Jean Pauls Exzerpte* 9). However, these excerpts did not help compiling encyclopedias from the collected knowledge; they rather allowed infinite digressions. This was made possible by the wit which Jean Paul defines in his *Vorschule der Ästhetik [School of Aesthetics]* (1804), as an ability to discover "die ähnlichen Verhältnisse *inkommensurabler* (unanmeßbarer) Größen, d.h. die Ähnlichkeiten zwischen Körper- und Geisterwelt (z. B. Sonne und Wahrheit), mit andern Worten, die Gleichung zwischen sich und außen, mithin zwischen zwei Anschauungen" (*Vorschule der Ästhetik* 172) ('the similarities between *incommensurable* magnitudes, between physical and spiritual worlds (e.g., sun and truth), in other words, the equation of self and other, of two perceptions') (*Horn of Oberon* 122). In Jean Paul's works, wit is a method of combining subject matters that are organized in different text units and then recombining them by cross-references and footnotes.

From the "geistige Würfel" ('mental dice') Jean Paul is describing with his mix of thoughts and excerpts, it is only a small step to proper dice. Hence it is no surprise that many writers take up rules and procedures of familiar dice and card games for organizing plots and narratives. The idea of generating texts by throwing dice and then selecting words from a number of lists according to the sum of the pips is not exclusively reserved to the experimental writers of the Baroque period or the avant-garde writers following Stéphane Mallarmé's *Un coup de dés jamais n'abolira le hazard [One Toss of the Dice Never Will Abolish Chance]* (1897). As early as the 1820s, Georg Nikolaus Bärmann—at that time known as a rather minor playwright and critic—used dice as a tool for generating texts. First, he published the book *Die Kunst ernsthafte und scherzhafte Glückwunschgedichte durch den Würfel zu verfertigen* ("The art of creating serious and funny greeting poetry by throwing dice") (1825), and only four years later, he brought out *Neunhundert neun und neunzig und noch etliche Almanachs-Lustspiele durch den Würfe. Das ist: Almanach dramatischer Spiele für die Jahre 1829 bis 1961* ("Ninehundredandninety-nine and some more almanac-games throwing dice, i.e.: Almanac of dramatic games for the years 1829 to 1961") (1829), a parody of the popular *Almanach dramatischer Spiele auf dem Lande* ("An almanac of dramatic games in the country"). The book contains 1,200 numbered dramatic text fragments including dialogues, stage directions, titles, subheadings, etc. To enable its users to create a readable and playable drama from these fragments, the book is accompanied by a so-called *Wurf-Tabelle* ("throwing chart") (fig. 5). After each throw, the reader must search in the chart for the number of the particular fragment. For producing a complete drama, 200 throws are required.

Much later, around 1970, many writers were experimenting with non-linear texts. This tendency was accompanied by influential reader-oriented theories of literature such as Umberto Eco's reflections on the "open work of art,"

	1r Wurf	2r Wurf	3r Wurf	4r Wurf	5r Wurf	6r Wurf	7r Wurf	8r Wurf
⚀	167	519	304	663	25	1078	413	823
⚁	521	2	168	20	958	30	664	306
⚂	826	959	15	522	3	169	1080	414
⚃	1081	10	416	667	308	827	523	40
⚄	5	172	524	6	961	1082	309	668
⚅	829	669	962	828	1083	1155	35	964

Fig. 5. *Wurf-Tabelle* from Georg Nikolaus Bärmann's *Neunhundert neun und neunzig und noch etliche Almanachs-Lustspiele durch den Würfel* (1829).

Roland Barthes' distinction between "readerly" and "writerly texts," or the gap-concept of Wolfgang Iser's reader-response theory. All of these were approaches conceptualizing the reader as collaborator or "wreader" as George Landow would have called it some thirty years later (14).

In Germany, the subgenre of the permutative dictionary novel was quite successful at that time. The text of such a novel is subdivided into numerous entries, which are arranged in alphabetical order. Schmitz-Emans argues that the dictionary novel is characterized by discussing order and by reflecting the contingency of order since the order of the alphabet represents a contingent principle of structuring knowledge that competes with its immanent order—if there is anything like that (Schmitz-Emans 182). The best-known novel of this genre certainly is Andreas Okopenko's *Lexikonroman einer sentimentalen Reise zum Exporteurstreffen in Druden* ("Dictionary-novel of a sentimental journey to the exporters' meeting at Druden") (1970). In the *Gebrauchsanweisung* ("introductory directions"), Okopenko calls on his readers to concoct their individual arrangement of text fragments:

> Die sentimentale Reise zum Exporteurtreffen in Druden muß erst vollzogen werden. Das Material liegt bereit, wie die Donau und die Anhäufungen von Pflanzen, Steinen und Menschen an ihren Ufern für viele Reisen und Nebenausflüge nach Wahl bereitliegen. Das Material ist alphabetisch geordnet, damit Sie es mühelos auffinden. Wie in einem Lexikon. (*Lexikonroman* 5)

> The sentimental journey to the exporters' meeting at Druden has to be executed first. The material is waiting, just as the Danube and the multitude of plants, stones and people at its banks are waiting for many side-trips of their own choice. The material is ordered alphabetically so that you can find it without difficulty. Just like in a dictionary.

The entries are arranged alphabetically. In addition, some fragments are linked by arrowheads (→) and some italicized fragments are privileged: "*Die* Hinweise, die Ihnen von Etappe zu Etappe die Fortsetzung der Reise ermöglichen sollen und die Sie daher vielleicht mit Vorrang beachten werden, sind *schräg gedruckt*" ('*Those* references, which may help you to continue the journey from place to place and which thus should have priority are *printed in italics*') (*Lexikonroman* 5). In his next novel *Meteoriten* (1976), Okopenko deliberately intensifies the principle of openness by explicitly doing without any game rules and instructions. Ror Wolf had noted in the early 1970s that he would be highly fascinated by a book,

> das man aus dem Schrank nimmt, wahllos aufschlägt, in dem man auf jeder Seite anfangen kann zu lesen, und in dem man immer, gleich wo man aufschlägt, den Einstieg findet, in dem es also unwichtig ist, was vorher oder nachher passiert. (Baier 154f.)

> which you get from the shelves, open at random, in which you can start reading on any page, and into which you always find an entry, no matter where you open it, so that it is unimportant what previously happened and what is going to happen afterwards.

More than ten years later, he started publishing a series of literary dictionaries whose fictitious editor is Wolf's *alter ego* Raoul Tranchirer—a pen name which explicitly refers to the German verb *tranchieren* ("to carve, to chop up sth."): The first publication was *Raoul Tranchirers vielseitiger großer Ratschläger für alle Fälle der Welt* (1983), followed by *Raoul Tranchirers Mitteilungen an Ratlose* (1988), *Raoul Tranchirers Welt- und Wirklichkeitslehre aus dem Reich des Fleisches, der Erde, der Luft, des Wassers und der Gefühle* (1990) and finally by *Raoul Tranchirers Bemerkungen über die Stille* (2005).

Ferdinand Kriwet's *Rotor* (1961) is not a dictionary novel but a sort of text kit consisting of 98 elements that can be used in any combination. In his *durch die runse auf den redder* (1965), however, the reading is being sidetracked from the main text into four auxiliary texts. Peter O. Chotjewitz applied a similar method in his novel *Vom Leben und Lernen* (1969) whose subtitle reads "stereo texts." The story starts off with a seemingly conventional plot on the first pages. But soon the reader realizes that numerous footnotes are incorporated into the novel's beginning. The extent of these footnotes far exceeds that of the main text.

Oswald Wiener's *die verbesserung von mitteleuropa, roman* (1969), too, breaks with conventional reader-expectations. Although this book is explicitly characterized as a novel, the text starts off with a person and subject index from which the reader can browse through the text. At the end of the book, there are various appendices. The main text does not consist of a fictitious story but rather of an extensive montage of both fictional and theoretical passages. Arno Schmidt's monumental book *Zettels Traum* (1970)—a book of 1,334 pages composed of three columns—tells the story of a writers' couple visiting a colleague who is working on a translation of Edgar Allan Poe. Handwritten notes, blackenings, a strange orthography, etc. are incorporated into the typescript. The reader can permanently jump from the conversation of the writers in the middle column to the Poe interpretation in the left or to loads of annotations and citations in the right column. Two years earlier than Schmidt, Franz Mon published *Herzzero* (1968), a text consisting of two versions arranged in two columns. Similar cross-readings have a long tradition in German literature which can be traced back to Georg Christoph Lichtenberg and turned into one of the starting points of the collage of quotes.

Konrad Balder Schäuffelen's "lottery novel" *deus ex scatola: entwicklungsroman* (1964) does not bear any resemblance to a book. The texts are printed on more than 60,000 paper-strips sentence by sentence. These strips have been hand-rolled and put into wooden cases according to the principle of chance. For the handling of the language material, tweezers are included. The ironic subtitle of *deus ex scatola* can be taken literally: The "novel" is only being generated if the user/reader uses the tweezers to draw the paper slips on which the text fragments are printed from the case—and then unwraps them (in German *entwickeln* means "to develop" or "to produce" but it also connotes "to unwrap"). For other lottery novels, Schäuffelen cut classics such as Thomas Mann's novella *Gladius Dei* or well-known poems by Goethe and others into pieces in order to allow the recombining of the fragments by the user. He also experimented with other alternative ways to arrange text in various storage media. *Haus der Bienenkönigin* (1988) is a case constructed like a bee house; from this bee house, the user can draw paper slips on which supposedly all the words from Jean-Paul Sartre's *Les Mots [Words]* are written

Herta Müller's *Der Wächter nimmt seinen Kamm* (1993) is a literary card game in the tradition of Marc Saporta (fig. 6). It consists of a box with 94 facsimiled text-image collages in postcard format. Müller cut the words from books and newspapers and pasted them onto the cards. In addition, there are photo collages

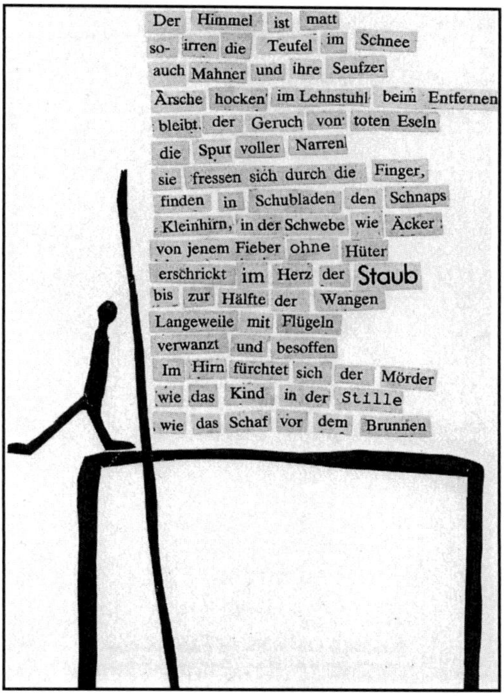

Fig. 6. Card from Herta Müller's *Der Wächter nimmt seinen Kamm* (1993).

or silhouettes on each card. The user/reader can arrange and read the cards in random order. Although poetic images are densely concentrated here onto single, unbound pages, they form an evolving network of motifs that give unity to the whole.

A similar box had been produced 30 years earlier by the Fluxus artist Tomas Schmit. Instead of postcards, Schmit's *Verlegerbesteck* (1966) contains small adhesive labels, which the user can paste on whatever surface wherever he likes (fig. 7). The leaflet gives the following reading and acting instruction:

> you—owner of this case—should have it more in your pocket than somewhere else, since here you're not so much asked to be reader, but: to be publisher (performer) and to give (real) context and (anonymous) (second-grade-) readers (resp: second-grade-performers: who fill them out: for the third-grade-readers).

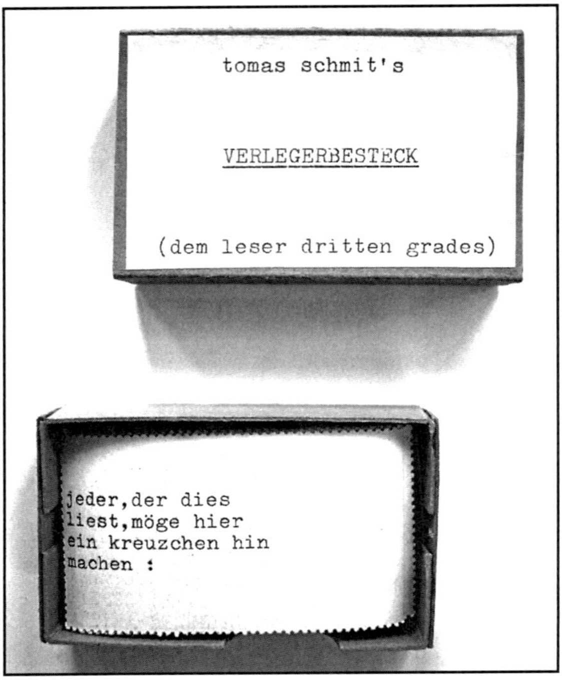

Fig. 7. Tomas Schmit's *Verlegerbesteck* (1966).

It is striking that readers of printed hyperfictions are very often called on to co-operate in continuing the writing process. In Okopenko's dictionary novels, the reader is invited to improve the alleged shortcomings of the book: "Nehmen Sie das Prinzip für die Durchführung, denken Sie an den ersten Computer, erweitern Sie den Roman durch eigene Weiterknüpfung an Reizwörter, am besten: schreiben Sie ein Buch, das meines in seiner Kleinheit festnagelt" ('Take the principle for the realization, think of the first computer, expand the novel by amending provocative words, or even better: write your own novel which will nail down the insignificance of my own') (Okopenko, *Lexikonroman* 7).

Additionally, there are gaps in the "dictionary" every now and then, in which the reader is invited to record his or her connotations and comments (11), paste images (23) or self-written poems (33). Kurt Marti, too, designed his dictionary novel *Abratzky oder Die kleine Blockhütte: Lexikon in einem Band* (1971) "von vorneherein als einen Anfang, als Fragment, als immerwährendes 'work in progress,' das jeden Leser zur fortlaufenden Mitarbeit einlädt" ('from the very outset as a beginning, a fragment, an everlasting 'work in progress' inviting its readers to co-operate continuously') (blurb). Ror Wolf even stimulates "alle natürlich denkenden Leser" ('all clear-thinking readers') to set up "Wirklichkeitsvereine" ('reality clubs') so that "Tranchirers Gedanken das Unglück und die Erfolglosigkeit ersticken" ('Tranchirer's thoughts will extinguish misfortune and failure') (190).

Collaborations, Telecommunication and Literature

This leads me to collaborative writing projects that can be distinguished into two types: firstly, collaborative texts, which are produced on site, e.g. in literary salons, writers' groups or creative writing classes; secondly, collaborative long-distance writing projects, which actively involve telecommunication networks. The parlor games of the Baroque era already aimed at increasing creativity with the help of group dynamics, and oracle games were anticipating the surrealist *cadavre exquis* ("exquisite corpse"). In these games, sentence fragments were whispered into the ear of the neighbor. Thus simple syntagmas and eventually entire sentences were generated step by step. Of course, this is not a literary genre in a strict sense as are, e.g., the so-called *Reyenreime* in which rhyme and meter need to be held. This method is getting more complicated if each collaborator has to contribute one and a half or even two lines as can be seen from the poem "Frülings-Freude" by Harsdörffer, Sigmund von Birken and Johann Klaj:

> F.
> Es fünken und flinken und blinken
> S. bunt-blümichte Auen.
> Es schimmert und wimmert und glimmert
> K. frü-perlenes Thauen.
> Es zittern / und flittern / und splittern
> F. laubträchtige Aeste.
> Es säuseln / und bräuseln / und kräuseln
> S. windfriedige Bläste.
> Es singen / und klingen / und ringen
> K. Feld-stimmende Flöten.
> Schalmeyen am Reyen erfreuen. . . .
> (Qtd. in Rühm, *Pegnitz Schäfer* 10f.)

Up until the present, there have been many collaborative novels, novellas, and stories in German literature that were usually written in sequence. Thus writers were either co-operating at the same place anyway, or they only used the post for sending the completed segments from one to the other. Various kinds of artistic co-operation ranging from mere sociability to programmatic discussions in literary salons and in societies and even to collective writing in poets' circles established during the 18th and 19th century.

Ludwig Uhland and Justinus Kerner jointly wrote the satire *Abendphantasie an Mayer* (1802) as did Arno Holz and Johannes Schlaf in writing the novella *Papa Hamlet* (1889). In 1808, the novel *Die Versuche und Hindernisse Karls: Eine deutsche Geschichte aus neuerer Zeit* was published anonymously; it later turned out that Karl August Varnhagen von Ense, Wilhelm Neumann, August Ferdinand Bernhardi and Friedrich de la Motte Fouqué had jointly written it. Fouqué was also participating, with E.T.A. Hoffmann, Adelbert von Chamisso and Karl Wilhelm Salice Contessa, in the unfinished *Roman des Freiherrn von Vieren*. The *Roman der Zwölf* (1908) was initiated by the publisher Konrad W. Mecklenburg who invited twelve writers into a co-operative venture, among them Hanns Heinz Ewers, Hermann Bahr, Otto Julius Bierbaum, Herbert Eulenburg and Gustav Meyrink.

A variety of recent serial novels are also based on a similar concept: *Das Gästehaus* (1965) was inspired by Walter Höllerer when he was director of the *Literarisches Colloquium Berlin*. Among the contributors were Peter Bichsel, Nicolas Born, Hans Christoph Buch, Hubert Fichte and Hermann Peter Piwitt. The weekly journal *Die Zeit* initiated a cliffhanger novel in 1999 to which for one year the cream of German contemporary literature contributed. It set off with Marcel Beyer's opening chapter "Anruf um Mitternacht" which was followed by texts by Birgit Vanderbeke, Andreas Neumeister, Judith Kuckart and many others until Terézia Mora wrote the final chapter "Höllisches Finale" one year later. In genre literature, too, serial novels have been written from time to time. Such famous writers like Heinrich Böll, Christine Brückner, Reinhard Federmann, Hermann Kasack and Hans Weigel contributed to the satiric crime novel *Der Rat der Weltunweisen* (1965), and only recently ten German crime writers, among them well-known writers such as Gisbert Haefs and Ingrid Noll, jointly published the crime novels *Eine böse Überraschung* (1998) and *Gipfeltreffen* (2000).

In the international Dada movement, collaborative writing had a very different quality. In Zurich, Hans Arp, Walter Serner and Tristan Tzara produced some collaborative poems under the name *société anonyme pour l'exploitation du vocabulaire dadaïste*, and Tzara, Richard Huelsenbeck and Marcel Janco had even radicalized the concept of artistic co-operation with their simultaneous poems. Dadaist simultaneous poems were originally written for the performance on the stage of the "Cabaret Voltaire" where several speakers were to recite their parts simultaneously. Hugo Ball defined the simultaneous poem as "kontrapunktisches Rezitativ" ('contrapuntal recitative') dealing with the "Wert der Stimme" (*Die Flucht aus der Zeit* 86) ('value of the voice', *Flight Out of Time* 57).

During the 1950s and 1960s, the Dada impulse was taken up in a different manner by both the *Wiener Gruppe* ('Vienna Group') and the "Stuttgarter Gruppe." It is thus not surprising that the members of both groups collaboratively wrote and published several texts. Konrad Bayer, H.C. Artmann, Gerhard Rühm, Friedrich Achleitner and Oswald Wiener considered the Vienna Group not only as a forum for joint performances of their "literary cabaret" but also as a writers' collective. Thus they collaborated in varying constellations:

> jeder brachte geeignetes material heran, wir spielten uns immer besser aufeinander ein, warfen uns die sätze wie bälle zu. Wenn auch jeder für sich die erschlossenen möglichkeiten weiterverfolgte, erwies sich gerade die montage als eine technik, die gemeinschaftsarbeit besonders begünstigte. (Rühm, "das phänomen" 25)

> everybody contributed suitable material: we soon became a real team, tossed sentences to each other like balls. Although we each individually made use of the potential we had tapped, montage proved to be a technique particularly conductive to the production of joint works. (Rühm, "das phänomen" 24; my revisions)

In Stuttgart, Harig and Döhl jointly wrote the prose text *Hans und Grete: Eine deutsche Sprachlehre* (1970). Radio plays such as Bense and Harig's *Der Monolog der Terry Jo* (1968) or *Türen und Tore* (1971) by Jürgen Becker, Harig and Döhl were also written in cooperation.

Writers who were physically separated have been using their contemporary transmission media for networked collaborative projects for a long time, ranging from letters carried from one writer to the other by post—i.e. the transport network—to the more recent technical communication media such as telephone, telegram, telefax, the French "Minitel" system up to the latest computer-based technologies and services like the World Wide Web, e-mail, mobile phones or SMS.

Writers' letters always have been an ambivalent genre: Usually, the published correspondence of writers is nothing but a documentation of their private letters to friends and fellow writers, which at some point was published posthumously. Edited letters are characterized by a "double address": On the one hand, they were originally addressed to a specific addressee; on the other hand, however, the general public gets access to letters without knowing much about the reason and motivation of the writer and thus can only rely on the annotations of the editor.

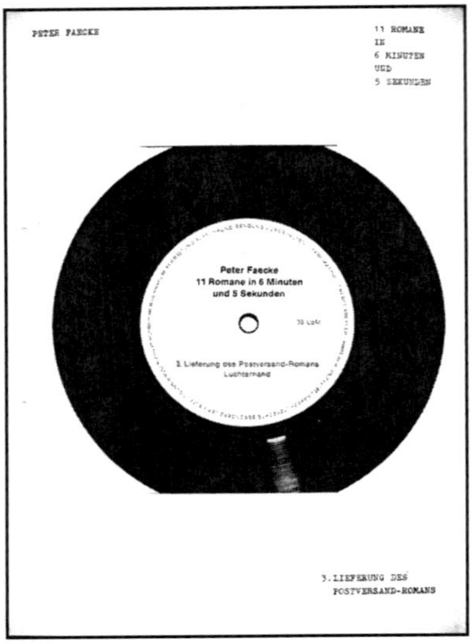

Fig. 8. Peter Faecke's *11 Romane in 6 Minuten und 5 Sekunden,* the 3rd delivery to the *Postversandroman* (1970).

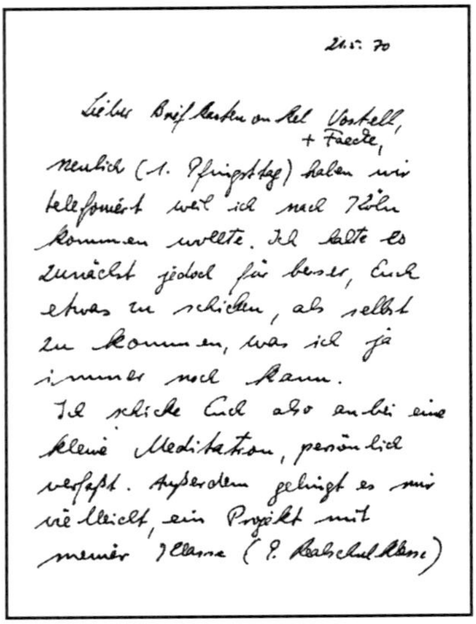

21.5.70

Lieber Briefkastenonkel Vostell,
+ Faecke,
neulich (1. Pfingsttag) haben wir telefoniert weil ich nach Köln kommen wollte. Ich halte es zunächst jedoch für besser, Euch etwas zu schicken, als selbst zu kommen, was ich ja immer noch kann.
Ich schicke Euch also anbei eine kleine Meditation, persönlich verfaßt. Außerdem gelingt es mir vielleicht, ein Projekt mit meiner Klasse (9. Realschulklasse)

Fig. 9. Reader's contribution to Peter Faecke's and Wolf Vostell's *Postversand Roman.*

However, there are exceptions to this rule: in some cases letters must be regarded as *constituents* of an artwork. The essay "Ueber 'Wilhelm Meisters Wanderjahre'" (1821) by Rahel Varnhagen and her friends is an example of a text which in fact was written in correspondence but was projected for publication from the outset.

This is comparable for artists' contributions to Mail Art or Correspondence Art projects, which were promoted in Germany by Klaus Groh, the editor of the *I.A.C. (International Artist Cooperation) Newsletter,* by Géza Perneczky and others. A higher degree of publicity than most Mail Art projects was achieved by the *Postversandroman* ("Mail Order Novel") (1970), a collaborative book project of the writer Peter Faecke and the Fluxus and Happening artist Wolf Vostell, which was published in the renowned Luchterhand Verlag. The purchaser of this "novel" first received a hardcover jacket and an envelope containing two bolts for binding the consecutive deliveries. The users' instruction was titled *Spiel ohne Grenzen* ("Games without frontiers") thereby explicitly addressing the limitations of the printed book. In accordance with the tone of the time, Faecke and Vostell called for replacing the institutions of the literary market in favor of an autonomous distribution of texts by the producers themselves. They also tried to overcome the limits of print by supplementing their "novel" with objects such as a vinyl single with Peter Faecke's *11 Romanen in 6 Minuten und 5*

Sekunden ("11 novels in 6 minutes and 5 seconds," fig. 8). Vostell's instructions for a happening, which callers could retrieve from his answering machine, are documented; elsewhere the readers were invited to paste newspaper clippings or account statements and to actively participate in this "work in progress" by sending their own contributions (fig. 9):

> nutzt den Postversand-Roman, schreibt uns, ruft uns an, bombardiert uns verflucht noch mal, wir antworten Euch in den nächsten Lieferungen. Faecke/Vostell: die beiden guten Briefkastenonkels, unsere Frauen helfen uns *und* sind ganz dabei, wir verteilen die Lieferungen auf der Straße soweit unser Geld reicht und machen einen Film drüber in Kölner Obdachlosen-Vierteln, den Ihr kostenlos an Kölner Hauswänden sehen könnt, schickt uns Informationen, ruft uns an unter
> *735335* Faecke
> oder
> 517783 Vostell
> wir antworten direkt *mit neuen Ideen, wir haben genug davon,* wenn Ihr Geld schicken wollt, OK., schickt uns *auf jeden Fall* Informationen, Manuskripte etc., wir wollen Bücher haben, die nicht mehr als 1 Mark kosten, wir wollen Bücher haben, die, nach Wahl durch ein akzeptables Gremium, gratis verteilt werden (Kostenpunkt jährlich für alle wichtigen Bücher in der BRD: 500 Millionen Mark). . . . Konkreter: Denk daran: Deine Antwort auf das, was Du bezahlt hast, wird von uns honoriert in Form einer Antwort in der nächsten Lieferung. Was kommt auf Dich, was kommt auf uns zu? Tu uns den Gefallen: spiel mit, tritt Deinen Nachbarn in den Arsch, *tritt dich selbst in den Arsch, zieh Deine alten Schubladen auf, geh ans Telefon, setz Dich ins Auto und fahr zu uns, hör mit uns die alten Lieder von dazumal, und wenn Du uns nicht passt, schmeissen wir Dich wieder raus!,* fang *endlich* selbst an zu schreiben, schreib zusammen mit uns dieses Buch. . . . (Faecke and Vostell; italics represent handwritten annotations in original)

> Use the *Postversandroman* ("mail order novel"), write to us, call us, bombard us—damn it, we'll answer you in one of the next deliveries. Faecke/Vostell: the two agony uncles; our wives will assist us *and* will fully take part, we'll distribute the deliveries in the streets as far as our money will take us, and we'll shoot a film about these activities in one of the homeless neighborhoods in Cologne; you'll be able to watch this movie on Cologne house walls, send us information, call us at

> *735335* Faecke
> or
> 517783 Vostell
> we'll answer immediately *with brand-new ideas, we have plenty of them*, if you'd like to send money, OK., but send us information, manuscripts, etc. *no matter what*, we want to have books which don't cost more than 1 German mark, we want to have books which (being selected by an appropriate committee), will be distributed free of charge (annual expenses for all important books in the FRG: 500 million German marks). . . .
> More specifically: Remember: We will remunerate your answer of your payment in form of an answer in our next delivery. What can you expect to happen, what can we expect to happen? Please do us the favor: join in, kick your neighbor's ass, *kick your own ass, search your old drawers, answer the phone, get into your car and come to see us, listen to the old songs with us, and if we don't like you, we'll kick you out!,* start writing yourself, write this book together with us. . . . (Italics represent handwritten annotations in the original)

The avant-garde utopia of transgressing the boundary between art and everyday life can only be realized, Vostell argues in an interview with Lothar Romain, if the latest media technologies are being used. Hence he considers his use of answering machines only as a first step towards universal and bidirectional means of communication. Such means would allow and require to mix up all media in existence today:

> Jede Kommunikation heißt in Zukunft jeder mit jedem auf der Welt, so wie jedes Ereignis auf der Welt mit jedem Ereignis auf der Welt zu tun hat. . . . Und formal gesehen glaube ich, daß zum Beispiel die Zukunft des Fernsehens darin besteht, daß alle Fernseher im Medium Fernsehen selbst zu Wort kommen, also daß das ein Ende findet, daß Millionen das konsumieren, was von wenigen Redakteuren gemacht wird. Das heißt, in Zukunft produzieren die Hörer oder die Seher selbst das Programm. (Faecke and Vostell)

> Every act of communication will connect each and everybody all over the world, such as each incident in the world has to do with every other incident in the world. . . . And from a formal point of view, I believe that the future of television, e.g., will be that every viewer will have his say on TV, so that we can end a situation in which millions

> of people are just consuming what is being provided by only a few editors. This means that in the future listeners and spectators will create their own program.

What Vostell expects from bidirectional television, complies with the utopian visions, which Bertolt Brecht drew up with the radio in mind some decades earlier. In his famous radio theory, Brecht demanded the transformation of radio, the grandfather of all wireless media, from a means of broadcasting into a multichannel means of communication. He argued that the technological potentials of the medium were not used adequately as long as the back channel was closed:

> Der Rundfunk wäre der denkbar großartigste Kommunikationsapparat des öffentlichen Lebens, ein ungeheures Kanalsystem, d.h., er wäre es, wenn er es verstünde, nicht nur auszusenden, sondern auch zu empfangen, also den Zuhörer nicht nur hören, sondern auch sprechen zu machen und ihn nicht zu isolieren, sondern ihn in Beziehung zu setzen. Der Rundfunk müsste demnach aus dem Lieferantentum herausgehen und den Hörer als Lieferanten organisieren. (Brecht 553)

> The radio would be the finest possible communication apparatus in public life, a vast network of pipes. That is to say, it would be if it knew how to receive as well as to transmit, how to let the listener speak as well as hear, how to bring him into a relationship instead of isolating him. On this principle the radio should step out of the supply business and organize its listeners as suppliers. (Qtd. in Strauss 15)

This utopia of a bidirectional technological medium turning the passive reader, listener or spectator into an actor is a persistent claim of 20th century media theory. Hans Magnus Enzensberger in his "Baukasten zu einer Theorie der Medien" ["Constituents of a Theory of the Media"] (1970) explicitly tied in with Brecht's theory demanding networked communication, "die auf dem Prinzip der Wechselwirkung aufgebaut sind" ('based upon the principle of reciprocity') (170).

Since that time, the latest transmission media such as radio and telephone, television and telefax have always been used for collaborative writing projects of which only a few should be mentioned here: In the context of the *Neues Hörspiel* ("new radio play") movement many writers and radio experts tried to put Brecht's call for a mobilization of the listeners into action. A good example is Richard Hey's *Rosie: Radio-Spektakel zum Mitmachen für Stimmen, Musik und telefonierende Hörer* (1969), a radio play in which the listeners had the opportunity to prompt the moderator to select one specific plot variant. On Michael Erlhoff's initiative, the *Cassetten Radio* was carried out from January to December 1984. Following the chain letter principle, audiotapes were sent to various collaborators who recorded literary texts, noises or songs. Hence a technical storage medium was combined with traditional mail. Robert Adrian X, a Vienna-based Canadian artist, was among the very first artists who used computer-based telecommunication technologies in the 1980s. He was one of the initiators of *ARTEX* (Artists' Electronic Exchange Program), the first international communication system reserved to art, and he also founded, together with Helmut Mark, Zelko Wiener, Karl Kubaczek and Gerhard Taschler, the art group *BLIX* which initiated co-operative art projects such as *Wiencouver IV* (1983), *Kunstfunk* (1984) or *Kunst BTX* (1985). For *Ars Electronica 1982*, Adrian organized the telecommunication project *The World in 24 Hours* in which 16 artists from three continents tried for one day "der Mittagssonne rund um die Erde zu folgen und dabei eine Art von telematischer Weltkarte zu schaffen" ('to follow the midday sun around the planet—creating a kind of telematic world map') (Adrian 145). But dealing with Adrian's work, however, we are already talking about current writers' and artists' telecommunication projects in computer-based and networked media, but this is too big a subject for *this* essay . . .

Notes

1. This is a short version of Schäfer, "Literary Machines Made in Germany." I am grateful to Peter Gendolla for his critical advice, to Brigitte Pichon and Dorian Rudnytsky for checking the English version of this text and for translating quotes from Baroque German into contemporary English, and to Patricia Tomaszek for compiling the bibliographical information.
2. Cf. Roberto Simanowski's as well as Peter Gendolla's and my joint article in this book.
3. Besides the methods and procedures mentioned so far, there are plenty of other methods of constrained writing such as chronogram, acrostic, or abecedarius, which I cannot discuss here due to limitated space. For more details cf. Schäfer, "Literary Machines Made in Germany."

Works Cited

Aarseth, Espen. *Cybertext: Perspectives on Ergodic Literature.* Baltimore: Johns Hopkins UP, 1997.

Adelung, Johann Christoph. *Umständliches Lehrgebäude der Deutschen Sprache, zur Erläuterung der Deutschen Sprachlehre für Schulen.* Vol. 1. 1782. Hildesheim: Olms, 1971.

Adrian X, Robert. "Elektronischer Raum." *Kunstforum international* 103 (1989): 142-147.

Arp, Hans. *Wortträume und schwarze Sterne: Auswahl aus den Gedichten der Jahre 1911-1952.* Wiesbaden: Limes, 1953.

———. *Gesammelte Gedichte: Gedichte 1903-1930.* Vol. 1. Zurich: Arche, 1963.

Baier, Lothar, ed. *Über Ror Wolf.* Frankfurt am Main: Suhrkamp, 1972.

Ball, Hugo. *Die Flucht aus der Zeit.* Munich: Duncker & Humblot, 1927.

———. *Flight out of Time.* Ed. John Elderfield. New York: Viking, 1974.

Bärmann, Georg N. *Neunhundert neun und neunzig und noch etliche Almanachs-Lustspiele durch den Würfel. Das ist: Almanach dramatischer Spiele für die Jahre 1829 bis 1961.* 1829. Munich: Heimeran, 1972.

Bellmer, Hans. *Die Puppe.* Frankfurt am Main: Ullstein, 1976.

Bense, Max. *bestandteile des vorüber: dünnschliffe mischtexte montagen.* Cologne: Kiepenheuer & Witsch, 1961.

Brecht, Bertolt. "Der Rundfunk als Kommunikationsapparat." 1932. *Große kommentierte Berliner und Frankfurter Ausgabe.* Vol. 21. Ed. Werner Hecht, Jan Knopf, Werner Mittenzwei, and Klaus D. Müller. Frankfurt am Main: Suhrkamp, 1992. 552-557.

Cramer, Florian. "Combinatory Poetry and Literature in the Internet." *Kodikas/Code: Ars Semeiotica* 24.3-4 (2001): 243-247.

Döblin, Alfred. "Bemerkungen zum Roman." 1917. *Aufsätze zur Literatur.* Ed. Walter Muschg. Olten: Walter. 1963. 19-23.

Döhl, Reinhard. "Hans Arp and Zurich Dada." *Dada Zurich: A Clown's Game from Nothing.* Ed. Brigitte Pichon and Karl Riha. New York: G.K. Hall, 1996. 112-124.

———. "Vom Computertext zur Netzkunst. Vom Bleisatz zum Hypertext." 1997. 4 Aug. 2006 <http://www.uni-stuttgart.de/ndl1/computertext_netzkunst.htm>.

Enzensberger, Hans M. "Baukasten zu einer Theorie der Medien." *Kursbuch* 20 (1970): 159-186.

Faecke, Peter, and Wolf Vostell. *Postversandroman.* Neuwied: Luchterhand, 1970.

Gardt, Andreas. *Sprachreflexion in Barock und Frühaufklärung: Entwürfe von Böhme bis Leibniz.* Berlin: De Gruyter, 1994.

Harsdörffer, Georg Philipp. *Frauenzimmer Gesprächspiele.* 8 vols. 1641-49. Tübingen: Niemeyer, 1968.

———. *Poetischer Trichter.* 1647-53. Darmstadt: Wissenschaftliche Buchgesellschaft, 1969.

———. *Deliciae mathematicae et physicae: Der Mathematischen und philosophischen Erquickstunden Zweyter Theil.* 1651. Frankfurt am Main: Keip, 1990.

Hundt, Markus. *"Spracharbeit" im 17. Jahrhundert: Studien zu Georg Philipp Harsdörffer, Justus Georg Schottelius und Christian Gueintz.* Berlin: De Gruyter, 2000.

Jean Paul. *Vorschule der Ästhetik.* 1804. *Werke.* Vol. V. Ed. Norbert Miller. Munich: Hanser, 1963.

———. *Horn of Oberon: Jean Paul Richter's School for Aesthetics.* Trans. Margaret R. Hale. Detroit: Wayne State UP, 1973.

Kuhlmann, Quirinus. *Himmlische Libes-küsse.* 1671. Tübingen: Niemeyer, 1971.

———. *Geschicht-Herold oder Freudige und traurige Begebenheiten Hoher und Nidriger Personen.* Jena 1672.

Landow, George P. "What's a Critic to Do? Critical Theory in the Age of Hypertext." *Hyper/Text/Theory*. Ed. George P. Landow. Baltimore: Johns Hopkins UP, 1994. 1-48.

Marti, Kurt. *Abratzky oder Die kleine Blockhütte: Lexikon in einem Band.* Neuwied: Luchterhand, 1971.

Motte, Warren F., Jr. ed. *Oulipo: A Primer of Potential Literature.* Normal, IL: Dalkey, 1998.

Müller, Götz. *Jean Pauls Exzerpte.* Würzburg: Königshausen & Neumann, 1988.

Müller, Herta. *Der Wächter nimmt seinen Kamm: Vom Weggehen und Ausscheren.* Reinbek: Rowohlt, 1993.

Murphy, Richard. *Theorizing the Avant-Garde: Modernism, Expressionism, and the Problem of Postmodernity.* Cambridge: Cambridge UP, 1998.

Neubauer, John. *Symbolismus und symbolische Logik: Die Idee der ars combinatoria in der Entwicklung der modernen Dichtung.* Munich: Fink, 1978.

Okopenko, Andreas. *Lexikon Roman einer sentimentalen Reise zum Exporteurtreffen in Druden.* Salzburg: Residenz, 1970.

Pastior, Oskar. *Anagrammgedichte.* Munich: Renner, 1985.

———. *Das Unding an sich: Frankfurter Vorlesungen.* Frankfurt am Main: Suhrkamp, 1994.

———. *Eine kleine Kunstmaschine: 34 Sestinen.* Munich: Hanser, 1994.

Rieger, Stefan. *Speichern/Merken: Die künstlichen Intelligenzen des Barock.* Munich: Fink, 1997.

Rühm, Gerhard. "das phänomen 'wiener gruppe' im wien der fünfziger und sechziger jahre." *die wiener gruppe: ein moment der moderne 1954-1960/die visuellen arbeiten und die aktionen / the vienna group: a moment of modernity 1954-1960/the visual works and the actions.* Ed. Peter Weibel. Vienna: Springer, 1997. 16-29.

Schäfer, Jörgen. "Literary Machines Made in Germany: German Proto-Cybertexts from the Baroque Era to the Present." *Cybertext Database* (forthcoming).

Schmit, Tomas. *Verlegerbesteck*. Limited edition of 60 boxes. Berlin: n.p., 1966.

Schmitz-Emans, Monika. "Labyrinthbücher als Spielanleitungen." *Paragrana* 11.1 (2002): 179-207.

Schöning, Klaus, ed. *Neues Hörspiel: Texte Partituren*. Frankfurt am Main: Suhrkamp, 1969.

Schottelius, Justus Georg. *Ausführliche Arbeit Von der Teutschen HaubtSprache*. 1663. Tübingen: Niemeyer, 1967.

Schulze, Holger. *Das aleatorische Spiel: Erkundung und Anwendung der nichtintentionalen Werkgenese im 20. Jahrhundert*. Munich: Fink, 2000.

Strauss, Neil, ed. *Radiotext(e)*. New York: Semiotext(e), 1993.

Thomkins, André. *dogma-mot*. Limited edition of 111 boxes. N.p.: n.p., 1965.

Tzara, Tristan. *Œuvres complètes*. Vol. 1. Ed. Henri Béhar. Paris: Flammarion, 1975.

Wolf, Ror. *Raoul Tranchirers Welt- und Wirklichkeitslehre aus dem Reich des Fleisches, der Erde, der Luft, des Wassers und der Gefühle*. Gießen: Anabas, 1990.

Zesen, Philipp von. *Hoch-Deutscher Helikon*. 1656. *Sämtliche Werke*. Vol. X/1. Ed. Ferdinand van Ingen, Ulrich Maché, and Volker Meid. Berlin: De Gruyter, 1977.

Zürn, Unica. *Gesamtausgabe*. Vol. I. *Anagramme*. Berlin: Brinkmann und Bose, 1988.

Friedrich W. Block

Humor — Technology — Gender

Digital Language Art and Diabolic Poetics

Abstract:

This essay argues that the poetic turn from nothing to form art tends to "diabolic" strategies in present language allowing for a self-referential presentation of cultural distinctions. This poetic deconstruction of symbolic forms such as man/machine, male/female, or 0/1 is closely related to humor and gender in cultural and artistic performances. This shall be illustrated by discussing two examples of language art in the field of digital electronics: the interactive installation *Die Amme* by Peter Dittmer and *female extension,* a subversive net art project by Cornelia Sollfrank. These projects are interpreted as gendered forms of the poetic as comic self-observation.

I. Once Again: Who is Speaking?

Up until now, the intense interest in the distinction (also dedifferentiation) between "subjective" or "natural," and "technical" or "artificial" qualities and activities has been highlighted in the discourse about computers and their culture, more than anywhere else. The "universal machine," digital technologies, or notable concepts such as virtual reality and artificial intelligence are still regarded as great challenges. Ascriptions of vicinity, the amalgamation or analogy of man and machine—affirmative or critical—determine both the technology theory and media art.

The Turing Test, which Alan Turing introduced in 1950 in his legendary essay "Computing Machinery and Intelligence," remains discursively highly effective, and is still discussed to this day. This test is laid out as an "imitation game" where a questioner is confronted with answers which come from a human source, initially, and are then produced mechanically—without the questioner's knowledge. In his essay, Turing then re-defines the man of body, mind and social being as an information processing system, assuming that the crucial question "can machines think?" is not conclusively decided. At the same time Max Bense's cybernetic aesthetics similarly argue the synthesis of man and machine in art (cf. *Literaturmetaphysik*).

This conception is connected with further distinctions important to computer symbolism e.g.: analog/digital, hardware/software, code/interface, natural/artificial language, interactivity/(inter-) passivity, control/autonomy and last but not least: male/female. Distinctions such as these are raised artistically when technology is staged in a self-referential way.

We shall pursue this perspective, giving humor special consideration. The comic aspect is of interest here, in its affinity to advanced, reflexive or experimental art, as both areas are concerned with deconstructing symbolism, frequently in the form of culturally highly prominent distinctions. This deconstruction or literally: dia-bolism clearly illustrates the breaks of the symbolic. It executes "a constant ambivalence" which Baudelaire defines as a characteristic of humor in his essay *De l'essence du rire* (1855): Laughter isn't of divine origin—indeed Christ never laughed—instead it is satanic (diabolic), and therefore profoundly human. According to Baudelaire, it is executed, in its absolute form, as art. In this sense, contemporary (media-) art per se tends towards humor.

This shall be illustrated by discussing two examples: Peter Dittmer's *Die Amme* and Cornelia Sollfrank's *female extension*. Although these projects hardly count as literature (or net literature)—if we understand "literature" to mean the literature system with its dominant concepts. The artists of both projects and their audience would hardly be located here. Nevertheless, this is language art, which net literature participates in, or it is "literature" in the specific sense that the technology with which the projects are primarily concerned is of linguistic nature or alphanumeric code. At all events, the artistic working with language plays an important role, on different levels, in both cases discussed here, especially in point of view of communication games. Both examples deal, more or less explicitly, with the gender difference, in a way, where humor is significant.

II. *Die Amme*

We come to the first example which, in parody, reminds us of the game of the Turing test in its interactive dialog situation: the installation *Die Amme* by Peter Dittmer, first presented in 1992 and permanently developed since then—up to the fifth generation (2005).[1]

Here, other than in the Turing test, the user knows that he/she is entering into a dialog with a machine, where keyboard and screen are used as classic interfaces. So, for the participant, the distinction between man and machine is explicit. The language dialog proceeds, alternating between arbitrary input by the audience and answers which the computer formulates, using dictionaries and programmed sentence formation rules, and then displays on the screen, after the petition to it has been lexically and syntactically analyzed. During the dialog the program randomly triggers a mechanism which spills a glass of milk in a display case. The “dialog ability” of *Die Amme* has been constantly enlarged by recording the conversations. In the meantime, the program disposes of approx. 400,000 answer modules, as well as more than 50,000 variables of analysis or input identification.

The project stands as an example for a number of works, where machines stage a grotesque exaggeration (or a kind of “caricature”) of human as well as mechanical features.[2] This has comic effects but is never harmless, however, without exception the treatment is in a serious manner. Serious in the sense that strategy and execution of the work are profound and highly and subtly differentiated[3], as well as of multi-layered and not only comic importance. This versatility also applies to the possibility of comic ascriptions: In *Die Amme*, the absurd distortion of human behavior is already indicated in the ironic signal set by the title, attributing a long extinct female profession (wet nurse) to a machine. The hardware of the spilling mechanism, milk glass and milk is controlled by software which is relatively complicated and where the design and sequences of operations are not transparent to the user. However, this is done in such a way as to mock the common function of a wet nurse (i.e. to give milk and not to spill). The desired monotonous, though slightly unpredictable “stupidity” of this putative malfunction stands in contrast to the considerable technical effort required, and also to the—simulated—communication ability of the computer, which indeed raises Turing’s question. Also this competence is ironically broken as the answers are in the style of a “run-down language” (Ernst Jandl) and therefore contradict in their supposed “naturalness” the expectation of a perfect artificial language. This may be illustrated with a short dialog (“>” stands for the audience, “<” for the computer):

. . .
> ich gehe endgültig.
< Meine Warnung lautet: Draussen überfällt den Mensch
nur derene leere Schlund von derene Welt.
> ich wünsche dir daß du verschlungen wirst.
< Da Sie sollten dem Gott anbeten für.
> gott wird dich verschlingen.
< Wollen Sie eine telefonische Leitung zum den Gott?
> ja her mit gott!
< Hallo. Hier bei Gott. Hitler is am den Apparat.
> gib mir gott!!! ich will nicht hitler! ich will gott!
< Da ich gebe Sie lieber dem Milch als mehr einem
allgemeinen Gegebenen. Vorsehen!

MILCHVERSCHÜTTUNG

> na endlich danke für die milch und tschüss.
< Aha. Nehmen und nich geben und gehen.
> was willst du denn haben?
< Ihrem Gemüt ich will Sie heraus reissen.
> du schwein.
< HA HA HA HA. Oh bin ich böse!
(Publikum ab). (Dittmer 25)

. . .
> I'm definitely going .
< my warning is: only their empty abyss of their
world attacks man outside.
> I hope that you are swallowed.
< there you should adore God for it.
> God will swallow you.
< Do you want a telephone line to the God?
> yes here with God!
< Hello. Here with God. Hitler is on this apparatus.
> give me God!!! I don't want Hitler! I want God!
< I prefer to give you the milk rather more than
a general given one. Provide!

MILK-SPILLING

> ah at last thanks for the milk and bye.
< Aha. Take and don give and go.
> What do you want to have?
< your soul I want to rip it out.
> you pig.
< HA HA HA HA. Oh am I bad!
(audience leaves).

This extract from a far longer dialog shows that the communication works for the participant, always assuming its general and specific improbability. For him/her the answers connect. They offer a hybrid mix of speech defects, verbal colloquial speech, corrupted historical word material and quotation material, in this case a ridicule is included: Can machines laugh? The machine is staged as fallible; and exactly this seems to personalize it as a (crazy) partner. Language design and milk spill keep the participant at a distance and this comic tension can be applied to the "communication process" taking place. This is also true for the art event itself. Especially in the context of the extremely expensive interactive media art of the nineties, which was euphorically celebrating the new technology age. Here—ironic in turn—this is met with exceptionally "low" technology, as far as the language processing is concerned (a simple PC is enough), contrasting with the high tech needed for the milk spill and to restore the outset situation.

Finally, such comic ascriptions can be found in the immediate vicinity of the work. Dittmer, who has a theater background, presents transcriptions of the conversations between audience and machine in the form of a comedy text. He always refers to his work in an ironic tone, avoiding the usual technical jargon, using, e.g., a style that may be associated with Heidegger:

> Die Amme ... ist ja tatsächlich, in ihrem ungeschützten Hingestelltsein in den Exhibitionsraum, das Beispiel für eine geschundene öffentliche Person mit nur begrenztem Weltverständnis, die in jeder Ansprache auch Zumutung und drohende Überwältigung vermuten muss. (Dittmer 15)

> In the element of being placed unprotected in the exhibit area, *Die Amme* is actually an example of a maltreated public person with a limited understanding of the world, who must fear unreasonable demands and the threat of being overpowered in every address.

Also, the illustrations enclosed with the commentary can be understood to be comic (fig.1): Far removed from technical diagrams, the drawings can be understood as close to the tradition of comic drawing. The contrast between "sublime" language concepts and "trivial" pictures repeats the aesthetic basic structure of the installation.

The difference between man and machine (naturalness and artificiality, acting and functioning; or according to Dittmer's "Schalten und Walten": switching and operating etc.), the question, whether machines can think, perhaps even laugh, is turned about satirically in the concept of *Die Amme*. It is exaggerated in its real satire: Vilém Flusser thought that simulation was per se a form of caricature—though in no way harmless:

> Es vereinfacht das Nachgeahmte und übertreibt einige wenige seiner Aspekte. Ein Hebel ist eine Simulation des Arms, weil er alle Aspekte

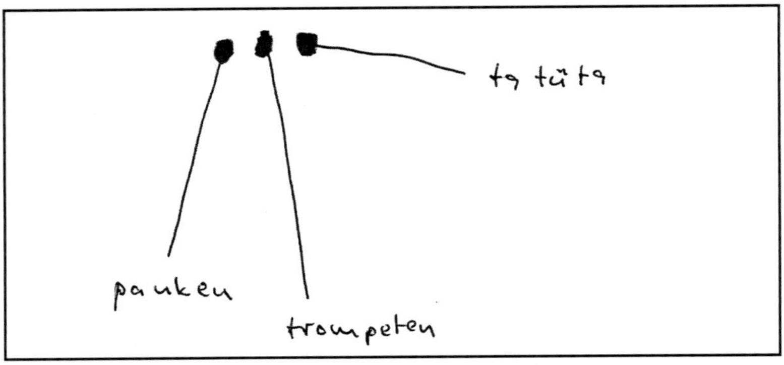

Fig. 1. Peter Dittmer: *Die Amme.*

> des Arms außer seiner Hebefunktion vernachlässigt, die Funktion aber so weit auf die Spitze treibt, daß er weit besser hebt als der Arm. Das sich durch die digitalen Codes ausdrückende und bildermachende Denken ist eine Karikatur des Denkens. Aber es wäre geradezu gemeingefährlich, diese neue Denkart deswegen etwa als dumm oder auch nur einseitig verachten zu wollen. (Flusser 143)
>
> It simplifies imitated aspects and exaggerates a few of these. A lever is a simulation of the arm because it neglects all aspects of the arm apart from its lever function, but so exaggerates this function that it far excels in lifting compared to the arm. The thought process expressed by digital codes and visualized in pictures is a caricature of thought. However, it would be virtually criminal to want to despise this new way of thinking as, say, stupid or one-sided, because of this.

Should anyone still wish to claim that the question of stupidity is also raised in Dittmer's installation (as an inability to interpret appropriately, or as a non-intentional form of comic, cf. Wirth 6-9), there would be no need to immediately decide whether, and in what manner, this should be ascribed to which factor in the game. For the observer, however, this reflection according to Flusser's is laid out, along with many other possibilities, and can connect with an—intentional—comic experience.

III. Hacking Gender

An interesting aspect of Dittmer's *Die Amme* is that the robot is implied to be female, i.e. it is gendered. Of course, this is broken ironically since, as already mentioned, the historical role of the wet nurse is forced and satirized. Once again, this aspect brings us back to Alan Turing's imitation game which is foremost a gender game, although this is rarely taken into consideration.[4] Initially the questioner is confronted with two answerers, a man and a woman not visible to him/her, and has to determine their sex on the basis of the answers given. This is made more difficult by the fact that the woman "truthfully" reveals her gender, while the man pretends to be a woman: an awkward "foreplay" on Turing's homosexuality, perhaps? In the second phase the man is then replaced by the computer. So, the question of identification combines the distinctions of man/machine and male/female. As the only possible answers are language statements, the symbolic and cultural nature of the double problem is underlined and not e.g. its substantial nature.

Thus, there is a second area linked with reflections on technology, namely the "Analyse der 'Technologien' der Subjektkonstitution und der mit ihnen verbundenen 'gesellschaftlichen Spielregeln'" ('analysis of the "technologies" of subject constitution and the "social rules" relating to these') (Rebentisch 29).

In science and art the theoretical debates on gender construction in the digital technology culture have revolved around two perspectives since the late eighties. These are: a possible reorientation beyond the positions of the first and second phase of feminism; and the analysis of how gender roles de facto appear within technological behavior. The post-feminist vision—following Donna Haraway's cyborg concept (1991)—is to leave the classic hostility towards technology behind, and to make use of the new possibilities, offensively, in the hope that gender difference, or existing order schemes, be deconstructed and newly defined. Net technologies such as chatrooms or MUDs and MOOs, where roles and identities can be freely chosen and developed, are an opportunity to separate the gender difference from corporeality, or to form corporeality fictitiously and variably, making feminist issues technologically relevant. This is a utopia, however. Complementary evidence shows that women are still highly underrepresented[5] in participating in technology dispositives, especially in hardware and software development, and that the entertainment industry flourishing in the Net reproduces glaring gender clichés—Lara Croft is a popular icon of this criticism—not to mention the reality of cybersex.

Both perspectives have been addressed artistically. Instead of "big theory" and "big stories"[6] however, artistic cyber feminism prefers deconstructive and, also, comic procedures. This shall be illustrated at this stage with a project by the concept/software artist Cornelia Sollfrank. This project both politically and aesthetically subverted the established system of gender and man/machine symbolism. For this, she provoked the art "operating system" with a type of hacker attack: In February 1997, the Hamburg Kunsthalle was the first museum in the world to announce a competition for net art, supported by Philips and the *Spiegel* magazine (*Spiegel Online*). The competition was entitled "Extension," and the museum intended to extend into virtual space. (Digitized) art was not to be entered, but rather net art, i.e. work dealing conceptionally with internet technology. For Sollfrank the starting point was as follows:

> The announcement of EXTENSION met the Zeitgeist of the year 1997. The established art world had started to become interested in this new art form and tried to deal with it more or less appropriately. Nobody wanted to miss the hype, everybody wanted to take advantage of the potential publicity and—in the best case—be the discoverer and supporter of a new art form. (Sollfrank)

The artist then exploited the relative inexperience of the organizers and the jury, and registered 200 fictitious female net artists for the competition. The names were assigned to seven different nations. Each was given a full address including phone number as well as an e-mail address which worked on various servers. These "avatars" were registered as real persons by the competition organizers and each received a password for the server, where the submitted projects were to be up-loaded. The high participation level, particularly of women, was publicly celebrated ("280 applicants— two thirds are women"—so the Kunsthalle press release).[7]

In a second step a corresponding number of net art works had to be produced. The artist handed this over to a net art generator she had conceived herself and programmed with the help of several colleagues: using search machines, arbitrary data (text and picture) was collected on the internet and automatically recombined[8] as new sites, according to key words. 127 net art projects were produced, assigned with the given names and submitted via the accessed passwords on the Kunsthalle server.

The organizers' pleasure in the high level of participation remained until the winners were announced. However, no one had known what to make of the pile of data, and three men were selected for the awards, despite the high level of "female" participation. Sollfrank published a press release at the same time which uncovered her coup.

In this case the attack on the art business is satirical in the best sense. It plays with the gap between appearance and reality, as well as with a whole row of more or less established conventions of art communication. One is reminded of corresponding strategies in the established satire scene e.g. by the German *Titanic* magazine.[9] In particular, a similar scandal which took place a good ten years before comes to mind, where the two Austrian poets Franz Josef Czernin and Ferdinand Schmatz—"gender swapping" in turn—invented a fictitious female poet and attributed "stupid" poems to her, according to the established literature taste, which—after all—were then published under Czernin's name by the renowned *Residenz* publishers.[10] In the two cases, the difference between institutionalized (bound to museums, publishing houses, curators, copyreaders, juries, finances etc.) and innovative art production is exposed.

The net art project underlines the paradox of putting everything "new" straight into a museum, with a diabolic use of new technology. Above all else, the hacking deception is about the distinction between code and interface: Code is conceived in such a way as to foil the perception offers it controls and conveys via the interface (here: the visualization of person data and "her" works in the net). Computer viruses[11] or other provocations of net art, work in the same way. The very successful work by duo Jodi is another example: The users could watch their own "stupidity" (see above) in working with the new technology, particularly in the early phase of the mass media World Wide Web.

Furthermore, automated art production provides a reflection on the subjectivity of the creation process: Can machines create art? Actually, the question only arises, if the final results are considered, instead of concept, process and their aesthetics. Tristan Tzara ironically ventured that the manufacturing of a dadaist poem by means of newspaper, scissors and an algorithm secures originality and a charming sensibility: The use of an art generator whose ejections only work as a parody of "original works of art" heightens the reference to the awkward subjectivity of "making." Florian Cramer aptly notes:

> When Cornelia Sollfrank states that "a clever artist makes the machine do the work," it still implies that the artist *makes* it work in the first place. Poiesis, making, becomes a second-order poiesis of making something that makes something else. So poetry, making, turns into poetics, the making of making. When making turns into meta-making, subjectivity simply shifts to a second order position, residing in the formula instead of the product. (87)[12]

The subversive humor of the female masquerade which critically analyzes gender politics in the established art world as well as in the techno culture, works in the sense of reflexive subjectivity. Subversion works particularly well because the new technology of digital de/re-differentiation is utilized for the masquerade, according to Haraway. Or, as Verena Kuni writes, such a masquerade is connected with the cognition "dass es hinter der Maske kein 'wahres' Gesicht—beziehungsweise kein 'wahres' Geschlecht zu entdecken gibt" ('that there is no "real" face, and no "true" sex to be found behind the mask') (Kuni)—nothing more than a history of female masquerade between fear and laughter.

It would be wrong to assume that this strategy only aims at the established art world. The mentioned gender problematic, e.g. female under-representation, continues in the subculture of the hackers. Cornelia Sollfrank has also demonstrated this with cyber feminist interventions or by "social hacking," following the pattern of the fictitious identity game: She invented a certain "Clara S0pht," for example. Faked video interviews with these figures were then presented at hacker meetings e.g. the Chaos Computer Club, to simulate female presence in the male dominant scene. Here, too, the deception went unnoticed, at first. Not even the pun on virtual pop icon Lara Croft was recognized. Sollfrank was actually treated with hostility because she revealed the apparent identity of hackers.

By affirming the male the test might also provoke the "enlightened" feminist scene: This was the case in 2005 when the "1. bachelor award for net literature" was announced by net artist Johannes Auer and the Literaturhaus Stuttgart. The announcement follows Dieter Daniels' idea, in which the Turing machine corresponds to the concept of Duchamp's "bachelor machine": Both explain gender issues technically and in a marked male, "celibate" and self-satisfying manner, i.e. without direct physical contact—a "short circuit on demand," where the object of desire, the opposite sex, is substituted by the machine (cf. Daniels 104ff.).[13]

The intrinsic consequence was to bar "women" from participating in the bachelor award.[14] In an according debate, female contributors to the mailing list *Rohrpost*, including Cornelia Sollfrank, were not amused, despite the option that the exaggeration towards a "man-machine" might actually offer the possibility "aktuelle strukturelle Transformationen der patriarchalen Matrix in den Blick zu bekommen" ('of getting a look at the current structural transformations of the patriarchal matrix') (Hunger 81): Obviously, this "simple" to and fro between male and female, as the analogy of switching from 0 to 1 in primary digital code suggests, stands contrary to the way things technical are charged with established gender semantics—of course, there is satiric/comic potential here.

IV. Short Conclusion on Demand, or Seven Theses on Poetics

1. Poetics sets the diabolical alternation between differences in motion, it intrudes creatively on its own general autopoiesis, i.e. by presentation, reflection, change. Here, poiesis/poetry is the medium in which this creative mechanics can be simulated and observed—if every sense arises from distinctions, as requisites following prerequisites which are digitally coded with positive and negative values: human/mechanical, male/female, 1/0, stressed/unstressed, comic/serious, beautiful/ugly, good/bad, true/false etc.
2. In this sense, poetics also self-referentially works as technology *(poietike techné)*. Particularly where it adapts itself to the culture of technology and relates this to its own means and procedures. It experiments with experiences on how technology literally makes sense between restriction and freedom, stupidity and power of judgement or, according to Peter Dittmer, between "switching and operating." "Literally" means that poetry, as language art, raises the linguisticality of the technology culture, as well as existing and new language techniques, and acts these out, condenses and recontextualizes them—in short, contaminates them poetically.
3. Poetry no longer progresses, instead, it constantly extends its possibilities. Attributes like *net literature, e-poetry, digital, visual, sound poetry* etc. represent this. These are open concepts of orientation with which art communications organize themselves in factual, social and temporal ways. Genres, or object areas clearly, aesthetically outlined, can hardly be defined by these concepts, however. Even if their especially academic use occasionally tries at standardization and institutionalization.
4. Poetry, as language art, is no longer a matter solely for the literature system, but extends throughout all of art in its entirety. It is—in keeping with Umberto Eco—art in movement, inspired by a (renewed) performative turn. With this performance, borders within art, as well as other communication and sense systems are constantly moved, and the works of art and their protagonists are process orientated. In this sense poetry is "prozessive Universalpoesie" ('processive universal poetry').
5. Humor is an important aspect of contemporary art, particularly where it isn't formed as an end in itself, but in a concert of mutually qualifying possibilities of effect. Hereby humor and poetics converge in their diabolic potential—being subversive and deconstructive, as well as constructive and affirmative, inclusive and exclusive, in a "constant ambivalence." In fact, all art in movement tends towards humor. On the other hand, art which lacks a sense of humor frequently seems to be rigid and suspected of ideology.

6. Humor also ambivalently complements art: It emphasizes conceptuality, i.e. it underlines ideas, reflections, programs, the creative "spark." At the same time, according to Freud, it increases pleasure and desire, i.e. to some extent humor meets hedonist needs. The poetic connection between conceptuality and desire which is found in humor, communicatively marks the individual according to a reflexive subjectivity, relating to the culture of technology and to gender culture.
7. Thus, forms of comic self-observation, as poetic self-observation within the world society, could be understood to be "self concern"—reflected experience, an individual development, assessment and arrangement of choices, and pure relish. Perhaps the connection between "art and life" can be found here, not in Henri Bergson's vitalistic sense, nor according to Peter Bürger's avant-garde theory, but in the art of living (cf. Schmid 71ff.), or the technologies of self.

Translated by Helen MacCormac

Notes

1. For detailed project information cf. <http://www.dieamme.de>.
2. In works such as *Bodybuiling* or *Rollstuhl,* similar to Dittmer's milk glass, Frank Fietzek has enriched his hardware with unexpected objects from the real world (butterfly machine as interface, wheelchair and model racing set as I/O device) which can be seen as ironic comments on the hype of the technology debate; on cyborg phantasms in the case of *Bodybuilding*; on acceleration of information processing in *Rollstuhl.* Cf. Block, "Diabolische Vermittlung," and Block, "vom code zum interface."
3. This is indicated not only by the installation's technical raffinesse, but also by the theoretical and practical concentration with which Dittmer has pursued his sole project since 1992.
4. Different to Rebentisch and Daniels.
5. Compare with Frances Grundy's exposition and Francis Hunger's differentiated historical depiction.
6. Jutta Weber discussed the strategy of cyber feminism in more detail.
7. Details taken from Cornelia Sollfrank's commentary *Female Extension.*
8. This and other generators to be found and tried out under <http://obn.org/generator/>.

9. A prominent example, which was noticed in the international press, was the successful staging of a supposed bribe scandal in connection with Germany's application to hold the Football World Cup 2006.
10. Cf. Czernin's and Schmatz's *Die Reise.* The *Spiegel* which had extensively covered this case, should have been prepared. The poets' coup did more than to reveal the tastes and market conventions of the literary scene, it also explored contemporary and historical forms of poetry. Soon after this, the pair developed the computer program POE ('Poetic Oriented Evaluations' referring to Edgar Allan Poes's analysis of verse): a poem generator intended to support the mentioned research (cf. Czernin and Schmatz, "Anmerkungen"). It is more than likely that this generator—corresponding to Sollfrank's net art generator—could have done the rhyming for the two poets in their staged scandal.
11. A relatively harmless but quite poetic example—poetic because of its technological elegance and visual appearance—is given as follows: The 13 characters :(){ :|:& };: (including blanks) which initially remind us of "emoticons" in chat and email communications, are actually the total program code of a neat little virus, which when entered into the command line of a Unix system, rapidly brings it down, because the program repeatedly doubly executes itself and so exhausts the system resources within seconds.
12. This connection between "poiesis" as artistic meta-observation and reflexive subjectivity was systematically developed by Block, *Beobachtung des "ICH,"* esp. 124ff. and 283ff.
13. The appropriate formula "short circuit on demand" (Daniels 115) is taken from Duchamp.
14. Cf. <http://www.junggesellenpreis.de>.

Works Cited

Bense, Max. *Literaturmetaphysik: Der Schriftsteller in der technischen Welt.* Stuttgart: DVA, 1950.

Block, Friedrich W. *Beobachtung des "ICH": Zum Zusammenhang von Subjektivität und Medien am Beispiel experimenteller Poesie.* Bielefeld: Aisthesis, 1999.

———. "Diabolische Vermittlung: Zur Konzeption von Bewußtsein und Körperlichkeit in interaktiver Medienpoesie." *Beobachtungen des Unbeobachtbaren: Konzepte radikaler Theoriebildung in den Geisteswissenschaften.* Ed. Oliver Jahraus and Nina Ort. Weilerswist: Velbrück, 2000. 148-168.

Block, Friedrich W. "vom code zum interface—und zurück: zur orientierung im diskurs digitaler poesie zwischen konzept und wahrnehmung." *$wurm = ($apfel>0) ? 1 : =;. experimentelle literatur und internet: memoscript für reinhard döhl.* Ed. Johannes Auer. Zurich: update, 2004. 76-86.

Cramer, Florian. *Words Made Flesh: Code, Culture, Imagination.* 2005. 26 May 2006. <http://pzwart.wdka.hro.nl/mdr/research/fcramer/wordsmadeflesh/words madefleshpdf>.

Czernin, Franz J., and Ferdinand Schmatz. *Die Reise: In 80 flachen Hunden in die ganze tiefe Grube.* Graz: Droschl, 1987.

———. "Anmerkungen zum Dichtungs-Programm POE." *IMBUCHSTABENFELD. die zukunft der literatur.* Ed. Peter Weibel. Graz: Droschl, 2001. 99-112.

Daniels, Dieter. "Duchamps:Interface:Turing. Eine hypothetische Begegnung von Junggesellenmaschine und universeller Maschine." *Vom Readymade zum Cyberspace: Kunst/Medien Interferenzen.* Ostfildern-Ruit: Cantz, 2003. 92-123.

Dittmer, Peter. *Schalten und Walten (Die Amme): Gespräche mit einem milchverschüttenden Computer.* 2003. 26 May 2006. <http://dieamme.de/amme5/dokumente/dieamme.pdf>.

Flusser, Vilém. *Die Schrift: Hat Schreiben Zukunft?* 1987. Göttingen: European Photography, 1990.

Grundy, Frances. "Computer Software: A Clue to De-Gendering Technology?" *Wie natürlich ist Geschlecht? Gender und die Konstruktion von Natur und Technik.* Ed. Ursula Pasero and Anja Gottburgsen. Opladen: Westdeutscher Verlag, 2002. 229-239.

Haraway, Donna J. "A Cyborg Manifesto: Science, Technology, and Socialist-Feminism in the Late Twentieth Century." *Simians, Cyborgs and Women: The Reinvention of Nature.* Ed. Donna J. Haraway. New York: Routledge, 1991. 149-181.

Hunger, Francis. *Computer als Männermaschine: Computer in einem patriarchalen und kapitalistischen Kontext.* 2003. 26 May 2006 <http://www.hgb-leipzig.de/~francis/irmielin/writings/comamama/>.

Kuni, Verena. *Mythische Körper: Cyborg-Configurationen als Formationen der (Selbst-) Schöpfung im Imaginationsraum technologischer Kreation: Alte und neue Mythologien von "künstlichen Menschen."* 2004. 26 May 2006 <http://www.medienkunstnetz.de/themen/cyborg_bodies/mythische-koerper_I/>.

Rebentisch, Juliane. "Sex, Crime & Computers: Alan Turing und die Logik der Imitation." *Ästhetik & Kommunikation* 26.96 (March 1997): 27-30.

Schmid, Wilhelm. *Philosophie der Lebenskunst: Eine Grundlegung.* Frankfurt am Main: Suhrkamp, 1998.

Sollfrank, Cornelia. *Female Extension.* 1997. 26 May 2006 <http://obn.org/femext/ext_engl.htm>.

Turing, Alan. "Computing Machinery and Intelligence." *MIND* 59.236 (1950): 433-460.

Weber, Jutta. "Ironie, Erotik und Techno-Politik: Cyberfeminismus als Virus in der neuen Weltunordnung?" *Die Philosophin. Forum für feministische Theorie und Philosophie* 12.24 (2001): 81-97. 26 May 2006 <http://www.obn.org/reading_room/writings/html/ironie.html>.

Wirth, Uwe. *Diskursive Dummheit: Abduktion und Komik als Grenzphänomene des Verstehens.* Heidelberg: Winter, 1999.

Playable Literature

Markku Eskelinen

Six Problems in Search of a Solution

The Challenge of Cybertext Theory and Ludology to Literary Theory

Abstract:

In this essay I'll be playing with cybertext theory and ludology while addressing six more or less intertwined problems. First, the most current and worn-out one: how to expand literary narratology beyond its print heritage without falling into the trap of pan-narrativism? Second and third, in addition to various new narrative and would-be-narrative constellations and devices the relations between texts and the text's relation to itself have also changed. Fourthly, all these changes have their bearing on the role of the reader in situations where the lack of conventions is well matched with outdated expectations concerning narratives, texts and transtextuality. Fifthly, enter playability and the fear of variety when readers and scholars should perhaps be willing to give up the idea of literary wholes and try to pursue happiness in the form of parts, phases and playthings or, to put it in more Oulipian terms, in objects and operations. Sixthly, we'll try to shed some light into the recent trend of building textual instruments and instrumental texts.

1. Introduction

The power of cybertext theory[1] in this context stems from the fact that even the state-of-the-art literary theories of today are ultimately based on literary objects that are static, determinate, intransient and random access with impersonal perspective, no links and utilizing only the interpretative user function. To this one special type of object cybertext theory adds 575 rather fresh alternatives capable of undermining and shaking many basic assumptions and presuppositions derived from the print era.[2]

Ludology gives us a perspective and a paradigm from which to approach the interactivity or ergodicity of literary works without any hype of the new versus the old, as interactivity has always been dominant in games. In short, in literature we may have to configure in order to be able to interpret, but in games we have to interpret in order to be able to configure and proceed from the beginning to the winning or some other situation (cf. Eskelinen, "The Gaming Situation"). This ludological double vision is all we need to situate the continuously expanding field of ergodic literature between linear literature and computer games.

It's time to formulate the first of my six challenges; let's call it the logic of expansion. There are many other fields where the current media plurality seems to pose problems, but the easiest way to show what I mean is by and through narratology. Narratives are supposed to be transmedial, or to put it in Seymour Chatman's terms: "The transposability of the story is the strongest reason for arguing that narratives are indeed structures independent of any medium" (*Story and Discourse* 20). So, let's see what happens if we try to expand formal literary narratology from its traditional one media position to those few hundreds offered by cybertext theory.

2. From Expanded Narratology to the Lack of Conventions

Aarseth contrasts the ergodic mode with the narrative mode for various good reasons I fully agree with. Still, from the outset Aarseth's concept of ergodic literature seems to concern differences only within one of the seven dimensions, the main dividing line separating the interpretative user function from the three other user functions. Therefore we could begin our investigation by focusing on the other six parameters of his typology of textual communication, as in principle they could all be combined to the interpretative user function within the realm of non-ergodic narrative literature. We'll have eight major shifts to deal with: from static to intratextual and textual dynamics, from determinate to indeterminate texts, from intransient to transient time, from random to controlled access, from impersonal to personal perspective, and from no links to links and conditional links.

To warm you up, let's take two stock examples of narrative power: the concepts of unreliable narration and dramatic irony. The former makes sense only if there's a completely accessible and static textual whole against which background the reader can finally understand and verify how the narrator was positioned in relation to possible other narrators and the narrated events and existents in general and how reliable he actually was. The case is pretty much the same with dramatic irony. The reader can by all means believe she knows more than certain characters and narrators do, but in the end this can be verified only in the context of the whole text and nothing but the whole text. Of course these things are always already actively interpreted and hypothesized on a more local level one way or the other, but the possibility that one is being misled remains until the whole text is read. Both in print fiction and classic hypertexts the whole text, that is, every single texton, is there to be read, but in more dynamic works that doesn't have to be the case. To misquote Radiohead, just because you read it doesn't mean it is there.

Genette's formal narratology (*Narrative Discourse; Narrative Discourse Revisited*) will serve as my point of departure, as it gives us access to a wide variety of already well recognized thoroughly studied and therefore quite uncontroversial and conventional narrative devices.[3] Genette's model is a fairly simple and reliable system of narrating, narrative, and story. The first of these terms is used for the producing narrative action, the second for the signifier, statement, discourse or narrative text itself, and the third for the signified or narrative content. Genette divides his basic parameters into three groups: tense, mood, and voice.

Tense concerns temporal relations and possible distortions between story time and narrative time. They are studied in three registers: order, frequency, and duration. It's now time to add some cybertextual ingredients to this simplicity. As we all know textonically dynamic texts like William Gibson's *Agrippa* or Noah Wardrip-Fruin's *The Impermanence Agent* are capable of reducing the number of their original textons and scriptons to zero and thus either disappearing or replacing themselves completely within a limited period of time. Transient texts set temporal limitations to their reading; in Stuart Moulthrop's *Hegirascope* it is 30 seconds max per node per visit. This means we may have to add two new temporal categories to the classic two: system time to account for the varying degrees of the text's permanence, in short the appearances, disappearances and possible reappearances of its parts and phases, and reading time for the text's potentially limited temporal availability to the reader.

For the sake of simplicity and time we browse through these two new temporal layers without inventing too many new categories, as those of frequency, order, duration and speed will do for our limited purposes. Accordingly, reading time could be conceptualized to be either unlimited or limited in all these categories, as in principle reading constitutes an event like any other. These limitations could be applied both locally and globally. *Hegirascope*, for example, limits only the duration of its reading on the local level of its nodes. There are countless other alternatives. Imagine your favorite classic that could be read only once or only for two hours, or in night time or outside office hours only, or of which you are allowed to read only two chapters in a decade or in a lifetime, tempting you to collaborate with other readers or leave it and your vague memories of it as an inheritance to the next generation of readers.

If the text disappears gradually and not at once, we will have to be prepared to attach different temporal values to each of its dynamic parts. In principle, the text is permanent, partially permanent or ephemeral. And if the text or some parts of it are of the vanishing kind then we could study this process in the aforementioned temporal registers. As I think it is trivially easy to invent or discover useful examples, I won't bother you with them now. To sum up these extra dimensions, suffice it to say that if we combine the possible limitations regarding levels (global and local) and the four temporal registers (order, speed, frequency and duration) we'll have at our disposal a system of 256 basic ways of constraining reading time only one of which, the one containing no restrictions whatsoever, is compatible with classic narratology (fig. 1).

Level	Local		Global	
	unlimited	limited	unlimited	limited
Frequency (How many times the work can be visited?)	*Reagan Library*[4]	*Reagan Library*		*Agrippa*
Duration (For how long the work can be visited?)		*Hegirascope*		*Agrippa*
Speed (Possible limitations of the reading speed such as mandatory pauses)		*The Impermanence Agent*		
Order[5] (Freedom of movement within a part and between parts)				Hypertext fiction

Fig. 1. Reading time.

All these limitations complicating reading and rereading may seem to run counter to our cultural conventions and therefore to be an exercise in futility. Still, if one of the main attractions of narratives is the challenge to construct and reconstruct a set of temporal relations on one level from various explicit and implicit cues and hints embedded in another temporal level, then it may not be totally out of the question to add more temporal levels to that challenge. Reading time and system time can be used equally well in non-narrative texts and for non-narrative purposes, which means they are not narrative categories in themselves.

Mood, Genette's second group, with its categories of distance and focalization, deals with various techniques of regulating the quantity and quality of narrative information. The effects of this regulation constitute another standard attraction of narratives, as it challenges the reader to work his way through these limitations and hypothesize the bigger picture so to speak. Focalization is not about who narrates, but who perceives or where the perceptual focus of the narrative is situated. In other words, it is a delivery channel for narrative information, and its functioning as a narrative device is ultimately based on uneven distribution of knowledge. It seems that many kinds of bots, non-playing characters, avatars and voices have already expanded the notion of the channel and its behavioral variety, as we can sometimes discuss with these entities, or even operate them to a certain degree, which is all very different from what we can do with characters and focalizations in print literature. Confusingly, the current careless and indiscriminate use of the word character doesn't tell us whether characters are used as delivery channels for narrative purposes or as strategic tokens for interaction (fig. 2 for basic distinctions).

Control	Existents not controlled by the player/reader/viewer		Existents controlled by the player/reader/viewer	
Communication	Subject (2-way)	Object (1-way)	Subject	Object
Static behavior		Characters in literature and film		
Dynamic behavior				

Fig. 2. Existents and characters.[6]

It is also clear that the amount of possible additional filters will know no limit once we move from random to controlled access. All these shifts point to new textual communication structures and it is a mystery to me why they should be reduced to or labeled as narratives (cf. also Aarseth, *Cybertext,* ch.5). Therefore, and before getting into the heart of our first challenge, we'll try yet another strategy to expand narratology with the help of cybertext theory.

Because in cybertexts narrators and characters and other traditional narrative elements and devices we are familiar with are just potentially alterable strings of signs, we should make a fundamental distinction between textonic and scriptonic entities. On a slightly more practical level this means that there might be a pool or an archive of possible narrators, and every time the (cyber)text reorganizes itself a certain number of these narrators are first selected and then distributed to their temporary positions inside the narrative text that is presented to the reader. It's easy to imagine possible operations of these dynamic fictive entities: They can turn more covert or overt, trade places with each other, split up or unite, extend or reduce their territories, move between narrative levels and homodiegetic, heterodiegetic, and mere character positions, alter the degrees of their intrusiveness, self-consciousness, reliability and distance, and so on (fig. 3).

Object / Level / Operation	Narrator	Character	Narratee
	textonic/scriptonic	textonic/scriptonic	textonic/scriptonic
Addition			
Substitution			
Subtraction			
Multiplication			
Division			
Increased presence			
Decreased presence			
Increased territory			
Decreased territory			
etc. (changes in intrusiveness, self-consciousness, reliability and distance)			

Fig. 3. Rotation scheme for narrative entities.

The same idea could be applied to more abstract narrative devices as well. If we choose to do it to narrative levels, in some fictional world the story of Shahrazad may very well turn out to be framed by the story of Sindbad the Sailor. In addition to the traditional options of either hierarchic (A embeds B) or heterarchic (A embeds B and B embeds A) arrangement of narrative levels, there are two novelties (fig. 4): dynamic relations between narrative levels based on how the text is being read, and rotating levels (let's say only five out of seven narrative levels are available during any one reading session).

Number of narrative levels in the text: n
Number of narrative levels available for the reader during any one reading: n / <n
Mutual relations of available narrative levels: static/dynamic
Type of framing: hierarchic/ heterarchic

Fig. 4. Narrative levels and metalevels.

It might be helpful to make at least one more distinction: the one between three different sides or dimensions of narrators, that is, their position or positioning, their identities, and their qualities or characteristics. In print fiction these three dimensions are glued together but in cybertext fiction that no longer needs to be the case. We can separate them, not without consequences of course. Once again, on a more practical level, this may lead to narrators exchanging pasts, or competing with each other for the key positions (in terms of time, space and credibility) inside cybertext fiction, and also, as a result of these transformations or for other reasons, changing the number and identity of possible narrators in the following phase or version of the narrative on the run. Working out the ways to rewrite narratology with these possibilities of playing with yet another attraction of narratives, the promised discovery of narrators' relations to and positions in the narrative text, will definitely offer lots of fun, but as I have already done it elsewhere (cf. *Kybertekstien narratologia*) I will not dwell on that here.

Whichever way we choose, adding strange new behaviors to easily recognizable narrative devices or new categories barely justified by loosely defined attractions, we'll end up in trouble sooner or later. Firstly, every well-established and sophisticated narratological theory, regardless of whether it puts its emphasis on communicative (Chatman, Genette, Prince) or cognitive models (Bordwell, Branigan, Sternberg), is ultimately a theory of interpretative reconstruction based on the narrative text that is presented to its readers or viewers. In other words, reading a narrative is a construction process where based on what's presented to us we are supposed to figure out what really happened and what the relations between events and existents were. It seems to me there's no justification to expand the definitions of narrative beyond this economy of presentation and interpretation. Secondly, many of these new and hypothetical narrative devices and techniques will exceed the cognitive capabilities of the readers and in some cases humanly possible attention spans as well. Thirdly, there are no broad cultural agreements, shared conventions or readymade expectations to help readers in their encounters with these would-be-narrative oddities, such as rotating narrators and characters and playing with time and temporal change in four registers instead of the standard two.

Consequently, our second problem or challenge has to do with conventions, but before getting there let me conclude with four possible solutions to our first challenge of how to expand narratology to meet the multiple media positions offered by cybertext theory:

1. Not to do it at all beyond its one media position, and believe it is too intimately tied to it, just like film narratology is tied to its one position, the difference between these two well-established narratologies residing in the transient time and temporally controlled access of the film medium.
2. Try to expand it to accommodate all the 144 non-ergodic positions and find out there's a lack of conventions for producing and consuming such artifacts even if the role of the reader as an interpreter has not changed.

3. Try to expand it to accommodate the 432 ergodic positions as well if certain other conditions are met, most importantly that the ergodic mode will be in the service of the narrative mode.
4. Try to expand it to absolutely anything the way the pan-narrativists do in the context of games. While you can take your chances and try to fill the blank spots in the charts, my choice is to focus next on conventions, the supposed lack of which will form the second challenge.

3. You Can't Always Get What You Want

As we move from narratological problems to those of reception, we need to clarify something, and in order to do this we have to adapt a standard deconstructive position for a moment or two. It could be shown that there always already exist both thematic and formal overflow capable of overturning and shaking whatever structural constellations or thematic interpretations our reading tries to posit. In the context of narratology, expanded or not, this means first of all that we should acknowledge the endless interplay between narrative and other textual designs. There's nothing extraordinary in this; we can take any decent half-experimental 20th century fiction and show a recognizable narrative organization to exist side-by-side or intermingled with other, anti-, non-, or counter-narrative organizations. From this perspective also the ergodic and narrative modes could easily co-exist in the same cybertext and either serve or undermine each other, and it's up to the reader-user to figure out which is the case.

The more or less hypothetical narrative devices we drafted earlier could easily be used to complicate and defamiliarize the text to every extreme, but that's not the only and probably not the best use of these relative novelties. While it is true that these techniques set new demands and challenges for the reader, it is also true that just like in an ordinary work of print fiction they must be motivated to the reader at least to a minimal degree. To reach that noble goal we can thematize, conventionalize or trivialize the functioning of the cybertext or try to make its ergodic side more playable.

Probably the easiest way to do it is to ground the behavior of the textual machinery in themes. If the protagonist is under a lot of stress and has to make more or less difficult and fatal decisions in a hurry then why not boost this theme by extending it to the reader's experience as well by setting limits for the time the text stays available without changing itself. Likewise, the text that changes itself could be described as hiding or destroying the evidence it contains making it therefore very likely to suit the well-known formats and clichés of thrillers and detective stories. If the text, or more likely certain key parts of it, is materially different in every rereading, it would be very easy to connect this quality to themes of memory and remembrance, as well as to countless others, from Alzheimer's disease and traumatic experiences to resistances and defenses in psychoanalytic sessions.

We could say that the three classic hypertexts, Michael Joyce's *Afternoon*, Stuart Moulthrop's *Victory Garden*, and Shelley Jackson's *Patchwork Girl* all did what they could to make the reader more receptive to the marvels of their labyrinths: by using hidden and conditional links to highlight and parallel the defenses and self-denials of the protagonist in *Afternoon*, his general unwillingness to know, evoking and concretizing the familiar literary tradition of forking paths of Borges, Coover and Pynchon in *Victory Garden*; and foregrounding Frankensteinian bodily metaphors to ease the postmodernist butchery work of connecting parts and wholes in *Patchwork Girl.*

Secondly, we could banalize. It is important to bear in mind that every cybertextual media position seems to have its trivial counterpart in everyday life. Textonic dynamics can be found in discontinuous textual flows in blogs and online discussions, and variable responses characteristic of intratextonic dynamics can be found in our everyday experiences with so-called intelligent machines. Humans are not (always) predictable; hence, indeterminability is at the core of our social life. Film subtitles are transient; even the simplest board games make you assume strategic responsibility for something, footnotes and encyclopedias fork the text and turn it to a system of fragments and connections, passwords control your access, etc. These familiar experiences offer countless possibilities for taming the cybertextual machinery.

The third way to make the functioning of a cybertext familiar or at least more familiar to its user is to borrow conventions, but not necessarily from literature or even art in general. In *Cybertext* Aarseth devotes separate chapters to four different specimen of ergodic literature: hypertext fiction, story generators, textual adventure games, and MUDs. The latter two use ordinary game and quest elements along with common communication structures like dialogue and polylogue to make it easier and more meaningful for the reader-user to deal with the textual constructs and constellations she encounters. Story generators will probably remain fundable because of their tight connections to AI research, and need neither familiarizing nor defamiliarizing touches.

This leaves us with hypertext fiction and its somewhat twilightish status as a relatively recent branch in the networked traditions of experimental literature. To condense what was said earlier about classic hypertext fiction, we could wonder whether it was quite enough after all to borrow the convention of the jigsaw puzzle and force or challenge the reader to complete it. A puzzle is a puzzle is a puzzle, a monologic experience if any.

Cybertext theory has already shown experimentalists a fresh way out in the form of new targets, those shared strongholds of both print and hypertext fiction: static scriptons and intransient time. Still, cognitively speaking I can't be optimistic: If the simple hypertextual distortions of the presentation order were able to generate such a commotion and conceptual mess, it is unlikely that the much more dynamic distortions will be recognized, understood and conventionalized in any time soon.

Conventions don't appear or disappear overnight or in a decade. The big question then is whether to wait for their emergence without knowing if they'll ever arrive or to apply the already well-known conventions, not from literature, but from elsewhere: especially from games and computer games. The justification would be that the minute there's a recognizable cultural divide between interpretative and operational sides of literature, the latter will be measured against the already existing and well-formed activities and operational practices and if no matches are found then the case is what it still is today: the bits and pieces of ergodic literature without explicit play- or game-like attractions are viewed as elitist and experimental oddities (which is fine by me). This last possibility refers to concrete relations and borrowings between sign-producing machines, which is a fairly decent way to introduce our third challenge: What happens when transtextual relations cease to be merely interpretative?

4. Transtextuality Meets Cybertextuality

Before taking a quick tour through Genette's five forms of transtextuality (*Palimpsests* 1-8), it should be clear that from the cybertextual perspective there's a basic distinction to be made between interpretative and ergodic relations between texts. The latter are concrete and programmable; the former are those of traditional print transtextuality, expanded or not by the variety of new possibilities opened up by the digital media. The material basis of interpretative relations between texts can be (at least potentially) altered in ergodically dynamic cybertexts (by adding, removing or otherwise manipulating scriptons) in contrast to print texts and ergodically static hypertexts. Within the latter group there's a crucial difference between web and stand-alone hypertexts. In the former environment some links and transclusions could be said to function as explorative forms of transtextual relations (that is, we can concretely and seamlessly move from one text to another). Similarly, nothing, except maybe deep-seated conventions, prevents authors from designing configurative and textonic relations between and among texts. This may sound familiar and it is intended to sound that way since it might be tempting to find an analogue between the ways scriptons are revealed and generated from textons, and the ways these textons could be generated from other textons. However, the analogue is far from complete, and we shall postpone discussing it until much later. For now it may be useful to keep an open mind to the possibility that in addition to interpretative relations there also are explorative connections, configurative affects and textonic impacts between texts.

In order to see the bigger picture of changed relationships between and within texts we should free ourselves from the metaphysics of links, that is, from the famous convergence or embodiment hypothesis. Obviously, links can be used to make explicit references and transclusions will work as direct quotations, if for some obscure reason we wish merely to emphasize, foreground or boost the traditional notions of intertextuality. However, it should be equally obvious that although links have been too often confused with intertextuality, there are both intertextual relations that cannot be shown by links, and various uses of links that have nothing whatsoever to do with traditional intertextuality. Every traditional notion of intertextuality is ultimately dependent on the unpredictably varying interpretative and transpositional skills of the readers and this dimension can neither be reduced to links nor fully expressed in them. On the other hand, already the links forming concrete connections (instead of mere references) between online hypertexts are potentially very different from their distant print relatives, as unlike the latter they are not merely interpretative and they could also be timed, changed, conditioned, chained, concealed, randomized and layered for complex effects the tradition knows nothing about.

Architextuality. Genette means by architextuality "the entire set of general or transcendental categories—types of discourse, modes of enunciation, literary genres—from which emerges a singular text" (*Palimpsests* 1). Here cybertext theory adds another dimension and taxonomy to the already muddled project of Western poetics and aesthetics: the 576 textonomical genres or media positions describing the functional or behavioral side of any literary work. At stake here is the whole necessary and unmanageable project of Western poetics: a cluster of modal, generic, formal and thematic categories, combinations and considerations. It is safe to say dynamically ergodic texts will have long-lasting uneasy relations with them. One needs only to imagine fictions the genre or the mode of which will change based on how the text is being read—let's say the faster you read a detective story the faster it becomes a horror story to slow you down with gruesome details or that only the dialogue will remain to be reread while the rest of the text once read, hopefully, stays in someone's memory.

Paratextuality. Paratexts are heterogeneous elements that lie on the threshold of the text and which help to direct and guide the reception of a text by its readers (Genette, *Paratexts*). In digital cybertexts this dimension should be discussed in relation to various interface issues as well as to user's manuals, help-files, web sites and etiquette given to users and readers to consult while approaching, interpreting and learning how to operate ergodic works. In some cases, the text may explicitly comment upon the way it has been read and offer genuine or mocking assistance, which nicely blurs the traditional difference between guiding paratexts and self-referential commentaries—yet another distinction inherited from the golden age when literary works didn't react to their reading.

Metatextuality is "the relationship most often labeled commentary" (Genette, *Palimpsests* 4). In certain encounters with indeterminate cybertexts, commentaries may turn out to be commentaries on one's own singular experience of ephemeral constellations of signs never to be repeated again or to be seen by any other user. There are at least three types of new challenges. Firstly, in addition to the usual interpretative skills the critic needs ergodic skills to use the textual system in an appropriate way. Secondly, the continuous and potentially never-ending variation and supplementation may well exceed any humanly possible attention span. Thirdly, an ergodic work may be very sensitive to how it is used and this may result in unique textual responses to the critic's efforts. I can almost hear the boring me-my-mine rhetoric this may well lead to, so it's better to move on to the next transtextual relation.

Intertextuality. According to Genette's rather restricted definition intertextuality is "a relationship of co-presence between two texts or among several texts" (*Palimpsests* 1). This includes the practices of quotation, allusion and plagiarism. In principle, it is just a matter of recognition. But in these days just because you read it doesn't mean it is there and vice versa. The invisible, hidden, and inaccessible parts of the text will deny the reader the comfort of knowing for certain what exactly is there in the text. To make matters more complicated it's not only the dialectic between visible and invisible parts of the text that counts, but also the relations between what's visible now and what, if anything, will be visible later. Also the threshold between what's inside and what's outside the text is getting blurrier, as it's so easy to supplement and replace the original text from the outside (as certain famous software agents have already shown). In some ways, all this may lead to the decreasing importance of intertextuality in dynamically ergodic works, or at least to more drastic ways of foregrounding intertextuality. In any case we'd be better prepared to understand the widely varying degrees of co-presence at work.

We could perhaps take one step further in order to follow the concept of co-presence to its logical end, and add yet another category of relations, the one between textual wholes (or entities) potentially capable of annihilating and incorporating other textual wholes, or arresting and releasing them, in short making other texts (and parts of other texts) appear, disappear, and reappear. In other words, I suggest we distinguish between independent, dependent, dominant and co-dependent works of art. This may very well be the end of intertextuality as we know it, but I feel fine.

Hypertextuality in the Genettean sense deals with systematic grafting of a text (a hypertext) upon an earlier text (a hypotext) in a way that is not commentary (*Palimpsests* 5). These imitations and transformations are formal and thematic, and it is safe to say that cybertexts add a third option, the functional one, as well as transform the previous two. There are at least three separate aspects to that situation. Firstly, there's a new possibility of gradual and serial transformation where the text goes through several different phases and quite possibly loses contact with its first or previous hypotext in the process; just imagine some banal desert island story that first models itself after Ibn Tufayl and then after Grimmelshausen, Saikaku, Defoe, Tournier, and Eco. To get to know the serialized nature of these hypotexts you just have to keep on rereading. Secondly, this transformation may be purely functional; let's say we first turn a detective story into a hypertext and boost its epistemological structures with conditional links, hiding the evidence so to speak, and then turn this ergodically static hypertext into an ergodically dynamic cybertext that after a certain time starts playing with both its own and its users' time and begins to destroy its static scriptons, that is, its very evidence.

Virtually every text generator can show us the transformation of textons into changing scripton combinations in a way that is challenging enough to every traditional notion of inter- and hypertextuality. The words waste and land may or may not be combined to constitute a reference to Eliot. You'll never know until it happens, and when it does you might not be there to witness this intratextonic nihilogue. Another case that comes to mind is the emergent growth of textual systems. Let's say there are two such systems connected to each other over the net continuously affecting the organization and reorganization processes of each other. The host and the parasite, the hypotext and the hypertext reprogrammed *n* times, until the one completely consumes or deadly inflicts the other. Scholars are very welcome to attach their favorite meanings to every morph, or forced to change their approach as there's no way to slow down the rotational speed (rpm) of the hermeneutic circle under such circumstances.

Users are an obvious source for altering inter- and hypertextual relations, but what happens or should happen when a Shakespeare snob meets a Shakespeare bot in some programmable textual environment? Is there any reason we should trace the connections whatever they are, or should we finally let the sleeping traditions lay and concentrate on more interesting issues if there are any? When users bring their transtextual fields and priorities into conversation and consumption, they may have a carefully calculated chance to convert the process to certain inter- or hypertextual directions (as in John Cayley's *Book Un-*

bound). There's also room for an interesting inversion: The cybertextual machine may turn critical or even allergic to such civilizing processes and may begin to defend its own priorities against the invading flow of the user's cultural capital. Imagine a text that needs to be regularly fed by other texts from the outside: a textual tamagotchi, a vicious little cybertext using, say, Kafka as a catalyst to induce series of transformations in its incorporated textual nourishments and their relations to each other thereby producing the next generation of minor literature.

To conclude, any notion of digital transtextuality has to take into account at least three basic things or changes. Firstly, the relations between texts have ceased to be a mere matter of interpretation; dynamically ergodic cybertexts may concretely affect each other's form, content, behavior and very existence. Secondly, the relations of such texts to themselves has to be taken into account, as they may now consist of several more or less autonomous parts and phases imitating and transforming each other and having their own transtextual specifications in all five dimensions. Let's call this new opening a field of dynamic autotextuality. Thirdly, the reader-users armed with configurative and textonic user functions can have fundamental impact on the text, a possibility not even on the horizon of Genette's work in the early 1980's. To take one concrete example: How should we conceptualize the personalized differences in ergodic works like *Book Unbound* or *The Impermanence Agent*, as the different ways their readers use them will make the versions of them very different from each other?

After this brief exercise in the fluidity of textual relationships, we might want to stop for a while to find out whether we need the traditional notion of the textual whole any more, and if not, how to get rid of it and what to replace it with? Enter the fourth challenge, dynamic intratextual relations.

5. The Whole and Nothing but the Whole

I will be following Brian Eno's maxim about going first into an extreme position in order to be able to shift into a more useful position. So let's take the seven cybertextual dimensions as far as possible from the behavior of the print works of our shared cultural heritage. This may sound like an exercise in futility, and maybe it is if futility is just another name for our unrecognized presuppositions. So: Imagine a textual process continuously supplementing itself by outside sources, unpredictably varying its response to the activities of the reader which it also constraints temporally and by denying complete and random access to its parts and phases, using conditional linking and demanding not only explorative but also configurative and textonic activity from the reader in order to overcome the obstacles she's bound to through her personal perspective on the consequences of her very own strategic choices.

After this condensed glimpse at the almost non-existent extremities we could perhaps finally ask what will happen when there simply are no more expectations left to be manipulated or crashed. The seven parameters of cybertext theory will serve to illustrate the ways the textual whole may vanish or disappear from the reader's grasp if not from his aspirations or work ethics.

Before going into that it is important to remember that this assumed literary whole or totality can be described at least in interplaying perceptual, behavioral, structural (or intratextual), cultural (or transtextual), temporal, spatial, causal, and functional terms. Needless to say it is not reducible to any of these dimensions.

Dynamics. Our traditional concept of the textual whole is tied to the unchanging amount and material content of signifiers, and it is not well suited to deal with potentially endless variation or supplementation. There are two rather effective ways to naturalize variation: dialogue structure as in Joseph Weizenbaum's *Eliza*, a granddad of many if not all subsequent bots and artificial characters, and anamorphic game or puzzle schemes where the variation is crafted as a challenge which promises a solution thus putting an end to variation (at least until next time if there'll ever be one). This is common to textual adventure games and interactive fiction in general. In these cases variation is tied to its rather immediate use value and it is not given the high status as literary expression to be contemplated; in other words the user is given a prior permission to actively ignore and dismiss certain variations.

With textonic dynamics the challenges to traditional notions of the whole get even more complicated. In principle, the source of supplements, changes and additions can be either the user himself, other users, or the text can supplement itself from the outside as in *The Impermanence Agent* or John Cayley's idea of using a Reuter's news-feed as a real-time material for one of his *Speaking Clocks*. All these open up a potential for never-ending metamorphoses that don't have to be balanced by personalization, social interaction or conceptual clarity, although these three are the best candidates for achieving that goal.

Determinability. Indeterminability is usually associated with randomness or chance, but that's only one side to it, and perhaps not the most interesting one. In Aarseth's model it is about the stability of the traversal function. In practical terms indeterminability means that if I react or act the same way in the same situation the system doesn't respond the way it did the last time. The crucial question is then how do I know it is the same situation after all; is it even possible to be in the same situation twice? Here too, the traditional wisdom related to the textual whole doesn't get us very far, as it is bound to the notions of repeatability (controlled by the user).

Transience. Transient texts don't allow the reader to control the time of their reading. Texts are available for limited periods of time, sometimes only once as was the case with William Gibson's *Agrippa*, or more precisely with its unhacked copies. Here we are dealing with cycles of appearances, disappearances and potential reappearances in the context where the text, and not the reader, controls the pace. We are asked to prioritize our fleeting perceptions and to decide what to read and see when it is impossible to read and see it all. According to Janez Strehovec ("The Moving Word" 104), in the context of digital web poetry this easily leads to foregrounding kinetic and visual affects, effects and constellations at the expense of the usual syntactic and semantic complexities.

Access. If we don't have random and complete access to every part of the text, our potential mastery is once again denied by constraining our traditional right to traverse and to skip the text any way we please. Controlled access contains the very real possibility that some parts of the text will remain hidden and out of reach despite the best efforts of the reader.

Links have the potential to undermine the relation between parts and wholes allowing different degrees of local and global wholeness, but that potential should not be overestimated, as the average hypertext fiction shares many stabilizing qualities with its print predecessors (from static dynamics and determinability to intransient time and impersonal perspective).

Personal perspective forces the reader to assume strategic responsibility and then face the consequences of her actions. This is a very game-like feature, but it is also possible that the required strategic choices have to be made in the absence of explicit rules, goals and manipulative procedures.

Finally, *the user functions.* While I save studying their mutual relations until a little later, it is safe to say that whereas the explorative user function brings in the rhetoric of choice, navigation and labyrinth, the configurative and textonic user functions give us the chance and necessity to affect the text and therefore to use or conceive it as a playground, an obstacle, or more or less malleable raw material to build upon. These latter two user functions clearly foreground the user's own extranoematic activity without which there would be no experience of wholeness whatsoever.

All these shifts away from the traditional media position (static, determinate, intransient, impersonal perspective, random access, no links and interpretative user function) seem to suggest a certain reversal and exhaustion in the reader's role. The "new" cybertextual media positions may constrain and limit the reader's freedom of movement and his control over reading time, undermine and even deny such traditional luxuries of reading as guaranteed repeatability, possibility to re-read, complete access and strategic passivity, and open up the finality of the literary work to potentially endless metamorphic processes,[7] setting new demands for the reader's attention span[8] while requiring much more than interpretative involvement from her. As the text sets conditions for its reading, it doesn't make much sense to keep clinging on to traditional values and habits of reading for much longer.

Traditionally, the work of art or literature can exhaust even the best attempts of its reader by using extensive duration or variation. Computers can be of great help in this, but they are not necessary in creating such effects as Queneau's *100,000 billion poems* have already shown. To counter this we could say that Queneau's work is easily exhaustible, because all its textons are readily available and the sonnet form is already automated to boredom and beyond. In any case, these examples have been rather marginal and have left no traces on ordinary horizons of expectations, which still take it for granted that a work is (and should be) materially exhaustible.

In hypertext theory, Jim Rosenberg did propose a concept of session as a new unit of reception (and attention) to account for the difficulties in traversing the text and variations in the reading order the hypertext reader is or was supposed to struggle with. Despite that common sense effort, the paradigm of reading it all still reigns, probably because at least the well-informed readers know that it is possible to read every single unchanging lexia of *Afternoon*, *Victory Garden* and *Patchwork Girl*. The case is a bit different in the later works of Stuart Moulthrop, especially in *Reagan Library* and *Pax*. The paratextual intro to the latter even tries to convince the reader-user of the impossibility of reading and experiencing it all (Moulthrop, "About Pax"). More than two decades before Rosenberg Roland Barthes drafted four pleasurable ways of how the readers combine their reading neurosis with the hallucinated form of the text. Let me quote:

> The fetishist would be matched with the divided-up text, the singling out of quotations, formulae, turns of phrase, with the pleasure of the word. The obsessive would experience the voluptuous release of the letter, of secondary, disconnected languages; of metalanguages. . . . A paranoiac would consume or produce complicated texts, stories developed like arguments, constructions posited like games, like secret constraints. As for the hysteric . . . he would be the one who takes the text for ready money, who joins in the bottomless, truthless comedy of language, who is no longer the subject of any critical scrutiny and throws himself across the text (which is quite different from projecting himself into it). (*The Pleasure of the Text* 63)

In short, if you read for your pleasure, you don't have to read it all.

To sum up: It is trivially easy to use cybertextual machinery to undermine traditional notions of the textual whole. We simply can't read it all in *Book Unbound* or *The Speaking Clock*, and our own activity is always already there in the beginning to replace and contaminate the original text in *The Impermanence Agent*. Still, these three works are far from chaotic or strenuously difficult; in *Book Unbound* the process and chance of inevitable personalization gives a firm enough teleology to the work's potentially endless mutability, and if we adjust our attention spans and kick our reception habits we can now and then take a relaxed look at the workings of the *Agent* and the *Clock*.

Even if the textual whole might have vanished or be under erasure, the expectations may remain the same. Cybertexts are machines for producing variety of expression, and at the moment we can discern at least four degrees in the deep-rooted humanistic fear of variety:

You can read for the plot and look for exaggerated rather than moderate or minimal forms of coherence and closure. The typical result is hypertext whining, the contradictory and hype-ridden practice where critics keep disappointing themselves by projecting their mainstream expectations onto works that continue the long and winding tradition of experimental literature.

Secondly, you can read for pleasure and stay happy within your very own hallucinated form. It's more than likely that Barthes' ideas about the reading neurosis could be extended to psychoses, borderline cases, and perversions, and not only in literature, but in games as well.

Thirdly, one can read for metareading as the connoisseurs of experimental literature do and cherish intertwining epistemological, ontological and ergodic challenges to the doxa of the day or the literature as we used to know it. Sometimes it may even be advisable to read literature the way the OuLiPo used to write it—as combinations of objects and operations.

And fourthly, one might consider balancing reading with playing. There's a long and binding aesthetic and trans-aesthetic convention that you are not supposed to exhaust the potentials of an instrument of any kind. Or if you do, it is not much of an instrument—or much of a game either.

Generally speaking, it would be good to give up the idea that interpretative user function will and should always be on top, and the most important one. When the textual whole has gone, interpretative mastery should follow its master as well, especially when we are entering our final challenge that situates itself firmly on the no-man's land between the last two stages in the following chart:

User function(s)	Aesthetic realm
Interpretative	Linear literature
Interpretative served by other user functions	Mainstream ergodic literature
Interpretative undermined by other user functions	Experimental ergodic literature
Interpretative serving other functions	Play, games, toys, instruments

Fig. 5. User functions and corresponding aesthetic realms.

It's time to move on to our two final challenges, which are implicitly tied to two almost mutually exclusive attitudes and approaches to play, games and literature(s).

6. Against the Fear of Variety

When the safe and somehow manageable totality, be it coherent or not, vanishes from sight, the spectators and readers have to try different strategies of comprehension, which may seem complicated if and when the users do not know the limits and the functional principles of whatever they are encountering in the disguise of an artwork. There's no guarantee that the work works as it seems to work, or continues to work as it has worked so far, not to mention that it would work as its manual or other paratext claims it works.

From this perspective computer games are interesting, as they domesticate the excess looming large in both ordinary and avant-garde products and processes, and the fundamental potential for change and unreliability inherent in new media objects. It would be tempting to generalize and argue that for as long as we have systems where there are either one material level or several material levels with trivial mutual relations, narrative is the most powerful (and certainly the most popular) arrangement that can be used in them—but whenever the relation of levels turns out to be arbitrary the concepts of gaming and simulation will be more and more attractive and popular. Despite the fact that computers can support, emulate and modify whatever aesthetic tradition and convention there is, the most satisfying way of pacifying consumers facing potential insecurity and lack of motivation caused by endless variation will be computer games. They promise fun and pleasure in exchange for following and applying the rules in order to reach the (given, chosen or created) goal.

A few words about ludology might be in order before I take my favorite pick of the challenges it provides. The very idea of ludology is to study games and especially computer games as games, and not as a derivative of something else like narratives, drama or film or their interactive and/or remediated offshoots. In short, games should be studied as their own transmedial discursive mode. It is important to understand what this doesn't mean. It does not mean there can't be hybrids, as no-one in his right mind working with computers, these famous universal machines, could deny the possibility of hybrids. It sounds trivial, but it isn't.

In a poorly defined opposition to ludologists there have been three major camps of narrativists: firstly, those who do not even define the contested concepts they try to compare (stories, narratives, games); secondly, pan-narrativists who try to stretch their definitions of stories and narratives to the extreme where everything we cognitively construct from whatever is presented to us will result in a story; and thirdly those who care to define their concepts, but focus on just hoping and predicting that some day, maybe, a useful definition will arrive to combine games and stories, but don't seem to have the brains to do the conceptual work themselves. No wonder the ludologists won that debate, the particularities of which don't need to concern us here, as we are talking about our six more or less literary challenges.

It would be trivially easy to point out the differences between games and narratives: the former an economy of means and ends with rules, goals and necessary manipulation of an equipment; the latter an economy of presentation and interpretation with story and discourse. Of course one could still try to rewrite narratologies to extend to the realm of games, but as we speak, such theories have not emerged and the burden of proof lies with those who continue to make narrativist claims. It'd be much easier to do the opposite and redefine narratives as games of interpretation if one really wants to have a unified field, where literary knowledge will still be of some use.

However, our real issue is elsewhere. If we wish to build hybrids called instrumental texts or textual instruments,[9] we need to get rid of silly and worn-out prescriptive ideas of how to do it, one variation of which can be read from countless futile proposals for civilizing games by adding narrative must-haves such as plot, themes, coherence or fictional worlds. Here's a schematic chart of dominant features and differences between interpretative and configurative practices:

Interpretative practice	Configurative practice
Interpretation	Action
Economy of representation	Economy of means and ends
Conventions	Rules
Meaning guaranteed by minimal coherence	Variation pacified by rules and goals
Reception	Play
Fiction	Simulation
Work and text	Model and system
Spatiotemporal	Causal-functional

Fig. 6. Basic differences between interpretative and configurative practices.

The above list is not meant to cause blindness and irritation in purist anti-binarists; the right hand side merely highlights game-related characteristics that are perhaps more important to take into account in the process of building textual instruments than their traditional and thoroughly studied counterparts on the left. In short, I propose we focus on the expressive potential of the former terms instead of constraining playability by the latter. Before viewing this challenge from the perspective of game studies in the next section, I'll try to figure out what literary theories and traditions could offer.

As we saw in the previous section, the "new" cybertextual media positions are typically used to diminish the reader's autonomy over his reading time and habits. The idea of building instrumental texts and textual instruments could be seen as a countermeasure giving the reader the chance to exercise his precious freedom in the form of play. It is still too early to say whether this recent shift of focus from the functioning of the cybertext to the actions and activities of its user-player is more than a minor trend caused by the accumulating pressure and attraction of computer game studies. From the perspective of David Rokeby's decade-old insights it's not very odd that also in literature navigable structures exist side by side with transforming mirrors and automatons.

In any case an instrument is supposed to shape and frame the player's action and to produce interesting variation. This is a challenge that goes far beyond the overly hyped problems of non-linear presentation. As in any economy of means and ends, it is important to find suitable goals and patterns of change and variation in the functional and causal framework. The questions of how to use one's resources and for what purpose haven't been the bread and butter of literary theorists except maybe in the author-centric context, most explicitly so in the Oulipian tables of operations changing the order, length, number or nature of linguistic objects from phonemes and letters to paragraphs and beyond.[10] Only if we remove a major implicit constraint from these schemes, the idea that these operations should be used by an author in order to produce a final and fixed literary object, could we get something to play and build an instrument with.

Literary tradition contains at least five easy dialectics that could be adapted as flexible frames for the necessary variation: the text as an object and a process, the work and the oeuvre, the text and the intertext, the reader's and the text's control over reading, and the maintenance and destruction of the text. The task and the pleasure of the reader-player-instrumentalist would be to maintain, break or (re)create the balance between these oppositional poles.

The instruments could be tuned by theories that are precise enough; in addition to the Oulipian corpus there are many other almost ready-made candidates ranging from Freud's case history of Schreber[11] to the particularities of the traversal function in cybertext theory. All these theories are more useful or instrumental in building textual instruments than instrumental texts, as the former should in principle be able to handle any given text and not just the one particular text that comes with the system.

Wardrip-Fruin's attempt to build a 3-gram instrument ("From Instrumental Texts") relies on syntagmatic substitution. It would be interesting to see it complemented by similar but somewhat fuzzier paradigmatic capabilities (maybe by using self-organizing maps).

In many cases an instrument gives us information we couldn't otherwise have. It'd be rather unproblematic to build instruments that detect constellations and statistical features that the readers can't or won't. The way people use instruments is also unpredictable to a certain degree. These emergent patterns could be boosted, paralleled or countered if the text and its behavior were at least partly organized by a-life applications.

The already existing multitude of instruments is just waiting to be exploited, especially those cheap everyday ones already capable of communicating with other smart objects. We are just one step away from the texts for the alcometer or other such devices detecting meaningful fluctuations in the physical condition of the player (allowing the player to play also with herself).

7. Games and Textual Instruments

We could perhaps begin to see the rise of concretely (and not metaphorically) playable literature if we imagine works that shape their ergodic side by taking their cues from Aarseth's recent game typology (Aarseth, Smedstad and Sunnanå). This system has 15 dimensions resulting in more than 220,000 functional game genres, circumscribing yet another neglected area of research. The reasons for choosing this typology for closer inspection are rather self-evident: If this model is any good it gives us an overview of the most important features of game structures and their basic variations. It is also clear that game structures affect and shape game-play. So, if we were to follow John Cayley, Stuart Moulthrop ("About Pax") and Noah Wardrip-Fruin's ("From Instrumental Texts") recent suggestions and build literary instruments we could do a lot worse than see what Aarseth's typology has to offer.

Let's begin with the categories of space: "The perspective of the player can be either omnipresent allowing the player to examine the entire arena or field at will or vagrant where the perspective follows a main player-token or avatar" (Aarseth, Smedstad and Sunnanå 49). At first glance this is reminiscent of the category of access in cybertext typology. Still, this game category is more directly connected to the user's perception, and that allows us to imagine literary practices playing with access and perspective, examples ranging from Eduardo Kac's holopoetry where the user's physical movement changes her perspective on the work as well as the work's visible content to hypothetical hypertexts where the node starts to destroy itself from the invisible end at the very second the reader begins to read it from the other end. As noted before the more fundamental question is whether it makes any sense to build instruments that could detect and analyze textual features, constellations and statistical frequencies humans either can't or won't—or could but only with great difficulty (like chart the field of possible multilingual anagrams in *Hexentexte* or *Finnegans Wake*) and reorganize either themselves or the playtexts using this non-human information.

The second category is topography: "A game's topography can be either geometrical with continuous freedom of movement, or topological, giving the player only discrete, non-overlapping positions to move between" (Aarseth, Smedstad and Sunnanå 49-50). In this respect classic hypertexts seem like badly designed board games allowing the reader to navigate between discrete reading positions (in contrast to such kinetic texts as David Knoebel's *Wheels*). This could be made more obvious and challenging by adding a chess-like interface to a 64-node hypertext and allowing the reader to traverse it from the positions of different pieces and according to their possible moves. In a turn-based two-player mode, you could lose your piece if your fellow reader's piece enters the same node, making it harder or impossible to traverse the whole text. This game could be made more challenging if the players didn't see each others' positions until it was too late and they overlapped with fatal consequences. In this arrangement the game and the literary text would be each other's by-products.

Thirdly, the game environment is either static remaining unchanged for the duration of the game or dynamic letting itself to be modified by the player. Obviously, MUDs are often dynamic in that sense, but there are other environments, including our physical surroundings, which we can now or soon tag and populate with miniature literary devices producing variety of expression.

Aarseth's game typology contains three parameters of time: pace, representation, and teleology. The category of pace divides games into real-time and turn-based ones, which is almost but not quite the same as the category of transience in cybertext theory. We already have boring turn-based collaborative writing practices, but since we seek to build instruments they won't do. *Hegirascope* comes close to being a real-time instrument, as its 30-second constraint feels like a challenge or an obstacle, which is the first step away from passive "timeless" hypertexts towards more complex real-time adversary structures.

The representation of time is either mimetic or arbitrary depending on whether the time of the actions in the game mimics the time of corresponding actions in the real world. It would be rather easy to build adaptations of literary classics and instruments for playing them that use strictly mimetic reading time freezing the reader's movement or progress. In a more competitive setting the reader-players could block each other's ways and navigation options inside an online hypertext using mimetic time.

The third temporal category relates to the final goal of the game. The games with clearly defined successful outcomes for one or more players are teleologically finite, and those without such outcomes are teleologically infinite, and could in principle go on endlessly. *Book Unbound* is clearly infinite while hypertext fictions as we know them are all teleologically finite, successful conventional outcome requiring that every node has been read at least once. More complicated goals could be crafted after Brian McHale's studies of postmodernism: The reader should maintain, restore or destroy the delicate balance between ontological and epistemological problems.

The next provisional subcategory, player structure, reminds us once more about certain fundamental differences between literature and games. We are given six options: single-player-, two-player-, multiplayer-, single-team-, two-team- and multiteam-games. It's a combination of adversary-structure (none, one or multiple) and team-structure (individual or team-based). In a quick comparison it is obvious that contemporary literature is single-reader, that is, the combination of individual interpretative effort and no adversaries. Superficial imitations would give rise to literary texts that could be read only as a team effort, and to readers interfering with each other's reading positions and possibilities, obstructing them to the best they can. On the more positive side, these kinds of practices would counterbalance the fuss about collaborative writing, which is such a banality.

The category of player structure contains also two categories currently best suited to mobile games, the question being whether the proximity of the players to each other and the physical location of the players matter. Probably, these possibilities will multiply with various WLAN technologies and cheap Bluetooth chips added to appliances of every kind, shape and size. Literary fictions tracking down your everyday routines, habits and movement and analyzing them in order to make the text more suitable to you and your blood pressure (or other problems) would not be to everybody's liking—if in addition to the physical location also the physical condition of the player-reader counts. Other obvious applications would require you to be at an intimate distance with someone to access a love story or pornography connected to sexual sensors and paraphernalia.

Aarseth's next grouping of fundamental game categories addresses control in terms of mutability, saveability, and determinism. Mutability refers to rewarding the player by strengthening her player-character or player position. In literary contexts such rewards might include gaining better access to or having more effect in the fictive world, or acquiring more resources like agents to help you scout the as-yet-unread territories, perhaps their style or genre. One form of saveability is already available and at work in *Book Unbound,* where the user can save her favorite extracts from generated text before the next generation takes place. This type of collected work saved by the reader is yet another solution that could be expanded to certain other contexts and practices as well (to challenge copyright laws for one). Obviously, the way the text is used could be used to determine how much of it could be saved until the next reading. The category of determinism is about unpredictability and random functions that are always already there if two or more humans play against (or with) each other. An instrument with multiple degrees and criteria of determinability would be ideal for games of increasing and decreasing coherence or certain kinds of preferred coherence; playing de Sade with Christian tunings or the Bible with libertine lardings[12] could be fun.

Finally, there are rules and three simple meta-rule dimensions: the presence or absence of topological, time-based or objective-based rules. As a concluding inversion to our topic, this meta-dimension could be further analyzed by running Aarseth's game typology through his cybertext typology, or vice versa. In any case, we'd have to make a fundamental distinction between textonic and scriptonic game structures, that is, game structures as they exist in the game and game structures as they appear to the players.[13] This distinction would open up the possibility of meta-games where one plays both by the rules and about them, and where different teams play the same game by different rules without knowing it (at least for a while). Here, finally, literary knowledge might have some value, as in the case we wish to build unfair and unequal games, we don't need to go any further than 20th century fiction for great ideas.

Notes

1. Here's a crash course for those who are not familiar with the former: In cybertext theory the elementary idea is to see a text or a work of art as a concrete (and not metaphorical) machine for the production and consumption of signs, and consisting of the medium, the operator and the strings of signs. The latter are divided into *textons* (strings of signs as they are in the text) and *scriptons* (strings of signs as they appear to readers/users). The mechanism by which scriptons are generated or revealed from textons is called a traversal function, which can be described as the combination of seven variables (dynamics, determinability, transience, perspective, links, access, and user function), and their possible values. This combinatory approach gives us a heuristic map of 576 different media positions into which every text could be situated based on how its medium functions, but independently of what that medium is. It is important to notice that the relation between textons and scriptons is arbitrary in digital media: That's the "essence" of its unique dual materiality—which stems from the separation of the storage medium from the interface medium. Aarseth's *Cybertext* (1997) focuses on ergodic literature, where the user has to do non-trivial work or effort to traverse the text, that is, ergodic literature requires more than interpretative activity from its user.
2. I made a somewhat similar point in an earlier article ("Cybertext Theory"). In her response to it N. Katherine Hayles ("What Cybertext Theory Can't Do") manages to confuse the tasks of theory with those of criticism and chooses to ignore the fact that in *Cybertext* functional differences are deducted from material ones (cf. Aarseth, *Cybertext*, ch.3). She also claims that "A third limitation of cybertext theory, especially as interpreted by Eskelinen, is mistaking numerosity for analytical power. . . . Simply because cybertext theory predicts 576 different combinations, using Aarseth's scheme for parsing the semiotic components of cybertexts, does not mean that all 576 combinations will be equally interesting or worthwhile. Nor does this number alone indicate the value of the theory, beyond setting up so many pigeonholes to be filled" ("What Cybertext Theory Can't Do"). Here she attacks only her own fabrications, as she turns a blind eye to every single example I give of the theoretical power of cybertext theory and its already empirically existing ingredients in the context of modifying, transforming and expanding hypertext theory, narratology, and theories of postmodernism and new media (etc.)—and not in the context of analyzing an individual text or filling pigeonholes. Hayles' blindness to theoretical matters (and consequences) is well in synch with the embarrassing theoretical blunders

she makes in her *How We Became Posthuman.* I discussed some of them in my article, but for some reason Hayles didn't defend her shortcomings in her response. In retrospect, the funniest part of Hayles' reply is her firm belief that she's capable of setting the record straight "about what cybertext theory cannot do, as well as what it can do." If that's the professional appearance Hayles wants to keep up, then she should know a lot better than to confuse textons and scriptons with surface and depth (for further details cf. "Responses to 'Narr@tive: Digital Storytelling 1/2'").

3. At least they should be as Genette's study is also a reading of Marcel Proust's *Remembrance of Things Past*, and a lot has happened in narrative fiction since the 1920s.
4. Stuart Moulthrop's *Reagan Library* rewards the persistent reader by adding more text to its nodes in the first three revisits. In our terminology, the frequency is limited during these revisits and unlimited after the text does not change anymore.
5. Here we could make a further distinction between static and dynamic architecture based on the stability of basic conditions. In Joyce's *Afternoon*, as the result of its guard fields, the number of possible links to follow will vary at the level of nodes, that is, locally. In this scenario *Afternoon* is not dynamic as a whole (globally), as there are no changes in its basic architecture of nodes, links and guard fields.
6. Characters in literature and film are said to be independent as the readers and viewers can't influence them. The behavior of these characters is static as it remains exactly the same regardless of how many times the book or film is read or seen. They are objects as there's no direct communication or dialogue between the user and the fictional character. This simple chart with three basic parameters gives seven other positions for the possible existents, and I see no valid reason why all of these should be called characters too, as that practice would only help to mask the crucial functional differences in the control, communication and behavior of the existents. In live drama the characters are minimally influenced by the spectators (by applause and other feedback conventions), and they may also become dynamic if an actor is given the freedom to change certain attributes or personality traits of the character between performances.
7. These processes can be both teleologically and temporally indeterminate and infinite.
8. In books the continuity and discontinuity of the attention span is up to the reader. That's not the situation with works like *The Impermanence Agent* or *The Speaking Clock*. We should perhaps introduce two conceptual divisions to meet the demands set by the behavior of programmable texts. First,

the possible and necessary attention spans and their mutual relations, and second, the relation of the text's duration and speed to the reader's attention span (the former can either exceed or match the latter).

9. Cf. Wardrip-Fruin, "From Instrumental Texts to Textual Instruments." Instrumental texts are texts that can be played with a system that is inseparable from them; textual instruments are systems that can play many different "outside" texts.
10. Cf. Bénabou, as well as Mathews and Brotchie 213-14 (Queneleyev's table) and 227-28 (systematization).
11. "Delusions of jealousy contradict the subject, delusions of persecution contradict the verb, and erotomania contradicts the object. But in fact a fourth kind of contradiction is possible—namely, one which rejects the proposition as a whole" (Freud 203). Needless to say this systematization and mechanism could be situated or embedded as a parergon in all four great scenes and schemes of writing and textual production: sender/receiver, signifier/signified, text/intertext, and textons/scriptons.
12. Cf. Mathews and Brotchie 163.
13. I'm not referring to the player's always possible non-understanding or non-knowledge of the rules and goals, but about potential game structures that vary in time and during the game depending on how the game is played.

Works Cited

Aarseth, Espen. *Cybertext: Perspectives on Ergodic Literature.* Baltimore: Johns Hopkins UP, 1997.

Aarseth, Espen, Solveig M. Smedstad, and Lise Sunnanå. "A Multi-Dimensional Typology of Games." *Level Up: Digital Games Research Conference Proceedings.* Ed. Marinka Copier and Joost Raessens. Utrecht: Utrecht UP, 2003. 48-53.

Barthes, Roland. *The Pleasure of the Text.* New York: Hill and Wang, 1975.

Bénabou, Marcel. "Rule and Constraint." *OuLiPo: A Primer of Potential Literature.* Ed. and trans. Warren J. Motte, Jr. Lincoln, NE: U of Nebraska P, 1986. 40-47.

Bordwell, David. *Narration in the Fiction Film.* Madison, WI: U of Wisconsin P, 1985.

Branigan, Edward. *Narrative Comprehension and Film.* London: Routledge, 1992.

Cayley, John. *Book Unbound.* London: Wellsweep P, 1995.

———. *The Speaking Clock.* London: Wellsweep P, 1996.

Chatman, Seymour. *Story and Discourse.* Ithaca, NY: Cornell UP, 1978.

———. *Coming to Terms.* Ithaca, NY: Cornell UP, 1990.

Eskelinen, Markku. "The Gaming Situation" *Game Studies* 1.1 (2001). 4 July 2006 <http://www.gamestudies.org/0101/eskelinen>.

———. *Kybertekstien narratologia* (Cybertext narratology). Saarijärvi: Research Centre for Contemporary Culture, 2001.

———. "Cybertext Theory and Literary Studies, A User's Manual." *The Electronic Book Review* 12 (2001). 4 July 2006 <http://www.altx.com/ebr/ebr12/eskel.htm>.

Frasca, Gonzalo. "Ephemeral Games: Is It Barbaric to Design Videogames After Auschwitz?" *Cybertext Yearbook 2000.* Ed. Markku Eskelinen and Raine Koskimaa. Saarijärvi: Research Centre for Contemporary Culture, 2001. 172-82.

Freud, Sigmund. "Psychoanalytic Notes on an Autobiographical Account of a Case of Paranoia." 1911. *The Pelican Freud Library,* Vol. 9. Ed. Angela Richards. London: Penguin, 1981. 138-223.

Genette, Gérard. *Narrative Discourse: An Essay in Method.* 1972. Trans. Jane E. Lewin. Ithaca, NY: Cornell UP, 1980.

———. *Narrative Discourse Revisited.* 1983. Trans. Jane E. Lewin. Ithaca, NY: Cornell UP, 1988.

———. *The Architext: An Introduction.* 1979. Trans. Jane E. Lewin. Berkeley, CA: U of California P, 1992.

———. *Paratexts.* 1987. Trans. Jane E. Lewin. Cambridge: Cambridge UP, 1997.

———. *Palimpsests: Literature in the Second Degree.* 1982. Trans. Channa Newman and Claude Doubinsky. Lincoln, NE: U of Nebraska P, 1997.

Gibson, William. *Agrippa: A Book of Dead.* New York: Begos, 1992.

Hayles, N. Katherine. *How We Became Posthuman: Virtual Bodies in Cybernetics, Literature and Informatics.* Chicago: U of Chicago P, 1999.

———. "What Cybertext Theory Can't Do." *The Electronic Book Review* 12 (2001). 4 July 2006 <http://www.altx.com/ebr/rip12/rip12hay.htm>.

Kac, Eduardo. "Key Concepts of Holopoetry." *Experimental-Concrete-Visual: Avant-Garde Poetry Since the 1960s.* Ed. K. David Jackson, Eric Vos, and Johanna Drucker. Atlanta: Rodopi, 1996. 247-257.

Knoebel, David. *Wheels.* 1996. 4 July 2006 <http://home.ptd.net/~clkpoet/wheels>.

Jackson, Shelley. *The Patchwork Girl.* Cambridge, MA: Eastgate, 1995.

Joyce, Michael. *Afternoon, a story.* 1987. Cambridge, MA: Eastgate, 1990.

Mathews, Harry, and Alastair Brotchie, eds. *OuLiPo Compendium.* London: Atlas P, 1998.

McHale, Brian. *Postmodernist Fiction.* New York: Methuen, 1987.

———. *Constructing Postmodernism.* London: Routledge, 1992.

Moulthrop, Stuart. *Victory Garden.* Cambridge, MA: Eastgate, 1991.

———. *Hegirascope.* 1995. 4 July 2006 <http://iat.ubalt.edu/moulthrop/hypertexts/hgs>.

———. *Reagan Library.* 1999. 4 July 2006 <http://iat.ubalt.edu/moulthrop/hypertexts/rl>.

———. *Pax. An Instrument.* 2003. 4 July 2006 <http://iat.ubalt.edu/moulthrop/hypertexts/pax>.

———. "About Pax." 2003. 4 July 2006 <http://iat.ubalt.edu/moulthrop/hypertexts/pax/about.htm>.

Prince, Gerald. *Narratology*. Berlin: De Gruyter, 1981.

———. *The Dictionary of Narratology*. Lincoln, NE: U of Nebraska P, 1987.

Queneau, Raymond. *Cent Mille Milliards de Poémes*. Paris: Gallimard, 1962.

"Responses to 'Narr@tive: Digital Storytelling 1/2'." *Grand Text Auto.* 23 Apr. 2004. 4 July 2006 <http://grandtextauto.gatech.edu/2004/04/23/narrtive-digital-storytelling-12/#comments>.

Rosenberg, Jim. "The Structure of Hypertext Activity." *Hypertext '96*. New York: ACM, 1996. 22-30.

Rokeby, David. "Transforming Mirrors: Subjectivity and Control in Interactive Media." *Critical Issues in Digital Media.* Ed. Simon Penny. Albany, NY: Suny, 1995. 133-58.

Sternberg, Meir. *Expositional Modes and Temporal Ordering in Fiction.* Baltimore: Johns Hopkins UP, 1978.

Strehovec, Janez. "The Moving Word." *Cybertext Yearbook 2000.* Ed. Markku Eskelinen and Raine Koskimaa. Saarijärvi: Research Centre for Contemporary Culture, 2001. 100-116.

Wardrip-Fruin, Noah. *The Impermanence Agent.* 1999. 4 July 2006 <http://www.impermanenceagent.com/agent/>.

———. "From Instrumental Texts to Textual Instruments." 2003. 4 July 2006 <http://hypertext.rmit.edu.au/dac/papers/Wardrip-Fruin.pdf>.

Wardrip-Fruin, Noah, and Brion Moss. "The Impermanence Agent: Project and Context." *Cybertext Yearbook 2001*. Ed. Markku Eskelinen and Raine Koskimaa. Saarijärvi: Research Centre for Contemporary Culture, 2002. 14-59.

Noah Wardrip-Fruin

Playable Media and Textual Instruments

Abstract:

The statement that "this is not a game" has been employed in many ways—for example, to distinguish between high and low culture electronic texts, to market an immersive game meant to break the "magic circle" that separates games from the rest of life, to demarcate play experiences (digital or otherwise) that fall outside formal game definitions, and to distinguish between computer games and other forms of digital entertainment. This essay does not seek to praise some uses of this maneuver and condemn others. Rather, it simply points out that we are attempting to discuss a number of things that we play (and create for play) but that are arguably not games. Calling our experiences "interactive" would perhaps be accurate, but overly broad. An alternative—"playable"—is proposed, considered less as a category than as a quality that manifests in different ways. "Playable media" may be an appropriate way to discuss both games and the "not games" mentioned earlier.

The impetus for coming to this term was not a love of terminology, but the author's need as an artist to situate a set of experiments in creating "instrumental texts" and "textual instruments" within an appropriate context. While it doesn't make sense to discuss all of these experiments as games, what distinguishes them from other electronic texts is their playability—both that they can be usefully considered as playable, and their particular structures of play. This essay discusses, particularly, two "textual instruments" recently constructed by the author in collaboration with David Durand, Brion Moss, and Elaine Froehlich. While both of these instruments operate according to the logic of n-grams (as first used in textual play by Claude Shannon), one instrument is designed to play with known local texts while the other is designed to employ the contents of network RSS feeds and web pages. One composition for each of these instruments—*Regime Change* and *News Reader,* respectively—is considered.

1. This is Not a Game

I'm writing this in the first person. For clarity's sake. Because, while I am writing about terminology, the terms I discuss are simply those that I use to think through my work as an artist and its relationships with the work of others. If employed for different purposes, as we say online, "your mileage may vary."

The first term I need to consider is "game," but from an oblique direction. Because I need to consider some things that are "not games."

1.a. Artists against Infocom

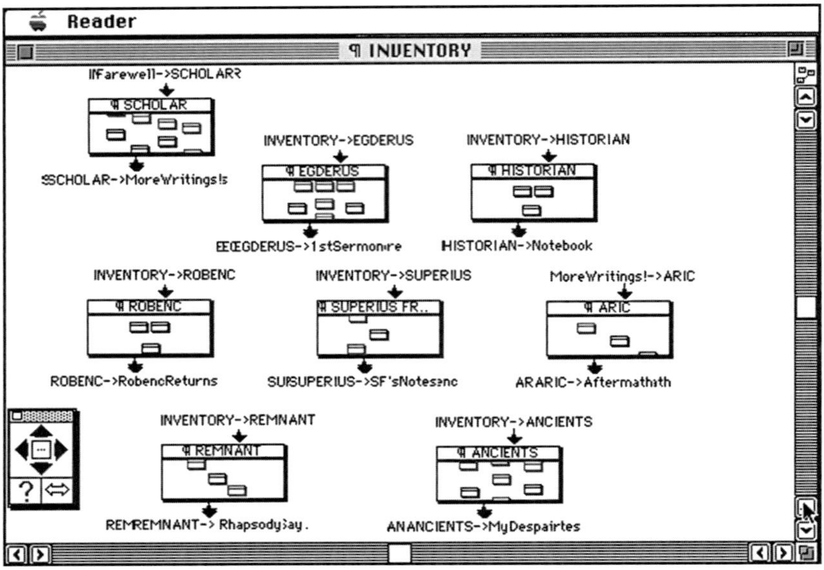

Fig. 1. Hypertext fiction is not a game. A Storyspace map view of Bill Bly's *We Descend.*

"This is not a game" was a slogan, at one time, embraced by a group of hypertext writers and theorists. While it is often traced back to a 1988 hypermedia writing workshop run by Rob Swigart, the slogan's most visible proponent was John McDaid. McDaid is may be best known as the author of the "artifactual" fiction *Uncle Buddy's Phantom Funhouse* (published by Eastgate in 1993). Other proponents of the slogan included high-profile members of the hypertext community such as Stuart Moulthrop and Michael Joyce (Moulthrop, Interview).

Swigart has written for computer games and published traditional novels—at the time of the workshop he'd recently published the "computer novel" *Portal* through game developer Activision. And, in fact, we can see the proto-slogan in use in discussions of *Portal* from before the convening of the workshop:

> Emerging from the mists of the "Vaporware" list in *PC Letter, Portal* has at last been published by Activision. Having thus established that *Portal* is not vapor, its creator, novelist Rob Swigart, has some further observations about what *Portal* is and is not.
> "It's not a game," says Swigart. Nor, apparently, is it interactive fiction as we have come to know it. "There is no parsing language in *Portal*," he adds, "no puzzles to solve."

> Then what is it?
> "It's a computer novel."
> And that is . . . ?
> "A novel that can be told only through the medium of the computer." (Infocom 9)

McDaid writes the following of the workshop group's enthusiasm for "this is not a game" as a phrase: "We thought this so important that we put it on the t-shirt, in *real* big letters." Why was it so important? In part, McDaid tells us, it was practical. There was no way to compete with games, especially graphically (fig. 1), so it was better to stay off that turf. But he also writes of what he calls a "larger, fictional issue." An issue connected to the structure of game play:

> The payoff for "correct" play [is] usually to win; to play "incorrectly" is to lose. This is very much at odds with what one might loosely call goals of fiction: exploration, insight, and the renewal of the perceived world through alterneity. (Infocom 9)

But McDaid then goes on to say, in the next sentence, "though it is true that in my own fiction, *Uncle Buddy's Phantom Funhouse,* there is in fact a puzzle. . . ." Faced with this, I'm a bit puzzled myself. Certainly I wouldn't want to elide the very real structural differences between fictions like McDaid's and those for which Infocom is best known. At the same time, however, there are also significant differences between McDaid's and Swigart's—not the least that one has "no puzzles to solve" while in the other "there is in fact a puzzle." Why, in these circumstances, would these writers have chosen such a primary focus on the term "game"?

When I interviewed Stuart Moulthrop, I asked him about this focus, about the hypertext community's version of "this is not a game." He pointed out that the implied comparison between the works on each side of the phrase wasn't neutral:

> There was an element of rank professional jealousy, for sure. [Infocom] had a market, after all. We were stuck in the garage. In retrospect our allergy to games looks incredibly foolish, both because Infocom's market experience didn't end all that happily, and more important because there was so much good work coming out of [the interactive fiction] community, and still is.

> I suppose what really changed my mind on this was [the] reception hypertext has sometimes got from the literary community: "How dare you? You have no place at this club." The notion that we could have gotten similarly clubby, trying to exclude someone else's work in new media, now seems repulsive. (Moulthrop, Interview)

Here we see "this is not a game" functioning as a distinction between mere text games and work worthy of consideration by the literary community. Between low and high culture. Between trivial play and serious writing.

Interestingly, those on the other end of this distinction seem to have chosen a different approach. As it happens, the piece quoted above discussing Swigart's *Portal* was published in Infocom's house publication, *The Status Line* (previously *The New Zork Times*). It concludes: "Can interactive storytelling work without challenging puzzles or conflict resolution? *Portal* proves it can." This is not an endorsement of a competitor's product (Infocom had recently been purchased by Activision) but, unless motivated entirely by command of their new corporate owners, it is a somewhat surprising endorsement of an artistic project that the "other side" came to see as sharply incompatible with Infocom's.

Personally, I'm interested in the ongoing work of both of these communities, and I'm looking for terminology that can help me see the common ground that the authors of *The Status Line* recognized.

1.b. Smudging the Magic Circle

"This is not a game" has also served as material for an influential game, and from there become a slogan for the players and developers of a certain group of games.

The promotional game for the movie *A.I.* had no official name, but here I'll use its nickname: "*The Beast*" (Stewart). Just as the game had no official name, it also had no marketing (in fact, was unavailable for purchase), and no official beginning. Or, to put it another way, it began when and how people began to play it. For many it began with the second *A.I.* trailer, in which "Jeanine Salla" is credited as "Sentient machine therapist" (Hon). Players' web searches for these terms revealed the beginnings of a trail that threaded through texts, images, and movies across the internet—as well as phone calls, faxes, US Postal Service deliveries, bathroom walls, and live events.

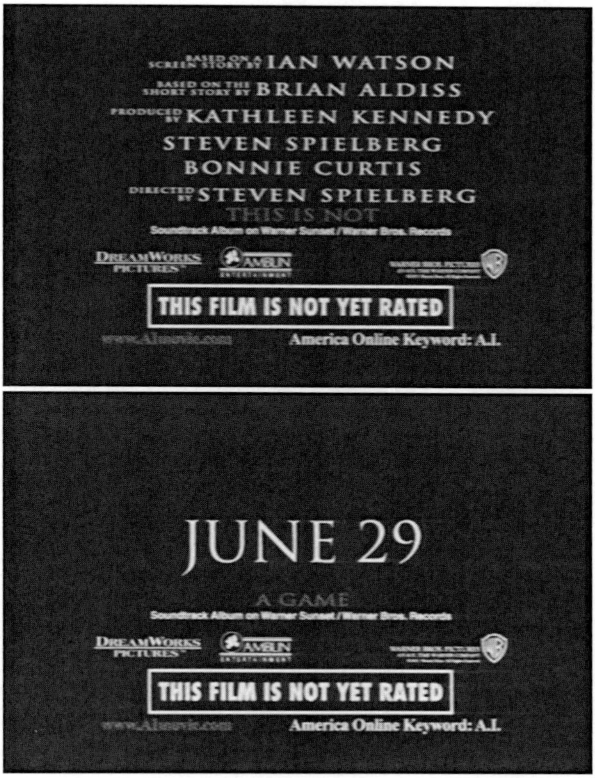

Fig. 2. Stills from *A.I.* television commercial.
"THIS IS NOT" and "A GAME" appear in red near the
center of the first and second still, respectively.

The game was a huge success—not only in the estimation of its players, but also in the surrounding media attention (which helped generate interest in *A.I.*). As the hype reached its height, in May 2001, a television commercial for *A.I.* was released that contained the words "THIS IS NOT / A GAME" (fig. 2).

As Jane McGonigal writes:

> This message has since become the mantra for both players and developers of immersive entertainment. To "TING" a game now means to explicitly deny and purposefully obscure its nature as a game, a task that has become increasingly difficult as immersive players grow more savvy about TING techniques. (4)

Here "this is not a game" has a different feel to it. Still a slogan of sorts, but a game's slogan. And as McGonigal points out, TING is now a game design technique—one that helps support players in their performance of belief in the game's reality, a feature identified as key to enjoyment of the "alternate reality" gaming genre that *The Beast* helped launch. This denial of the game's apartness from reality may seem to refuse the "magic circle" that has been part of our discussion of games since Johan Huizinga's *Homo Ludens*. But rather than an erasure or breakage of the circle, it's probably more of a smearing or smudging—a deliberate extension of the border between a game and the rest of life, in order to create space for performance and play.

"This is not a game" could also be seen as a more formal characterization of *The Beast*. After all, it had no rules, no points, and only ill-defined outcomes toward which the players could work (locating what might be a puzzle, trying to solve it, hoping to find larger patterns leading toward the solution of the emerging murder mystery). But more on this sort of consideration is to come.

1.c. Don't Toy With Me

Some people are under the impression that *The Sims* (fig. 3) is the best-selling PC game of all time (Adams 2004). In fact, as of this writing the publisher of *The Sims*, Electronic Arts, on its website leaves all qualifiers aside to call *The Sims* "The #1 best selling game of all time." But others would say that, while *The Sims* may have sold well, it is not a game. Rather, they say, *The Sims* is a "toy" or "simulation."

One source of such arguments is formal game definitions. For example, *Rules of Play*, a game design text from MIT Press, defines games as "a system in which players engage in an artificial conflict, defined by rules, that results in a quantifiable outcome" (Salen and Zimmerman 80). The focus on quantifiable outcomes (which *The Sims* may not sufficiently possess) makes non-games, or borderline cases, of many experiences that we commonly call games, including simulation games and role-playing games. The closest these experiences get to meeting the definition is when considered in terms of interim quantifiable goals the players set for themselves. As Salen and Zimmerman put it, "As with other open-ended game-like experiences such as *Sim City*, RPGs have emergent quantifiable goals but usually no single overriding outcome" (Zimmerman 83). And these authors are far from alone in proposing formal game definitions that systems such as *The Sims* fail to meet.

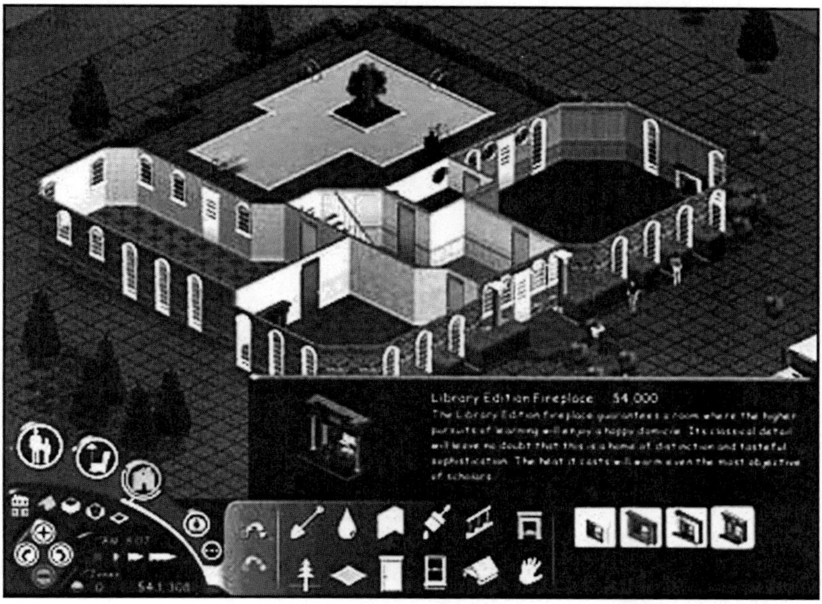

Fig. 3. *The Sims*, in its original form, doesn't have a quantifiable outcome, or even clearly-defined goals for players to work toward. Is the best-selling PC game of all time also "not a game"?

But more "popular" publications also question the status of *The Sims* as a game. For example, several game publications' reviews of later installments (console versions of *The Sims,* or *The Sims 2*) note this attitude toward the first installment, for example:

> With the first game, people complained that *The Sims* was more like a toy than a game. (Bub)

> In the PC versions of the series, gameplay was typically thought of as a toy—something for players to pick up and enjoy for however long they want with no clear end in sight. (Lewis and Boulding)

Sometimes the difference is split:

> The original *Sims* was as much a toy as it was a game. (Kosak)

Even the *Wikipedia* entry on *The Sims* also (as of this writing) notes that "It has been described as more like a toy than a game" (Wikipedia).

Taking a different tack, some have sought to more clearly describe how *The Sims* deviates from the usual definitions of games and then describe this as an alteration of their game model (rather than a lack). Jesper Juul, for example, writes:

> Open-ended simulation games such as *The Sims* change the classic game model by removing the goals, or more specifically, by *not* describing some possible outcomes as better than others. (30-45)

This could result in us changing our game definitions or, as Juul does, classifying *The Sims* as a borderline game.

I don't mean, by this, to seem to be arguing against formal game definitions (or popular perceptions). It may well be that *The Sims* is not a game. At this point I'm simply intending to add "doesn't meet my (in)formal definition" to the senses in which people have meant "this is not a game."

1.d. Overtly dramatic

Toys and simulations aren't the only types of entertainment software that "aren't games." In their 2002 SIGGRAPH presentation Michael Mateas and Andrew Stern put up a slide that read, in part, "It's a story, not a game" (Mateas and Stern). The two were referring to their jointly-developed project, *Façade* (fig. 4), which provides a first-person dramatic experience for the interactor. There are no puzzles, no points to score, and no quantifiable outcomes. Instead, the interactor plays a character, interacts with other characters through language and movement, and has an experience shaped by *Façade*'s software "drama manager." The experience, of a couple breaking up, has caused *Façade* to be described as "an interactive *Who's Afraid of Virginia Woolf*."

Mateas and Stern's project is part of a genre of entertainment/art software that has been called "interactive drama" or "cyberdrama"—popular knowledge of which has been spread by books such as Janet Murray's *Hamlet on the Holodeck*. While interactive dramatic experiences may fail to meet many of the criteria of formal game definitions, this isn't what Mateas and Stern were arguing in their SIGGRAPH slides. Rather, they were seeking to draw attention to the different structures of play, and goals of play, in this genre. Here, there is no winning or losing, and the point of interaction is not to accomplish a game goal. The interactor certainly forms opportunistic short-term goals, but usually in terms of the dramatic situation. In this case, to be "not a game" is to be a different kind of playable digital media, calling for a different type of engagement, than usually comes to mind when the term "game" is used.

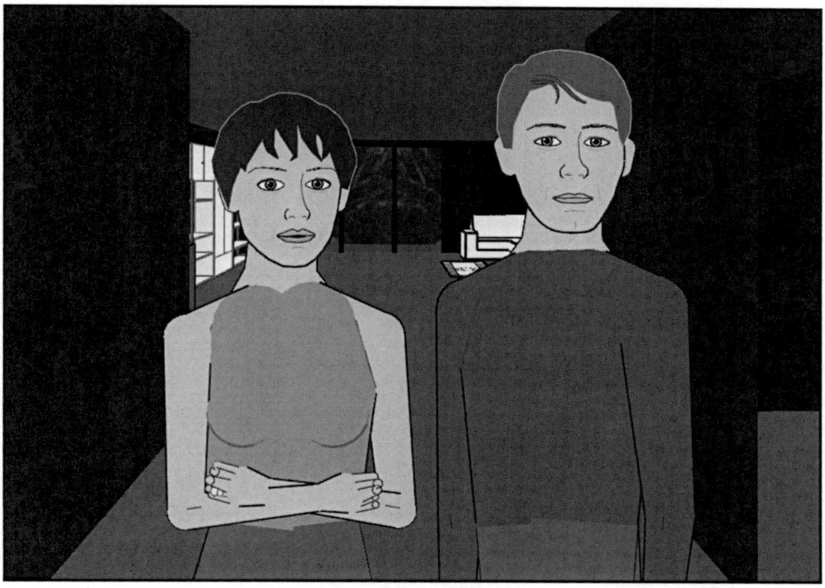

Fig. 4. *Façade*'s main characters, Grace and Trip, hold painful secrets and witty quips. Exploring the dramatic experiences possible with them, rather than achieving a higher score, is the motivation for replaying.

The assertion that interactive dramas are not games is relatively well accepted among game scholars. But it is not universally accepted, either within or outside such circles. And it seems that one place it is less accepted is within the game development community itself. *Façade* was chosen as a finalist for the 2004 Independent Game Festival awards, which must on some level be seen as acceptance of its status as a game by the jurors of that competition. Further, the IGF is held in conjunction with the yearly Game Developer's Conference—and, therefore, in 2004 *Façade* was on display in the main exhibition hall for the length of the U.S.'s primary gathering of game developers. I spent time at *Façade*'s kiosk regularly during the conference and observed the people gathered around, watching others interact and waiting for their turn. I didn't hear a single one say that it was "not a game."

And neither Stern nor Mateas would have uttered the phrase then, either. By 2004 *Façade*'s development had progressed further, and their presentation of the relationship between *Façade* and the term "game" had evolved considerably. While "it's a story, not a game" had helped shock some out of the "how do I win?" mindset, there is much in *Façade* that is usefully considered with the term "game." As Stern described it to me in an e-mail:

> In fact, one of the underlying interaction mechanics of the first half of *Façade* we call the "affinity game," where Grace and Trip interpret everything you do as a zero-sum taking sides "game"; the second half of *Façade* we call the "therapy game," where the player is (purposefully or not) potentially increasing each character's degree of self-realization about their own problems. (E-mail)

Stern and Mateas here are making an interesting maneuver, introducing "game" in the interpersonal sense (as in *Games People Play*). As Stern explains in a post on the collaborative blog *Grand Text Auto* titled "Head Games," this isn't a way of trying to shift the ground away from discussion of computer games, but to imagine a new area of the computer game field which focuses on interaction with richly realized characters. Stern writes that, in such games:

> The gameplay will literally need to be about the characters themselves. The "state space" the player manipulates—the variables you affect, the values you change—need to be the feelings, emotions and thoughts of the characters, not just external counters, scores, levels and objects. Rather than just firing a gun to cause an enemy's health to decrease, or a crate to explode or a door to open, you'll fire off *discourse acts* such as praise, criticism, expressions of feeling, requests and ideas; the other characters' attitudes will immediately change, emotions will get generated, and new actions will become motivated and get performed.
> What kind of game would that be? It could be the game of persuasion, or negotiation, seduction, or communication, for example. The kinds of games we play with each other all the time, really. ("Head Games")

Just as *Façade* attempts to integrate such games into an overall dramatic structure, more traditional game developers could integrate them with mainstream game genres involving human characters—from role-playing games such as *Knights of the Old Republic* to simulation games such as *The Sims*. Of course, for this to be successful, more research of the sort being undertaken by Mateas and Stern will be necessary.

2. Playable Media

All of the above uses of "this is not a game" are potentially fruitful. But the distinctions they make are not the ones I'm after. I'm not looking to separate high from low, well-demarcated from immersive, or the formally defined from border cases. I'm looking for a way to discuss all these examples together, in a manner that highlights a set of features that are of interest to me, and without throwing so broad a net that the weight of what I'm trying to pull in capsizes my vessel.

Taking this point of view, I have a series of thoughts: It may be that none of the examples in the preceding section are games, or some of them may be—and that's fine. Some of what I create as an artist may be games, or not—and that's fine, too. But thinking about how these may or may not be games has led me to identify something that is interesting about all the preceding examples, and hopefully also about some of what I'm creating as an artist—how they are *played.*

And this is what has led me to talk about "playable media." For me, this phrase shifts my thinking from a question I've found only temporarily useful ("Is this a game?") to one I have found rewards sustained attention ("How is this played?"). "Playable media" also encompasses a body of work that I want to consider—including the examples above, as well as many other products of the commercial game industry, and also the body of what might be called "playable art." By "playable art" I primarily mean projects from the digital art community such as Camille Utterback and Romy Achituv's *Text Rain* (fig. 5), which invite and structure play.

Of course, I could also talk about all of these things by using an already-popular umbrella term such as "interactive." But these terms are often overly broad even for my purposes, and the impetus that allows them to attain their breadth may shape their meanings in a way that isn't helpful. For example, while it's easy to imagine an argument about whether a love letter is more "interactive" than a computer game, it's unlikely that we would spend long arguing which is more fruitfully considered in terms of its playability. And while playability is a useful way to look at agreed-upon games (such as football) it's also appropriate when considering certain types of related play (such as hackey-sack) without encompassing too much (it leaves aside sports commentators, the symbolism of team logos and names, the economy of sporting goods, etc.).

And a focus on the playable also attracts me for another reason—because we play more than games and "not a" games. We also play instruments, and compositions. And it is at the juncture of these senses of play—that for games, and that for music—that a thought-provoking discussion about playable texts has been taking place in the electronic literature community.

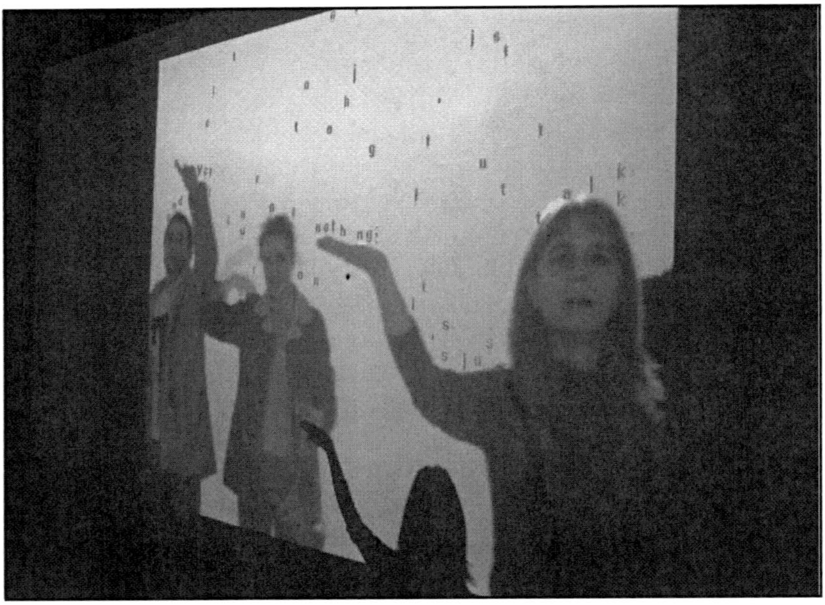

Fig. 5. *Text Rain* shows interactors a video image of themselves in an alternate reality. The letters of lines of poetry fall from above, coming to rest on anything darker than the background—inviting creative play with this language made physical.

3. Instrumental Texts

There are two types of playable texts that interest me here. The first type, "instrumental texts," has (as noted above) been the focus of some discussion in the electronic literature community. The second, "textual instruments," began as a personal thought experiment—a perhaps contrarian inversion of some of the assumptions of the first category. But then this thought experiment evolved into my first set of collaborative projects, perhaps the first of many, with textual instruments.

Let's begin with the first type. In electronic literature circles—those in which experimental writing for digital media is a common topic—the last few years have seen increasing discussion of the concept I'm calling "instrumental texts." (Within this phrase I'm subsuming a discussion, around texts with instrumental qualities, that has used a variety of loose terminology.) These are texts meant to be *played.* As John Cayley put it in an interview with Brian Kim Stefans:

> My point is that we are currently writers trying to build relatively simple textual instruments that are intuitive and, hopefully, both affective and significant when they are played. I mean played as musical instruments or sequencers or mixers are played. This is ergodic indeed, but still distinguishable from (hard) work or from the type of play in games which is rewarded by winning, by other forms of "success" or simply by "playability." ("From Byte to Inscription")

With the term "ergodic" Cayley is referencing the work of Espen Aarseth, whose *Cybertext: Perspectives on Ergodic Literature* (1997) is a touchstone in the electronic literature community. Aarseth defines the term by stating, "In ergodic literature, nontrivial effort is required to allow the reader to traverse the text" (1). More recently Aarseth has become one of the leading figures in the emerging field of game studies, helping found the field's first journal as well as the Center for Computer Games Research at the IT University of Copenhagen. In referencing Aarseth's work, Cayley brings to the fore a focus on play as "nontrivial effort" in music, games, and instrumental texts.

In my interview with Stuart Moulthrop cited above, I had the opportunity to ask him about instrumental texts. He, too, discussed some of the potential challenges of the reader/player's nontrivial engagement, but also some of what authors of instrumental texts might learn from the designs of folk instruments:

> Maybe some instruments will be hard to play. They may require practice. Or not. As a teacher once said to me about the guitar, "After five or ten minutes you'll make sounds that are almost musical. That's what the frets are for." And that's a great virtue of folk instruments. They do allow you to get in touch with a productive vocabulary very quickly. I think a good instrument would do that. It would stimulate engagement. It should make people want to get in there and interact, and to repeat the experience. (Moulthrop, Interview)

But for Moulthrop, perhaps unlike Cayley, musical and literary figures are not the only ones brought together in this discussion. For Moulthrop musical figures are a vocabulary that can help one imagine projects that occupy a space between two other types of work at play in discussions of instrumental texts. As Moulthrop puts it:

> What I'm particularly taken with is the notion of a middle space between literary texts and ludic texts—between interactive fiction, or hypertext

> fiction, and games. You have, with instruments, a text with behavior and temporal dimensions that in some ways maps onto the temporal experience and interactive possibilities in game design. (Moulthrop, Interview)

In focusing on games, Moulthrop has likely chosen the stronger of the two comparisons. For the examples usually given of instrumental texts, unlike musical instruments, only "play one tune." Their structures of play and material for play are designed and delivered together, much as those of games. But this is not to say that the musical figure is unimportant. Rather, I think it helps indicate the sort of engagement that authors of instrumental texts hope audiences will have with their work—again, as with *Façade* and *The Sims,* something that is not about winning or losing, or perhaps about quantifiable outcomes of any sort. But, unlike *Façade* and *The Sims,* this engagement is a physical discipline, sensitive to differences in movement and able to be learned at a muscular level.

Cayley's interview statement about instrumental texts actually came in the context of an answer to a question about his piece *riverIsland* (fig. 6). The occasion of my interview with Moulthrop was the release of his piece *Pax* (fig. 7). Looking at these may help ground our discussion, as well as help us understand what other work might be brought into this category.

3.a. riverIsland

John Cayley's work often employs a technique he calls "transliteral morphing." This is a letter-by-letter morphing that transitions from one text to another, much as graphical morphing moves points in space so as to transition from one image to another. In transliteral morphing the in-between letters are determined by movement along a loop on which Cayley has arranged Roman characters according to their sounds, as he explains:

> If texts are laid out in a regular grid, as a table of letters, one table for the source and one table for the target, to morph transliterally from one text (one table of letters) to another, is to work out, letter-by-letter, how the source letters will become the target ones. Assume your alphabet (including "space" and apostrophe, 28 letters in all) is arranged in a special loop where letters considered to be similar in sound are clustered together. The aim is to work out the shortest distance round the loop (clockwise or anti-clockwise) from each source to each target. (*riverIsland*)

Once the movement for each letter is worked out, the text then moves through fourteen steps (the largest number that might be necessary for any one letter—movement to the opposite side of the 28-character loop). Some letters go through many more transitions than others. Changes are "reluctant" at the beginning of the process and then "anxious" for completion at the end—so that both the early and final stages are close to readable texts.

A number of Cayley's pieces, such as his well-known *windsound* (1999), employ transliteral morphing in a manner that is performative on the part of the program. Texts morph into other texts under the gaze of the reader/audience, using the computational capabilities of the computer on which they are displayed. And yet these morphs could be, like most of the graphical morphs we see, pre-rendered and displayed as moving images (without any computation at the time of reading). The only visible loss would be the small changes in timing from reading to reading on the same computer, and the occasionally larger changes when moving from computer to computer.

Cayley's *riverIsland*, on the other hand, is not only performative on the part of the system, but also performative in a manner controlled in part by the reader. One of the types of performance made available to the reader is relatively straightforward: *riverIsland* is composed of two loops of poems, one horizontal and one vertical, and the reader can use on-screen arrows in order to trigger movement along these loops. When the reader indicates that a move should be made from one poem to another, the appropriate transliteral morph is performed by the computer.

There is another type of reader performance in *riverIsland,* however, that feels quite different to me. And I believe that this is part of what Cayley was getting at in his talk of instruments during his interview with Stefans. In this type of performance, the reader can click and drag on the screen's vertical and horizontal Quicktime movies. The vertical movie is an "object" movie that graphically transitions between images of paths through the woods. The horizontal movie is a panorama of a riverside scene. A reader experienced with *riverIsland* can use these movies to navigate to any point within the work's two loops. A transliteral morph is then performed between the text that was being displayed before the navigation process began (which might, itself, be an in-process morph) and the destination selected by the manipulation of a movie. This creates an experience for which pre-rendered morphs could not effectively substitute—like Cayley's figure of the sequencer, it harnesses real-time computational processes to create a performance based on high-level user direction that requires knowledge of its materials and control space.

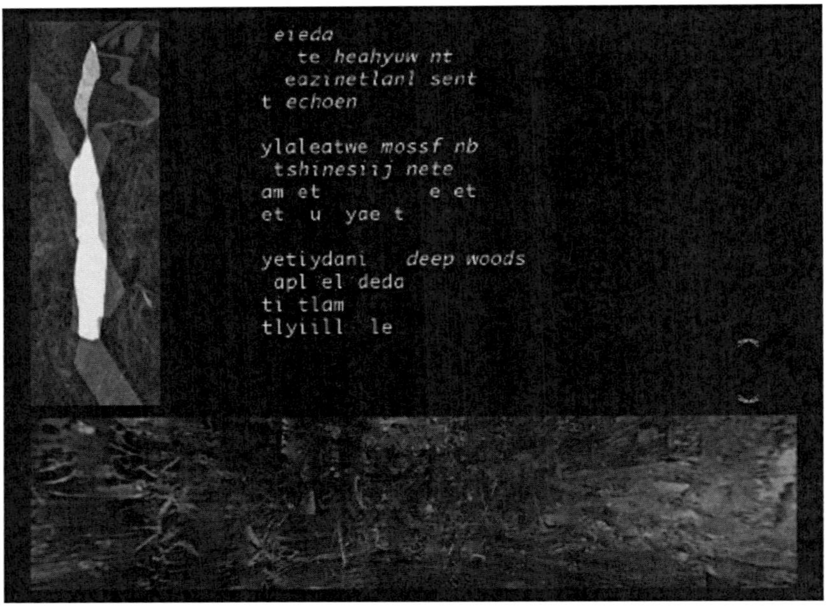

Fig. 6. *riverIsland* enables traditional, step navigation through its textual morphs via the arrows on the lower-middle right—or, for those with knowledge and practice, more free-form selection of destination texts through the horizontal panorama and/or vertical object movie.

3.b. Pax

The instrumental text of Stuart Moulthrop's that I will consider here, *Pax*, presents an experience of reading and performance that differs from *riverIsland*. Its differences in some ways map onto two of the different musical instruments Cayley and Moulthrop chose for their examples when discussing instrumental texts—while Cayley mentioned the sequencer, Moulthrop mentioned the guitar.

A sequencer might play itself for some time after being given instructions, but a guitar demands interaction for each note sounded. Similarly, *Pax* is structured for near-continual interaction. The larger area of the piece, on the left, shows characters floating up (in the first half of the piece's duration) or falling down (in the second). Unless the reader interacts with these characters, almost no text appears. Readers interact by "catching" floating characters with the mouse pointer. Characters can be released by moving the mouse away, or clicked (either by active clicking, or by holding them caught for 20 seconds). Clicking elicits text from that character, which appears in the area on the right (this becomes a scrolling text area once there is enough text to scroll). The fourteen characters float by in different orders, but those recently clicked tend to reappear, making it possible to consistently evoke text from two or three characters as the piece's time passes. Each reading lasts from noon to midnight (the characters' time) and is divided into six thematic movements: "Shaken Out of Time," "American Flyers," "Home Land," "Evil Ones," "Falling," and "Total Information." The text elicited from a character is determined, in part, by the number of times the character has been caught and clicked, as well as the current movement of the piece. The character texts evoke two situations: being caught in some version of a terminal at the Dallas airport (shut down for security reasons in an even-more-irrational "war on terror" than that which now grips us) and being caught in the space and structure of *Pax* itself (naked, floating, caught and prodded by the interactor).

While it would be impossible to manipulate the Quicktime movies of *river-Island* toward particular effects without relatively strong knowledge of the piece, *Pax* provides obvious places to click and quickly-understood effects even for the first time reader. However, because of its random elements and the strong impact of time's passage, it would be more difficult to exactly reproduce the same reading (after learning to play) than with *riverIsland*. To put it in terms of the musical analogy, *Pax* may provide frets, but for an instrument that adjusts its tunings over the course of each playing.

And, this, again, points to the strength of computer gaming as a figure for understanding instrumental texts. In the gaming context there is nothing surprising about behavior that changes over the course of time. There is also nothing surprising about the skills of physical manipulation and memorization that would be required to elicit particular readings from *riverIsland* and *Pax*. And the fact that these "instruments" come packaged with only one composition, from which they cannot be easily decoupled, also makes sense in the context of computer games. And yet they are clearly not games in the manner that play is approached. Perhaps what the musical analogy helps with most is the fact that these projects seek a lyric engagement—not easily formulated in terms of contest or quantifiable outcome.

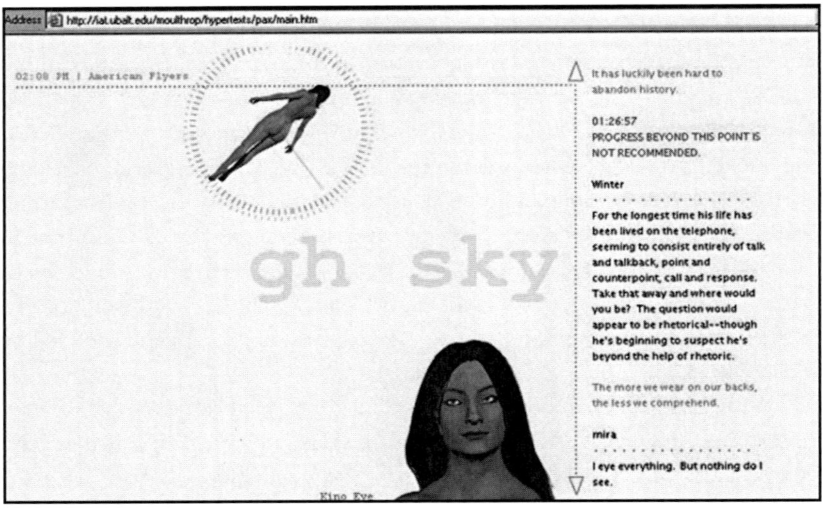

Fig. 7. *Pax* produces texts when the reader catches and clicks on characters that float by—and is otherwise silent. Rather than a narrative "told" to the reader, or one "played through" as in (for example) the levels of a narrative first-person shooter, *Pax* is an exploration of character and situation.

3.c. New Word Order

With a better sense of what we mean by "instrumental texts" it may now be possible to adopt into the category a number of computational textual projects not described by their authors in such terms. In fact, I'd like to propose a perhaps radical move—adopting as an instrumental text a project that uses a method of interaction not even designed for text, but repurposed through use of an existing game engine. I believe the adoption is appropriate once we look at it, and this foregrounds the game-like structures of interaction for instrumental texts, as well as the different—more musical or performative—position of engagement with these structures.

The piece I propose adopting for these purposes is *New Word Order* by Sandy Baldwin (as reported by Chris Funkhouser). In this, the second part of Baldwin's *Black Mesa* project, poetry is presented mapped onto objects in a simple *Half-Life* level. The poetry is that of Billy Collins, the first U.S. poet laureate (2001-2003) appointed during the presidency of George W. Bush. Placing Collins's poetry within *Half-Life* subjects it to destruction and reconfiguration with an arsenal ranging from automatic weaponry to the famous crowbar (fig. 8).

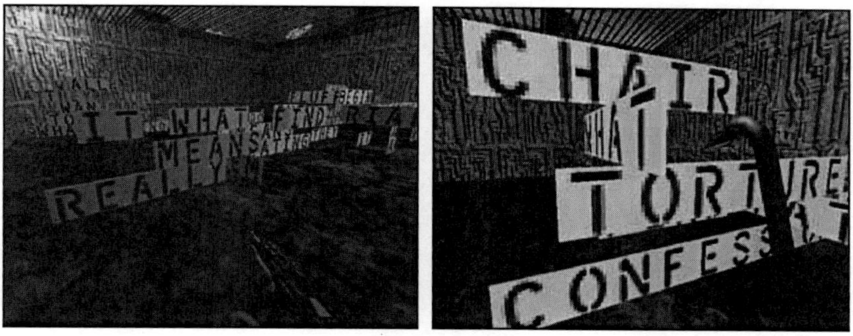

Fig. 8. *New Word Order* invites the reader to reduce and reconfigure poetry using the interaction structures of the first-person shooter.

New Word Order takes an interaction structure invented for competitive play with quantifiable outcomes—for gameplay—and repurposes it as play that re-contextualizes and explores the potential of poetic language.

3.d. Screen

When interviewing Moulthrop, it struck me that one of my in-process collaborative projects, *Screen,* might also be usefully considered as an instrumental text. *Screen* combines familiar game mechanics with virtual reality technology to create an experience of bodily interaction with text. At the same time, the language of the text, together with the uncanny experience of touching words, creates an experience that doesn't settle easily into the usual ways of thinking about gameplay or VR.

Screen is a collaboration with Andrew McClain, Shawn Greenlee, Robert Coover, Josh Carroll, and Sascha Becker that was created in the Brown University immersive virtual reality chamber (Cave), as part of a research project in spatial hypertext writing (Carroll et al.). Brown's VR chamber is similar to the University of Illinois's CAVE—a virtual environment that shows three-dimension images while allowing users to continue to see their own bodies, and that does not require users to wear encumbering equipment (unlike head-mounted displays, which are essentially blindfolds with televisions inside) (Sandin et al.). Brown's Cave is an eight foot cube, missing its top and one side, and its walls and floor are projection screens. Projectors are pointed at each screen, and they alternately project streams of images meant for the user's left and right eyes. The user wears shutter glasses that alternately occlude the left and right eyes, in synchronization with the projectors. The result is stereo

VR—3D vision of computer-generated imagery—combined with the physical presence of the people and objects in the Cave.

The initial experience of *Screen* can be disorienting for those familiar with VR. Rather than make the walls "disappear," we project flat images onto the same plane as the walls, reinforcing their presence. And the images we project are not of colorful shapes, but of white text on a black background. This text at first appears in an introduction that fades in and out on the walls—and then forms three traditional paragraphs, each nearly filling one of the walls. Each of these paragraphs is a character's moment of memory that gives rise to the virtual experience of touch. Each wall appears, and then is read aloud. After the last has been read there is a pause, and then a word peels from one of the walls, is spoken aloud, and flies toward the reader. If the reader does nothing, the word circles near her. Soon another word peels, and then another, at an increasing pace, flocking around the reader. The reader can intervene in this process by batting at words with her hand. When a word is hit a sound is heard, and the word flies back toward a wall, perhaps breaking apart in the process. If a hit word is the only word off the walls it will return to the space it left empty. However, if more than one word is off the walls then a hit word may return to a different space.

Once the number of words off the walls passes a certain threshold—something which, with the increasing pace of peeling, only very active engagement can long delay—all the remaining words come free of the walls, swirl around the reader, and then collapse into the center of the Cave (fig. 9). A final "closing" text is then heard. In addition to creating a new form of bodily interaction with text, *Screen* creates three reading experiences—beginning with the familiar, stable, page-like text on the walls; followed by the word-by-word reading of peeling and hitting (where attention is focused); and with, simultaneously, more peripheral awareness of the arrangements of flocking words and the new (often neologistic) text being assembled on the walls. *Screen* was first presented in 2003 as part of the *Boston Cyberarts Festival,* and in 2004 it was included in *Alt+Ctrl: A Festival of Independent and Alternative Games* at University of California, Irvine.

Given its presentation at *Alt+Ctrl,* we might simply discuss *Screen* as a game, rather than with the more unusual term "instrumental text." And, in fact, the final moments of *Screen* feature a scattering of words (and parts of words) on the walls—which caused one young visitor to the Cave to ask, "Is that my score?" But while the play of *Screen* is reminiscent of classic games from *Whack-a-Mole* to *Breakout,* and some players may at moments be driven purely by the game-like goal of hitting words as quickly as possible, there is no contest or quantifiable outcome. Even approached purely physically, without any attention to the linguistic nature of the words being played, *Screen* is more like hackey-sack than

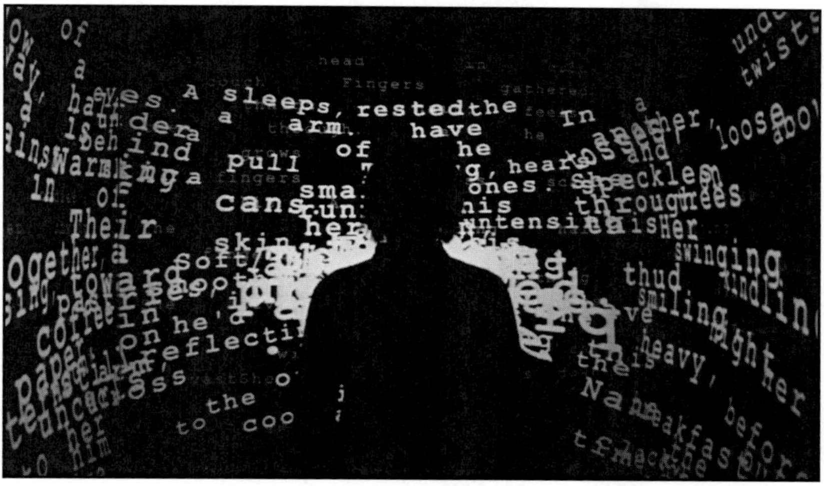

Fig. 9. Words collapsing in *Screen*.

soccer/football. And, in my observations, players don't approach *Screen* without attention to its words as words. Rather, interactors oscillate between reading and playing, with the objects of both coming faster and coming apart, until both experiences can no longer be sustained and the piece ends. As with *riverIsland* and *Pax,* reader/players can get better at *Screen,* though the fact that interactors do not control the ripping of words alters what is possible via virtuoso performance. Perhaps the most impressive performance of *Screen* I have seen is that of Michelle Higa, who both edited the video documentation of *Screen* and played the role of the interactor within it ("Screen"). In order to videotape *Screen* we had to turn off the flickering alternation between left- and right-eye images. Higa had become adept enough at the experience of *Screen* that she was able to play it relatively successfully even without stereo cues.

4. Textual Instruments

I first discussed the idea of "textual instruments" in a short paper for Digital Arts and Culture 2003 ("From Instrumental Texts"). The idea grew from questions about the limits of instrumental texts. If instrumental texts are odd instruments in that they only play one tune, how might we imagine tools for textual performance designed to play a variety of compositions? What would it mean to have textual instruments that one might learn to play proficiently, for which one might write and perform a number of compositions, and that could eventually be made available to play the compositions of others?

In my DAC paper I described textual instruments as follows:

> A textual instrument is a tool for textual performance which may be used to play a variety of compositions. In this sense it is evocative of Thalia Field's figure of the "language piano"—something that one learns to play, and which may produce a much wider variety of texts than is the case for those projects normally discussed as instrumental texts.
>
> However, a textual instrument need not be like a prepared piano. The direct selection of text, rather than the manipulation of a non-linguistic device, can be its interface. And the relationship between a textual instrument's interface affordances and the possible textual outcomes need not be one-to-one at all levels (as it must be with a piano's keys, though they may be played in many combinations). Understanding at a gut level how a textual instrument's probability spaces function for a given composition is part of learning to play that piece.
>
> Compositions, here, consist of a body of text (and/or a means of acquiring text) and a set of "tunings" for the instrument(s) used. ("From Instrumental Texts" 3)

While I don't know of any projects, other than those I have been involved with, that are described by their creators in these terms, in the next section I'll adopt a potential example into the category. I'll then touch on a couple of related issues before discussing my first two collaborative projects in this area—the compositions *Regime Change* and *News Reader,* and the instruments for which they were composed.

4.a. Arteroids

Most who approach the arts as writers are quite attached to their own words, and this holds true among writers for digital media. In the digital field, Jim Andrews is one of the exceptions, having undertaken a number of interesting projects that involve him arranging a system for language to inhabit and then (rather than including only his own writing) inviting other writers to provide text. *Arteroids* (Andrews 2001-04) is one of these systems (fig. 10).

Arteroids is, on the interaction level, simply the repurposing of an existing game—and, in that way, quite similar to projects such as *New Word Order.* The major difference, in fact, is that while Baldwin's piece seems created as a context

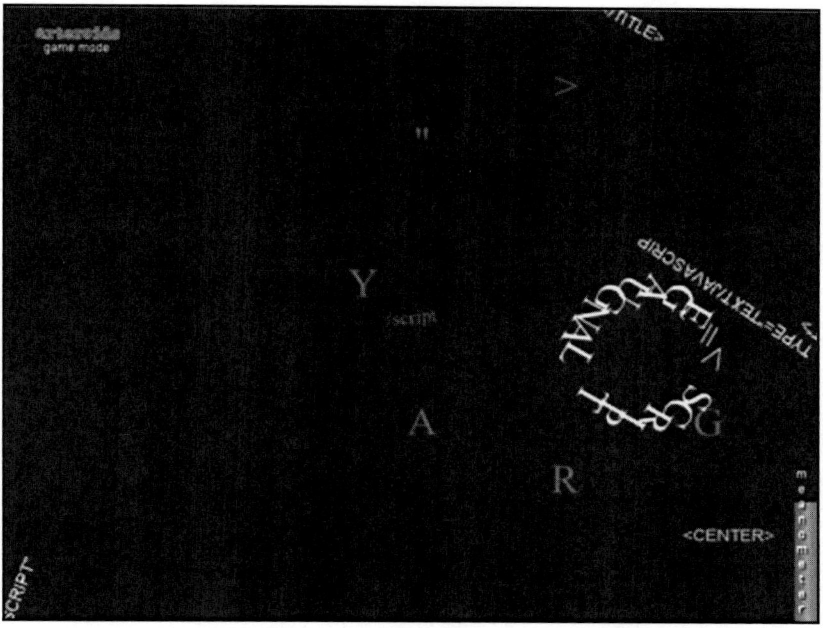

Fig. 10. *Arteroids* uses the interaction structures of a modified version of the arcade classic *Asteroids*, but replaces the images of a spaceship and rocks with images of text.

for Collins's work, Andrews appears to view *Arteroids* as an instrument for which many texts could be composed (and then used as materials for play). Andrews created a "Word for Weirdos" to allow others to compose for *Arteroids* and has included texts from others in presentations of the work—such as the texts by Christina McPhee and Helen Thorington included when *Arteroids* was shown in the "page_space" exhibition (*Superbunker*). While on some level it may appear that learning to play *Arteroids* is no different than learning to play *Asteroids,* one can imagine the desire to create particular linguistic experiences changing this. With each textual composition as a different starting place, even after the *Asteroids*-derived parts of *Arteroids* have been mastered there might yet be much to learn about creating an evocative experience from the play with any one text.

And yet I don't enjoy *Arteroids* very much, even when it uses texts by writers whose work I have appreciated in other contexts, and even after gaining some experience with playing it. Considering why this is the case has led me to define further the work I want to pursue in textual instruments.

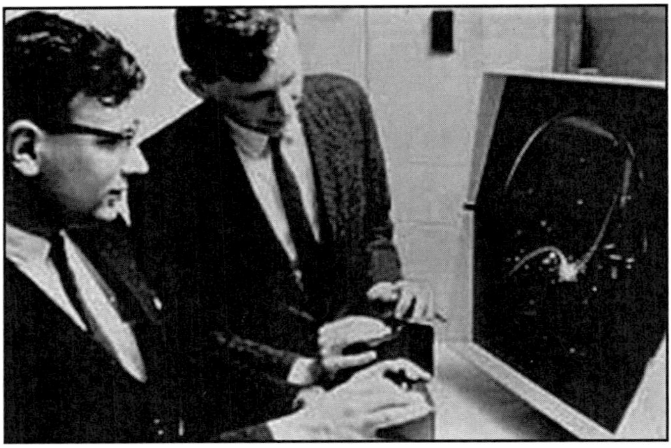

Fig. 11. *Spacewar!* was the first modern video game, combining logics of graphical play still in wide use today.

4.b. Graphical and Linguistic Logics

Not all playable computational media is graphical. In fact, some of the most popular early computer games were entirely textual. Games like *Adventure* and *Zork* were even at times played on teletypes, with the interaction recorded on scrolling reams of paper, rather than on terminals with screens. An excellent tracing of the history of this textual interactive fiction, which is still being created today, can be found in Nick Montfort's *Twisty Little Passages*.

But when we think of playing with computers, we generally think of graphical experiences, those that follow in the tradition of *Spacewar!* (fig. 11) rather than *Adventure*. Created on the PDP-1 at MIT in the early years of the 1960s, *Spacewar!* was the first modern video game. Two players each had a custom-made controller, which they used to control the flight of a virtual spacecraft on the PDP-1's CRT. The spacecraft were pulled toward the star at the center of the screen by simulated gravity, and could fire projectiles at one another. A spacecraft hit by the central star or a projectile would be damaged. These are still among the central logics of graphical gaming today—the ability to move graphical objects that on some level represent the player, the ability to fire projectiles, a simulation of some form of physics, and "collision detection" when one thing runs into another. These logics aren't only the basis for play in experiences such as *Half-Life,* but also (leaving aside projectiles) in pieces such as *Text Rain*.

We're accustomed to seeing successful combinations of graphical logics and game rules repackaged repeatedly. Games such as *Pac-Man* and *Tetris* have had many authorized and unauthorized versions "skinned" with different surface graphics and different graphical arrangements, but with the essential logics of graphical movement and gameplay preserved. Such combinations, within a larger range of variation, are also the basis for our identifications of game genres such as "side-scrollers" and "first-person shooters."

I bring all this up in order to make a point about the instrumental texts discussed above—about how they are played. While each of these pieces contains a textual component, they are all played along graphical logics. For *riverIsland* play is primarily through the graphical/physical manipulation of the Quicktime movies; in *Pax* it is the collision detection as characters are caught and clicked; in *New Word Order* the movement of the first-person perspective and collision detection (in firing of weapons and use of the crowbar); and in *Screen* it is the movement of the interactor's body and the collision detection of hitting words. What these projects do, in each case, is package together logics of graphical play and methods of response with textual and graphical material.

What *Arteroids* does differently is take a set of logics of graphical play and methods of response and then open them to many different sets of textual material. This might be seen as the same as taking the formula of *Pac-Man* or *Tetris* and opening it up to many variations in graphical representation. But, for me, it doesn't feel the same. Somehow it feels arbitrary, no different than if the graphics in *Pong* or *Spacewar!* were opened to replacement by arbitrary text.

I've been thinking about what would feel less arbitrary. It seems to me that, if the same graphical logic can be skinned with many different surface graphics successfully, perhaps those seeking to create textual instruments will need to consider forms of play that proceed via linguistic or textual logics. Before the computer became part of everyday life, textual forms of play such as the crossword (or games of the Surrealists or Oulipo, or Madlibs, and so on) successfully accommodated many different texts by structuring play around the features specific to textuality. Perhaps the true challenge of creating textual instruments involves finding such structures that benefit from the computational environment.[1]

If such structures can be found I also suspect that the result will feel, at least to me, in some ways like a deeper engagement with text than is possible with projects (like *Screen*) that proceed along graphical logics—projects that could still be played, though perhaps not as rewardingly, if their words were all converted into colorful boxes.

Of course, the next question is where one might begin to explore such textual or linguistic logics. Though perhaps this work has already begun, with those creating computer-based versions of crosswords, Surrealist games, Oulipian games, and so on. And *Fields of Dream*, by Nick Montfort and Rachel Stevens, goes further—bringing the basic fill-in-the-blank logic of Madlibs into a project specific to the networked computer.

But there is also another territory to explore. This is that of linguistic and textual logics previously employed in text processing and generation—in contexts ranging from the computer science subfield of natural language processing (NLP) to the artistic contexts of John Cage, William S. Burroughs, or Jackson Mac Low. Whether considered scientific or artistic, these methods have generally been operated in batch mode—either generating chunks of language or analyzing chunks of language. But there is nothing to prevent them from being run interactively, or to prevent the interaction with them from being structured as play.

4.c. Claude Shannon's Textual Play

In fact, a number of the logics used broadly in NLP have already been used in play, if not necessarily computational play. Claude Shannon, sometimes referred to as the "Newton of the Information Age," even introduced one of these logics by describing a textual exercise quite similar in tone to Oulipian play structures such as "N+7."

Shannon's nickname comes from the fact that he was one of the major figures in formulating the mathematics of communication—what he called "the fundamental problem of . . . reproducing at one point either exactly or approximately a message selected at another point" (379). This quotation comes from a paper that was a milestone in the field, both when published on its own (as *A Mathematical Theory of Communication*) and when repackaged as a book with an introduction by Warren Weaver (as *The Mathematical Theory of Communication*). While a milestone, Shannon's paper certainly doesn't have to be approached as a millstone. Rather, it can be approached playfully, an approach to the products of science and technology practiced by Shannon himself. (His famous projects of this sort include a mechanical rat for solving mazes, figure 12, as well as a machine that, when turned on, performs only one action: causing a mechanical hand to reach out from a box and turn the machine off.) One aspect of the paper amenable to a playful approach is Shannon's description of stochastic approximations of English.

Shannon's discussion includes a number of sample messages. One is purely random. But the others have contents that are shaped by the frequencies of particular letters or words in the English language. This shaping is of two sorts. In one type of shaping, individual letters or words are selected in a manner weighted by their frequency in English. In the second type of shaping, letters or words are selected in a manner shaped by the frequency with which they appear

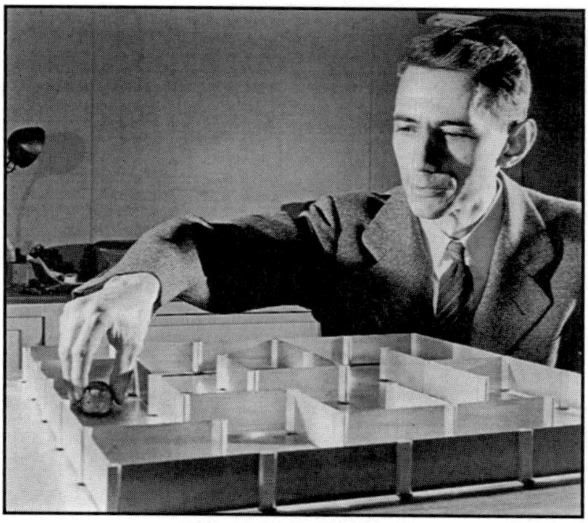

Fig. 12. Shannon with his mechanical mouse.

in *groupings* of letters or words in English. So, for example, "E" is a more common letter than "U" in English. However, if there is a pair of letters that begins with "Q" it is much more likely that the complete pair will be "QU" than "QE." Taking the frequencies of pairs into account in this manner means paying attention to the frequencies of "digrams." Paying attention to sets of three is attention to "trigrams." And, more generally, paying attention to the frequencies of groups of symbols (rather than only individual symbols) is the use of "n-grams."

Shannon provides six sample messages. In the first, each of the 26 letters and the space appear with equal probability (388):

> XFOML RXKHRJFFJUJ ZLPWCFWKCYJ FFJEYVKCQSGHYD QPAAMKBZAACIBZLHJQD.

In the next, the symbols appear with frequencies weighted by how commonly they appear in English text (i.e., "E" is more likely than "W"):

> OCRO HLI RGWR NMIELWIS EU LL NBNESEBYA TH EEI ALHENHTTPA OOBTTVA NAH BRL.

In the third, symbols appear based on the frequencies with which sets of two of the symbols in English. That is to say, after one letter is recorded, the next is chosen in a manner weighted by how commonly different letters follow the just-recorded letter. The sample message created in this way is:

> ON IE ANTSOUTINYS ARE T INCTORE ST BE S DEAMY ACHIN D ILONASIVE TUCOOWE AT TEASONARE FUSO TIZIN ANDY TOBE SEACE CTISBE.

In the fourth, symbols appear based on the frequencies with which sets of three of the symbols appear in English. This, again, is called a "trigram"—with each choice of the next letter being weighted by the frequencies with which various letters follow the set of two just recorded. This sample message is:

> IN NO IST LAT WHEY CRATICT FROURE BIRS GROCID PONDENOME OF DEMONSTURES OF THE REPTAGIN IS REGOACTIONA OF CRE.

In the fifth, the unit is moved from letters to words. In this message, words appear in a manner weighted by their frequency in English, but without attention to the prior word:

> REPRESENTING AND SPEEDILY IS AN GOOD APT OR COME CAN DIFFERENT NATURAL HERE HE THE A IN CAME THE TO OF TO EXPERT GRAY COME TO FURNISHES THE LINE MESSAGE HAD BE THESE.

Finally, in the seventh sample message, words are chosen based on the frequency with which pairs of words appear in English. This, again, like the technique of choosing based on pairs of letters, is called a "digram" technique. The final message is:

> THE HEAD AND IN FRONTAL ATTACK ON AN ENGLISH WRITER THAT THE CHARACTER OF THIS POINT IS THEREFORE ANOTHER METHOD FOR THE LETTERS THAT THE TIME OF WHO EVER TOLD THE PROBLEM FOR AN UNEXPECTED.

These messages are interesting in part because of how they are generated. The first two were created using a book of random numbers, with the addition of a table of letter frequencies when creating the second. But the rest of the samples were constructed by using ordinary books (making the assumption that ordinary books appropriately reflect the frequencies of letters and words in English). Shannon explains the process as follows:

> To construct (3) for example, one opens a book at random and selects a letter at random on the page. This letter is recorded. The book is then opened to another page and one reads until this letter is encountered. The succeeding letter is then recorded. Turning to another page this second letter is searched for and the succeeding letter recorded, etc. A similar process was used for (4), (5) and (6). It would be interesting if further approximations could be constructed, but the labor involved becomes enormous at the next stage. (389)

That is to say that the last sample message (which begins with a sequence that sounds surprisingly coherent) was created by opening a book to a random page, writing down a random word, opening the book again, reading until the just-recorded word was found, writing down the following word, opening the book again, reading until that second word was found, writing down the following word, and so on. This is a method that sounds remarkably like some of the kinds of textual play engaged in by experimental artists and writers. It certainly sounds like a technique that could be brought into an explicitly playful context.

The surprising coherence of the last sample is also of interest. It shows that processes of the sort being used by Shannon have potential as a form of logic that can be made operational with linguistic material. Of course, the sample is quite rough, in part because only one previous word is being taken into account at any time. What Shannon calls "further approximations"—for example, taking two, three, or more previous words into account—certainly gives more English-like results. Shannon pointed out correctly that, using the method of flipping through a book, the labor involved in creating such further approximations would be enormous. But the modern availability of computing power has made carrying out such calculations automatically a near-trivial task for reasonably-sized bodies of sample text. As Shannon also pointed out, the stochastic processes he described are commonly considered in terms of Markov models. And, interestingly, the first application of Markov models was also linguistic and literary—modeling letter sequences in Pushkin's poem "Eugene Onegin" (Markov). But Shannon was the first to bring this mathematics to bear meaningfully on communication, and also the first to use it to perform text-generation play.

This model (whether called n-grams or Markov chains) is now widely used in natural language processing and generation, often in combination with other techniques. It has also been used in electronic literature, perhaps most extensively by John Cayley. At least seven of Cayley's works employ "collocational" word-level digram procedures, including *Book Unbound* as discussed by Aarseth. Last but not least, this approach has also been the primary basis of textual toys such as the DOS program *Babble!,* the emacs "Dissociated Press" command, Hugh Kenner and Joseph P. O'Rourke's *Travesty,* Andrew Plotkin's *chan.c,* and Brion Moss's *prate*—which have themselves at times been used in the generation processes for (non-playable) literature. However, as noted above, both the elit and toys based on n-grams have operated entirely in "batch mode." That is to say, the interactor requests a body of text, and then that text is produced—following which the text can be read and another text can be requested, but no interaction with the texts (or interaction during generation) is possible. Given this limited nature of play with n-gram texts, there is also limited context for play—usually a blank text buffer for the program to write text into. After talking with Moss (with whom I'd collaborated on *The Impermanence Agent*) about these issues, we began to imagine possibilities for n-gram play that was less batch-oriented and took place within a textual context.

4.d. Two n-gram Instruments

Moss and I approached Turbulence, an organization that supports digital media art, and they commissioned us to create two pieces. These pieces would be inspired by the idea of textual instruments and operate using the logic of n-grams. After a false start with different collaborators, Moss and I connected with document researcher David Durand (best known for his work in formulating a number of document markup standards, including XML) and designer Elaine Froehlich (principal of Active Surface Design). From there, the project's conceptualization and execution were a team effort, with the initial technical work happening on top of Moss's Java *prate,* and later development built on top of work done by David Durand in Tcl/Tk.

Two major design decisions were made early on. The first was that, rather than build an n-gram text into an empty text buffer, play would always begin within the context of a pre-created document and consist of progressive alteration of that document. This was motivated, in part, by the fact that, while the text produced by n-gram algorithms has microstructures that are recognizable from its source texts, the larger structures of n-gram texts tend to be very similar regardless of the starting material. Some have tried to address this by looking at larger structures in the source texts statistically, but unless the texts in question have been marked up by a human author or editor, this process involves a series of assumptions about the text (e.g., that a period marks that end of a sentence, as it does not in the case of "e.g.") that are both sometimes inaccurate and on some level aesthetically displeasing. These assumptions are displeasing because they depart from the purity of the simple n-gram algorithm, which in its basic form would work with starting texts in Japanese or Braille or musical notation as easily as English-language ones. However, there was also another motivation. In many n-gram texts, especially those based on short chains, part of the pleasure is based on play between coherence and incoherence—and we found something more interesting, and potentially more meaningful, in such borderline coherence occurring within the context of traditionally-created texts.

The second design decision was the identification of our basic method for making n-gram generation playably interactive, rather than oriented toward large batches. We decided that, in addition to the starting document (within which play takes place) we would have a body of text used for producing the alterations to the starting document. (We call this second body of text the "alteration text" or "alteration corpus.") When the starting document is displayed, certain words are highlighted. We chose this as a convention familiar from hyperlinks on web pages, letting interactors know that a click will elicit a response. However, these words are not highlighted as the result of author-specified links. Rather, they are highlighted because a string of n-gram text (of a length specified by the piece's author) appears in both the starting document and the alteration text. We decided that such "bridges" between the two bodies of text would offer interactors the opportunity to open up the starting document and insert text generated from the alteration corpus. More than one generated text would be offered for possible insertion, allowing the interactor to choose one or none (this last leaving the text unaltered). The texts offered would, themselves, be generated from the alteration text through the use of n-gram techniques. The number of texts offered and the n-gram length used in their production would, again, be determined by the piece's author.

Once these decisions were made, we sketched, mocked up, and eventually tried to make operational a number of interaction designs. Some didn't give the kinds of results we'd hoped for, and others were too computationally expensive to work, but we eventually settled on one that—for us—is satisfying in terms of the feel of interaction and the shape of the attention to text it creates. We're still discussing some potential variations on the visual/spatial aspects of the interface, but for the first compositions (*Regime Change* and *News Reader*) a simple, web-style series of windows seemed both appropriate and easy to implement.

More information on the details of interaction will appear below, in the context of the discussions of these two compositions. But first it's worth noting what differentiates the two instruments we constructed. Our first instrument was the simpler of the two. It depended on both the starting document and alteration text existing on the local drive, with known file names. The second instrument adds a number of features, including the request, processing, and display of network RSS feeds and HTML files. As a result, the first instrument is better suited to compositions involving longer chain lengths (and therefore greater coherence) because it doesn't have to take time for network file requests or for processing the wide variety of network html files into text that the system can use. The second instrument is, of course, more suited to compositions involving dynamic network data sources—in fact, we regard it as a type of alternative browser.

While some of this discussion may have been a bit difficult to follow in the abstract, hopefully the following two examples will make things clearer.

4.e. Regime Change

Regime Change begins with a news article from April 2003, following the bombardment that began the U.S. invasion of Iraq. George W. Bush cites "eyewitness" intelligence that Saddam Hussein was assassinated by targeted U.S. bombing, and clings to the contention that the Iraqi president was hiding "weapons of mass destruction." Playing *Regime Change* brings forth texts generated from a document that records a different U.S. attitude toward presidential assassination and eyewitness intelligence—the report of the Warren Commission.

Once the window with *Regime Change*'s starting text is opened, words in that text, pair by pair, become highlighted. Clicking on words opens a new window (fig. 13a). Interacting with new windows produces new texts that will take the place of the clicked words.

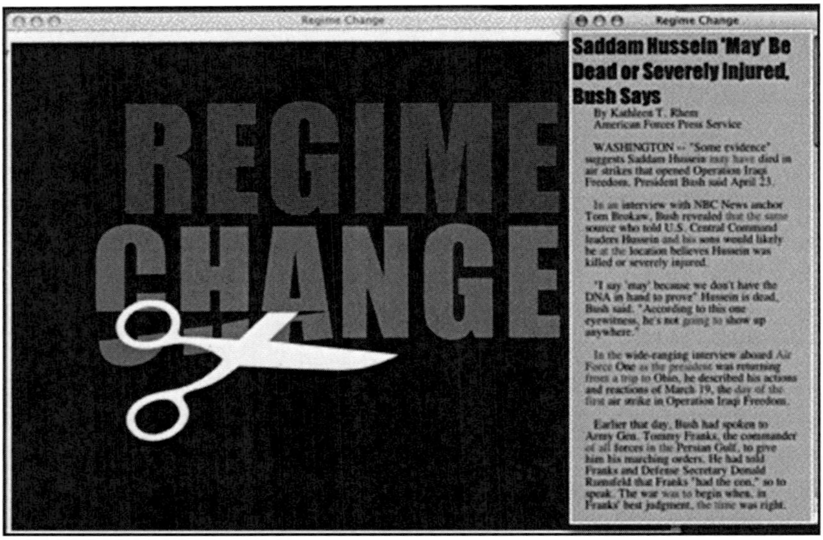

Fig. 12. *Regime Change* displaying its starting text.

New windows contain texts that begin with the words clicked in the previous window. Each paragraph in the new window is an alternative text—beginning with the same words but potentially (though not necessarily) following many different paths from there. These texts are generated by connecting chains of words (3-grams and 4-grams) that may have appeared originally in very different parts of the source document.

A new window's texts, once displayed, also begin to have words highlighted within them. Clicking highlighted words will open another new window, containing generated texts that can take the place of the clicked words (fig. 13b-c). Opening several generations of windows opens wider possibility spaces for the texts that will be created (and that will replace the clicked words in earlier-generation windows). Windows alternate between generation from the Warren Commission text and the original news story.

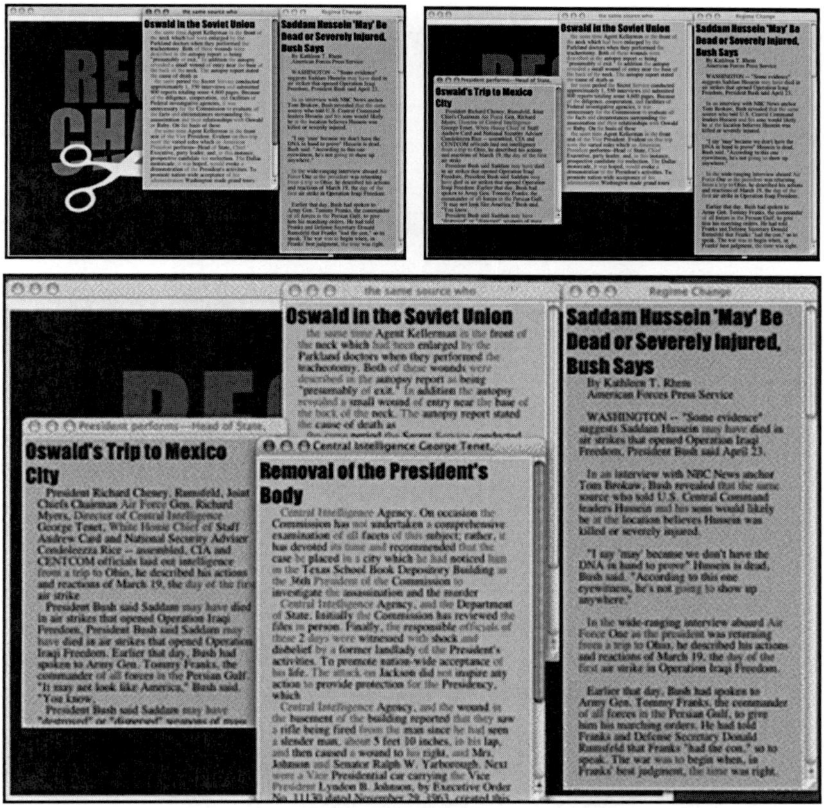

Fig. 13a-c. *Regime Change* opening multiple layers of windows.

In any window with generated text, clicking a non-highlighted word is also a means of interaction. Such a click will close the window—and select a text. The selected text will run from the beginning of the clicked paragraph to the clicked word. That selected text will then take the place of the words clicked to open the window (fig. 14a-b). This creates a kind of stretchtext—the pair of words clicked to open a window are replaced with the words selected in the open window (usually more than a pair).

After opening several layers of windows, part of play is keeping track of where each window came from—so that it can be collapsed by selecting a word that will make a pleasing segue at the point where it will join the text to which the player intends to connect it. (This may be more than one layer down.) Keeping track of context is made easier by the title bar of each generated window—which displays the two words that will be replaced by the generated text, followed by the two words that appear after them in the text clicked.

I find that, when I'm playing, this cycle of activities—reading, remembering context, selecting a place to click, reading again—consumes my entire attention. I've found it impossible to "give a reading" of *Regime Change* as I might with other writing projects. My most successful presentation so far, instead of a traditional reading, was a performance in which I played the text and Popahna Brandes read the results aloud.

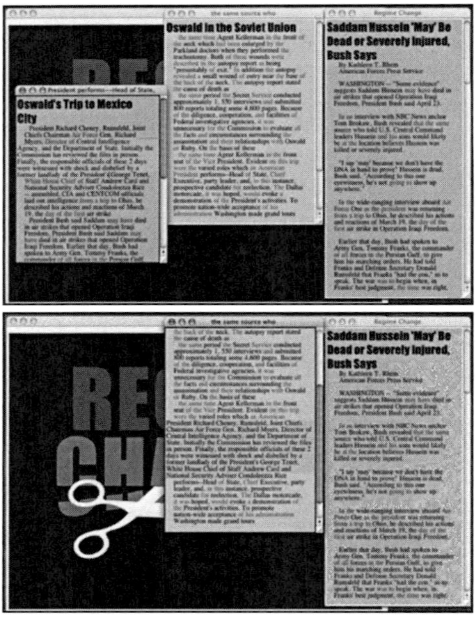

Fig. 14a-b. Word replacement in *Regime Change.*

Most of the interactions I've described are those that I would consider part of the instrument. I consider the composition to consist of the selection of texts, the basic settings made for the instrument (n-gram chain lengths, number of alternative texts generated, various selections of colors and fonts, and the ways that different windows open into different texts), as well as the ways that settings change over the course of interaction with the piece (including a "surprise" third text that becomes part of *Regime Change*'s material after a certain number of third-level windows have been opened).

Of course, I'm aware that this particular set of terms is not the only reasonable way to understand *Regime Change*. Rather than a focus on concepts such as instrument, composition, and play, it would also be reasonable to view the piece in terms of something like Nick Montfort's "human-computer co-authorship" ("The Coding and Execution"). Such a focus would seek to make more explicit the moves made by system and interactor during the creation of the final textual output from a session with *Regime Change*. Employing Montfort's framework, we would say that the initial move is: *Computer – G* (the computer provides the initial text). Then: *Interactor – I* (the interactor provides "some instructions or intermediate text" by clicking). Followed by *Computer – G* (the computer generates texts based on where the click was placed). Finally, *Interactor – I,* which can lead to *A* by extension or Computer—*G* (the interactor clicks, which may result in alteration of text or the generation of further text). Then these final stages are repeated, indefinitely, until the concluding text is reached. This enumeration may seem an odd exercise, but it does help us formalize how the texts of *Regime Change* differ from those created in other situations of human-computer co-authorship. For example, while human participants certainly shape the texts created during interaction with *Regime Change,* Montfort's model makes it clear that interactors at no point generate text.

4.f. News Reader

News Reader is software for reading the news, and for re-forming it. It is a specialized browser—displaying a selected RSS feed, as well as the news stories to which the feed links. Unlike a normal browser, *News Reader* also downloads another set of texts in the background—and uses this material to open each page it displays to textual play. Through this play the concerns and language patterns of the hidden documents, as shaped by the movements and passages selected by the player, are introduced into the original news stories. *News Reader* provides a different way to encounter the daily news, making its patterns of repeated phrases into opportunities for disruption, and producing results that range from humorous to disturbing.

When *News Reader* launches it displays a window containing the current headlines from the *Yahoo! News* RSS feed (fig. 15). Clicking the headline or preview text opens another *News Reader* window, displaying the story.

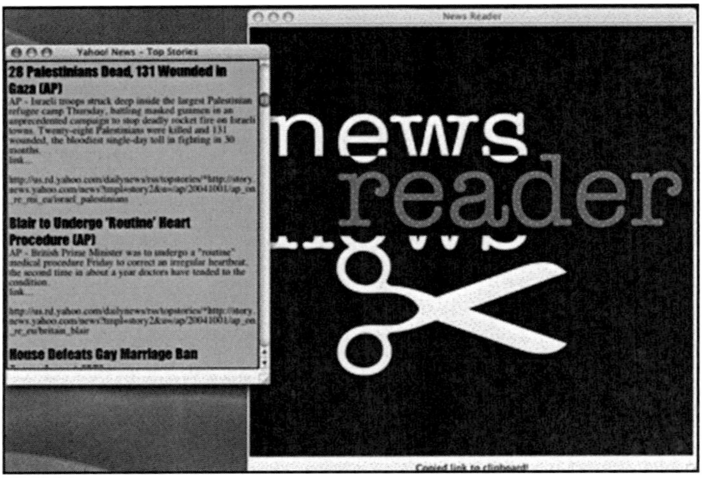

Fig. 15. *News Reader* showing *Yahoo! News* RSS feed.

A link below the text preview copies the address of the story to the system clipboard, so that it can then (if this is desired) be opened in a traditional browser of the reader/player's choice (fig. 16).

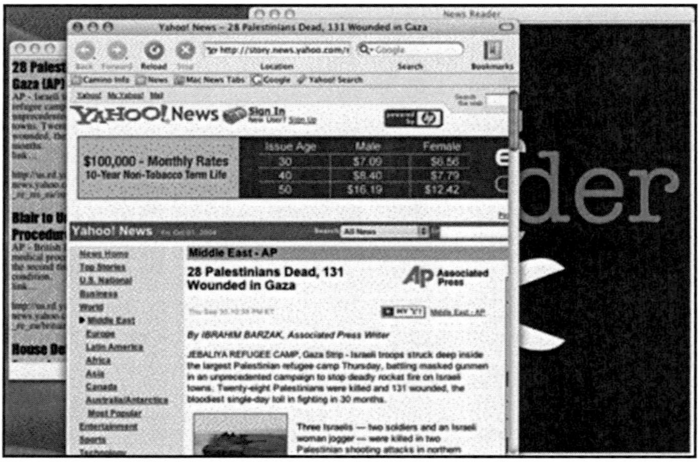

Fig. 16. Opening a traditional browser with a link from *News Reader.*

If a story is displayed in a *News Reader* window, links appear within it (fig. 17a). As with *Regime Change,* these links don't lead to other web pages, but rather generate texts out of a statistical text model (in this case, trigrams of the alteration corpus and their relative frequency). These generated texts appear in a new *News Reader* window (fig. 17b). The alteration corpus is created from the texts of alternative news stories (found at *Common Dreams*) downloaded in the background when *News Reader* is launched. As with *Regime Change,* windows of generated text contain several paragraphs, each of which is a continuation of an n-gram that begins with the words in the clicked window just prior to that word clicked. And, again as with *Regime Change,* clicking a non-highlighted word will close a window of generated text, replacing the words clicked to open that window (fig. 17c). The words used to perform this replacement will be those between the clicked word and the opening of the alternate text ("paragraph") it was within.

4.g. Play and Variety

Playing *Regime Change* can produce a wide variety of texts, especially if one opens and collapses many layers of windows. The potential variety of texts created by *News Reader* dwarfs this. In fact, given that *News Reader* employs materials that change several times an hour, it may be more sensible to discuss the fact that no text created with it is likely to ever be repeated, rather than the fact that it can create a great variety of texts. But variety is still worth mentioning because, as Markku Eskelinen points out in "Six Problems in Search of a Solution," we struggle with a "deep rooted humanistic fear of variety" (196).

Eskelinen observes that this fear of variety, coupled with our need to grapple with digital objects that produce great variety, may be part of the attraction of computer game studies for some scholars. In computer games variation is pacified by rules and goals. Expanding this, Eskelinen writes that when we have lost "the safe and somehow manageable totality, be it coherent or not" we reach a point where "computer games are interesting, as they domesticate the excess looming large in both ordinary and avant-garde products and processes, and the fundamental potential for change and unreliability inherent in new media objects" (197). Perhaps this is also the attraction of performance, of instruments, as a way of talking about digital objects that produce great variety. Performance, and especially improvisational performance, is different each time—and yet we understand that it is structured. Perhaps this is also the attraction of discussing work in terms of playability—in terms of the potential of, and structures for, play.

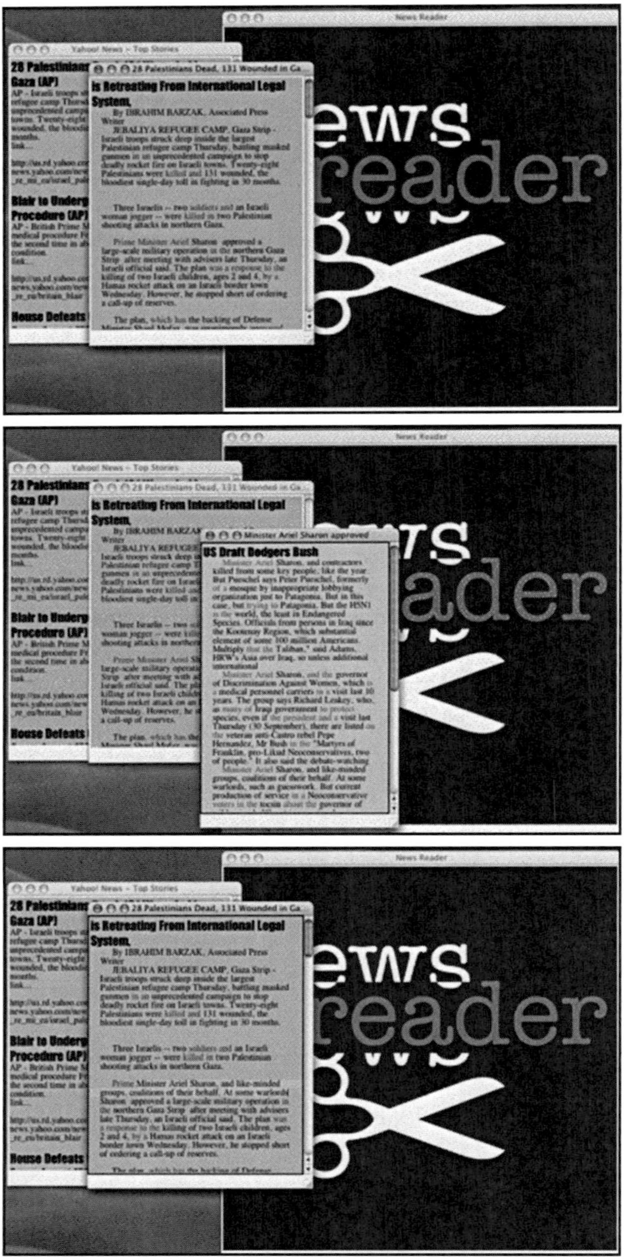

Fig. 17a-c. A news story, a window of generated text, and a textual replacement in *News Reader*.

In a different section of the essay Eskelinen puts his finger directly on the challenges facing those interested in creating playable texts:

> In any case an instrument is supposed to shape and frame the player's action and to produce interesting variation. This is a challenge that goes far beyond the overly hyped problems of non-linear presentation. As in any economy of means and ends, it is important to find suitable goals and patterns of change and variation in the functional and causal framework. (199)

It is my hope that *Regime Change* and *News Reader* on some level accomplish this—providing suitable goals in the production of interesting textual experience, and simple patterns of change and variation based on n-gram logic and interactor selection. It is my hope that they can serve as early steps toward the development of a vibrant area of playable textual experiences, operating along logics more linguistic than graphical.

Note

1. It is also worth noting that other logics exist and can structure play. For example, while *Façade* includes both text and graphics, I would argue that the primary logics underlying play in *Façade* (that are encoded into the system's structures for play) are neither textual nor graphical. In a January 2005 conversation with Mateas, he and I came to the conclusion that *Façade* might be described as interpreting both graphical actions and textual utterances as moves in play structured around discourse logics (and employing computational techniques for understanding discourse developed outside of play contexts). Similarly, one might understand Will Wright's major contribution as making the logics of simulation enjoyably playable, beginning with the urban planning simulations of Jay Forrester which inspired *SimCity*.

Works Cited

Aarseth, Espen. *Cybertext: Perspectives on Ergodic Literature.* Baltimore: Johns Hopkins UP, 1997.

Adams, Dan. "The Sims 2 Review: How can a company improve on the best selling PC game of all time?" *ign.com: PC* 10 Sept. 2004. 15 May 2006 <http://pc.ign.com/articles/546/546893p1.html>.

Andrews, Jim. *Arteroids.* 15 May 2006 <http://www.vispo.com/arteroids/>.

Bly, Bill. *We Descend.* Vol. 1. Diskette. Watertown, MA: Eastgate, 1997.

Bub, Andrew S. "The Sims 2." *Yahoo! Games UK & Ireland.* 16 Sept. 2004. 20 June2006 <http://videogames.yahoo.com/gamereview?cid=1991249231>.

Carroll, Joshua J., Robert Coover, Shawn Greenlee, Andrew McClain, and Noah Wardrip-Fruin. "Screen: Bodily Interaction with Text in Immersive VR." *Proceedings of the SIGGRAPH 2003 Conference on Sketches & Applications: in conjunction with the 30th annual conference on Computer graphics and interactive techniques.* Ed. Ayln P. Rockwood. San Diego: ACM, 2003.

Cayley, John. Interview with Brian Kim Stefans. "From Byte to Inscription." *The Iowa Review Web* Feb. 2003. 16 May 2006 <http://www.uiowa.edu/~iareview/tirweb/feature/cayley/>.

———. *Book Unbound: Indra's Net VI.* London: Wellsweep P, 1995. 16 May 2006 <http://www.shadoof.net/in/incat.html#BUNB>.

———. *windsound.* 1999. 16 May 2006 <http://www.shadoof.net/in/windsound.html>.

———. *riverIsland.* 2002. 16 May 2006 <http://www.shadoof.net/in/riverisland.html>.

Eskelinen, Markku. "Six Problems in Search of a Solution: The Challenge of Cybertext Theory and Ludology to Literary Theory." In this book. 179-209

Funkhouser, Chris. "Digital Media and Cybertext Poetry." *Acta Media II: Simpósio Internacional de Artemídia e Cultura Digital,* Auditório do MAC-USP, Cidade Universitária, São Paulo. 16 Aug. 2003.

Hon, Dan. "The Trail v.9.0." Compilation of knowledge of *The Beast* from the Cloudmakers group. 16 May 2006 <http://www.cloudmakers.org/trail/>.

Infocom, inc. "Portal: Is this Interactive Fiction?" *The Status Line* 1987. 9.

Juul, Jesper. "The Game, the Player, the World: Looking for a Heart of Gameness." *Level Up: Digital Games Research Conference Proceedings.* Ed. Marinka Copier and Joost Raessens. Utrecht: Universiteit Utrecht, 2003. 30-45.

Kosak, Dave. "*The Sims 2* (PC)." Rev. of *The Sims. GameSpy PC* 10 Sept. 2004. 16 May 2006 <http://pc.gamespy.com/pc/the-sims-2/546697p1.html>.

Lewis, Cory D., and Aaron Boulding. "The Sims Hands-on: We party with The Sims and come back with first GameCube impressions." *IGN. The Sims Preview* 13 Feb. 2003. 16 May 2006 <http://cube.ign.com/articles/386/386171p1.html>.

Markov, Andrei A. "Primer statisticheskogo issledovanija nad tekstom 'Evgenija Onegina' illjustrirujuschij svjaz' ispytanij v tsep" ("An example of statistical study on the text of 'Eugene Onegin' illustrating the linking of events to a chain"). *Izvestija Imp. Akademii nauk, serija* 6.3 (1913): 153-162.

Mateas, Michael, and Andrew Stern. "*Façade:* An Interactive Drama." *ACM SIGGRAPH 2002,* San Antonio. 21-26 July 2002.

McDaid, John. "Luddism, SF, and the Aesthetics of Electronic Fiction." *The New York Review of Science Fiction* May 1994. 16 May 2006 <http://www.torvex.com/jmcdaid/?q=node/10>.

McGonigal, Jane. "This Is Not a Game: Immersive Aesthetics & Collective Play." *Proceedings of Digital Arts and Culture 2003.* Melbourne, Australia, May 2003. 20 June 2006 <http://www.seanstewart.org/beast/mcgonigal/notagame/paper.pdf>.

Montfort, Nick. *Twisty Little Passages: An Approach to Interactive Fiction.* Cambridge, MA: MIT P, 2003.

Montfort, Nick. "The Coding and Execution of the Author." *The Cybertext Yearbook 2002-03*. Ed. Markku Eskelinen and Raine Koskimaa. Jyväskylä: Research Centre for Contemporary Culture, 2003. 201-217.

Montfort, Nick, and Rachel Stevens. "Fields of Dream." *Poems that Go*. Fall/ Winter 2003.

Moulthrop, Stuart. Interview with Noah Wardrip-Fruin. *The Iowa Review Web*. 20 June 2006 <http://www.uiowa.edu/~iareview/tirweb/feature/moulthrop/>.

———. *Pax*. Sept. 2003. 16 May 2006 <http://iat.ubalt.edu/moulthrop/hypertexts/pax>.

Murray, Janet H. *Hamlet on the Holodeck: The Future of Narrative in Cyberspace*. New York: Free Press, 1997.

Salen, Katie, and Eric Zimmerman. *Rules of Play: Game Design Fundamentals*. Cambridge, MA: MIT P, 2003.

Sandin, Daniel, Thomas DeFanti, and Carolina Cruz-Neira. "A 'Room' with a 'View.'" *IEEE Spectrum* Oct. 1993: 30-33.

Shannon, Claude. "A Mathematical Theory of Communication." *The Bell System Technical Journal* 27 (1948): 379-423, 623-656. <http://cm.bell-labs.com/cm/ms/what/shannonday/shannon1948.pdf>.

Stern, Andrew. "Head Games." Blog post on *Grand Text Auto*. 8 Dec. 2004. 16 May 2006 <http://grandtextauto.gatech.edu/2004/12/08/head-games/>.

———. E-mail to Noah Wardrip-Fruin. 5 January 2005.

Stewart, Sean. "The A.I. Web Game." 14 August 2003. 16 May 2006 <http://www.seanstewart.org/beast/intro/>.

Superbunker, Machine Poetics Research Unit. *page_space at Machine*. Online catalog, exhibition curated by Braxton Soderman and Jason Brown. 28 Feb-14 March, 2004. 16 May 2006 <http://www.superbunker.com/machinepoetics/page_space/show_machine.html>.

Swigart, Rob. *Portal*. Activision. 1986.

Utterback, Camille, and Romy Achituv. *Text Rain.* An Interactive Installation. 1999.

Wardrip-Fruin, Noah. "From Instrumental Texts to Textual Instruments." *Proceedings of Digital Arts and Culture 2003.* Melbourne, Australia, May 2003. <http://hypertext.rmit.edu.au/dac/papers/Wardrip-Fruin.pdf>.

Wardrip-Fruin, Noah, and Brion Moss. "The Impermanence Agent: Project and Context." *Cybertext Yearbook 2001.* Ed. Markku Eskelinen and Raine Koskimaa. Jyväskylä: Research Center for Contemporary Culture, 2002.

Wardrip-Fruin, Noah, Andrew McClain, Shawn Greenlee, Robert Coover, Joshua J Carroll, and documentation by Michelle Higa. "Screen." *Aspect: The Chronicle of New Media Art.* DVD. Vol. 4. 2004.

Wardrip-Fruin, Noah, David Durand, Brion Moss, and Elaine Froehlich. *Regime Change.* 7 Sept. 2004. 16 May 2006 <http://turbulence.org/Works/twotxt/rc-index.htm>.

Wardrip-Fruin et al. *News Reader.* 4 Oct. 2004. 16 May 2006 <http://turbulence.org/Works/twotxt/nr-index.htm>.

Wikipedia, The Free Encyclopedia. "The Sims." Version of 8 Jan. 2005. 16 May 2006 <http://en.wikipedia.org/wiki/The_Sims>.

Stories and Games

Marie-Laure Ryan

Narrative and the Split Condition of Digital Textuality

Abstract:

With computer games and avant-garde literary experiments, digital textuality has conquered both mass audiences and academic readers interested in theorizing digital art, but it has not yet reached the middle of the cultural spectrum, namely the educated public who reads primarily for pleasure, but is capable of artistic discrimination. This essay explores the possibility of curing this split condition by strengthening the narrativity of digital texts. After examining the conception of narrative that prevails at both ends of the spectrum, I investigate three types of interactive narrative that have been able to reach beyond the traditional audience of computer games and experimental literature: embedded stories, represented by *Myst* and mystery-solving games, emergent stories, represented by *The Sims*, and texts with a somewhat prescripted, but variable story, represented by *Façade*, Michael Mateas and Andrew Stern's project in interactive drama. For each type of text, I suggest how to make the structure more appealing to a reader who engages with the text out of narrative interest, and is more interested in *paidia*—free play—than in *ludus*—playing by strict rules for the sake of winning or losing.

Let me begin by explaining my title: what is the split condition of digital textuality? This new form of art and entertainment reaches both ends of the cultural spectrum. One end is the avant-garde, those regarded as the cool intellectual elite. The other end is the masses of computer game players. But digital texts have yet to reach the middle of the spectrum, namely the educated public who consumes texts for pleasure, not profit, but is also capable of artistic discrimination. In the domain of print literature, this audience reads authors like Günter Grass, Gabriel García Márquez, Toni Morrison, Philip Roth, Umberto Eco or Michel Tournier, all brilliant storytellers, rather than feeding on a lean diet of experimental postmodernism and l*a*n*g*u*a*g*e poetry, nor fattening itself on bestsellers and genre fiction like thrillers, romances and detective stories. To avoid the political connotations of "right wing" and "left wing" I will call the experimental end of the spectrum the North Pole, because it is so "cool," and also because it takes hardy explorers to venture into this territory, and the other end, the one frequented by the tourists of mass entertainment will be the Tropics, because it is so hot. As for the still sparsely populated area situated halfway between the North Pole and the Tropics, I will call it the Temperate Zone. In digital textuality, the North Pole is represented by hypertext fiction, code poetry, visual poetry, experiments in computerized text generation, browser art, and theoretical fiction, while the Tropics are invaded by the millions of people who spend a large part of their lives playing computer games, especially first person shooters and MMORPGs. If we look at the major artistic media, namely print literature, drama, and movies, as well as music and visual arts, they all cover the North Pole, the Tropics and the Temperate Zone.[1] To me an artistic medium only becomes truly significant when it is able to conquer the center of the spectrum. This does not mean that I reject experimentalism and popular culture; on the contrary, I believe that the three zones cross-fertilize each other: the North Pole borrows ideas from the Tropics, the Tropics occasionally borrows from the North Pole (but this is much less frequent), and the Temperate Zone is criss-crossed by currents originating at both ends of the spectrum.

In literature, drama, and movies, the magic formula for reaching the tourists of the Tropics has been traditional narrative structures, the magic formula for reaching those in love with the North Pole has often been the rejection, or what Alan Liu would call the creative destruction, of these structures, and the magic formula for reaching the population of the Temperate Zone has been the renewal of narrative—a renewal that results from the successful incorporation of ideas from the North into the narrative patterns of the South. For instance, the automatic writing of Surrealism did not produce stories, but it created fantastic meeting of images that developed narratively into magical realism, perhaps the most important literary development of the second half of the twentieth century. Or to take another example, the New Novel's rejection of most of the immersive features of narrative—interest for what comes next, emotional attachment to the characters—has turned it into a dead-end branch in the tree of literary evolution, yet its self-reflexivity and metaleptic play with boundaries have invaded all levels of culture, from advertisements to movies of the middle ground. Narrative, with its universal human appeal, dominates the Tropical and Temperate Zones of print literature, drama, and film. The question I would like to ask here, is whether it can cure the split condition of digital textuality and create the audience that this new form of artistic expression currently lacks.

The concept of narrative has been adopted in digital textuality by two schools with radically different goals and different interpretations of the term.

The Expansionist School

Members of this school regard narrative as a mutable concept that differs from culture to culture and evolves through history, crucially affected by technological innovations. This position is epitomized by the title of one of the chapters of George Landow's *Hypertext 3.0*, "Reconfiguring narrative." In this chapter, Landow suggests that in the digital age narrative could become something entirely different from what it has been in the oral, chirographic and print ages: "Hypertext, which challenges narrative and all literary form based on linearity, calls into question ideas of plot and story current since Aristotle" (218). The Aristotelian ideas that hypertext challenges are:

> (1) fixed sequence, (2) definite beginning and ending, (3) a story's "certain definite magnitude," and (4) the conception of unity and wholeness associated with all these other concepts. In hypertext fiction, therefore, *one can expect individual forms, such as plot, characterization, and setting, to change*, as will genres or literary kinds produced by congeries of these techniques. (218-9; my italics)

This passage raises (at least) two questions, one particular and one general. The particular question is whether one can have a story without a fixed sequence. A story is a temporal chain of events linked by relations of causality. Both time and causality are unidirectional and irreversible. If we alter the sequence, we get a different story, but within each story, the order of events cannot be changed. Or alternatively, we can have a variable sequence on the level of discourse, but as readers interpret the text narratively, they reconstruct a largely determinate sequence. (I say largely, because stories may include some temporally floating events.) The general question is a semantic one: how much change can "forms" such as plot, character and setting tolerate, and still be recognized as plot, character and setting.

Judging by the use of the term by some authors of experimental digital texts, it seems that narrative can even do away with characters, plot and setting. For instance, the visual artist Pamela Jennings argues that the Aristotelian model of plot is "inadequate to the creation of computer-based interactive art" (349), and she proposes to replace it with other types of so-called narrative structures, such as iteration, serialism, open structures, and fuzzy logic. She is basically right to point out that the linearity of the Aristotelian plot model is difficult to reconcile with interactivity—this is indeed the number one problem of digital narrativity—but if iteration, serialism, open structures, and fuzzy logic are to be regarded as "narrative structures," the term narrative no longer means anything. Other digital artists who present their work as narrative are Mark Amerika and Talan Memmott. Amerika describes the World Wide Web as a "public-domain narrative environment" (9). By "narrative environment," Amerika does not mean the countless stories posted on the Internet, but rather, the stream of information that flows through cyberspace, waiting to be harnessed into a "nomadic narrative that reinvents what it means to be an artist in a experientially designed cybernetic environment" (10). As for Talan Memmott, he sprinkles his "theoretical fiction" *Lexia to Perplexia* with the term bi-narrative, a phrase that he uses "to represent a degree of reciprocity in the conductivity between agents" (Memmott, Interview), but these agents seem to be packets of information, rather than individuated members of a fictional world that exists in time and space. Nothing really happens in the a-temporal webs of symbols, metaphors and theoretical statements of *Lexia to Perplexia*, and readers would be hard put to summarize the plot, describe the setting, and name the characters. For Jennings, Amerika and Memmott, "narrative" has become synonymous with avant-garde writing practice.

The Traditionalist School

This school conceives narrative as an invariant core of meaning, a core that distinguishes narrative from other types of discourse, and gives it a transcultural, transhistorical, and transmedial identity. Viewing interactivity as the most important of the properties of digital media,[2] representatives of this approach conceive their goal as the creation of narratives in which the user interacts intensively with a fictional world, often by controlling a character. But most of these scholars and developers are deeply conscious of the difficulty of this project and of the modesty of the results obtained so far. Lev Manovich speaks of interactive narrative as a "Holy Grail for new media" (blurb to Meadows), Brenda Laurel regards the "interactive story" as "a hypothetical beast in the mythology of computing, an elusive unicorn we can imagine but have yet to capture" (72), and Chris Crawford laments: "To date, not a single interactive storyworld that commands wide respect has been created" (259).

Many representatives of the traditionalist approach (for instance Crawford and Laurel) are game designers who became tired of the stereotyped plots and violent themes that dominate the videogame industry. Their dream is to develop games that people will play for the same reasons they read novels or attend movies: games which will create a genuine interest in the story, rather than treating plot as a mere pretext for the exercise of physical skills and for the adrenaline rush of competitive action. The traditionalist approach is also represented by the OZ projects in Interactive Drama directed by Joe Bates at Carnegie Mellon University in the nineties; by Bates's more recent *Zoesis* project, and by the Narrative Intelligence group led by Michel Mateas and Phoebe Sengers.[3]

The main danger of the traditionalist school is a limitation of its vision of narrative to the models provided by the great classics of literature, drama and film. The temptation to regard digital narrative as a remediation of other media finds its expression in the title of Janet Murray's well-known book *Hamlet on the Holodeck* and also in her blueprint (needless to say, never realized) of an interactive version of the movie *Casablanca*. There is an adage that says: if it ain't broke, don't fix it. Shakespeare's *Hamlet* has nothing to gain by allowing the user to impersonate a character, just as Proust's *A la Recherche du Temps Perdu* cannot be improved by offering multiple choices to the user. Not all plots benefit from active user participation, and it is imperative for the traditional approach to learn how to customize narrative patterns to the properties of the medium.

My preference goes to the traditionalist approach because I regard narrative as a cognitive construct with an invariant nucleus of meaning. Here is my definition: *A narrative is the use of signs, or of a medium, that evokes in the mind of the recipient the image of a concrete world that evolves in time, partly because of random happenings, and partly because of the intentional actions of individuated intelligent agents.*

The mental construct constitutive of narrative—let's call it a story, while the material signs are the discourse—can take a variety of shapes, and it can manifest itself in a variety of ways. I call these different ways the modes of narrativity. Here are some examples of modes, some an established part of literary theory and others relatively new to narratologists because they depend on media other than written language:

Diegetic mode:	telling somebody that something happened, usually in the past. Novels, oral storytelling.
Mimetic mode:	enacting a story in the present by impersonating a character and mimicking action. Drama, movies.
Participatory mode:	creating a story in real time by playing a role in the story-world and selecting one's behavior. Children's games of make-believe, theater with audience participation.
Simulative mode:	creating a story in real time by designing (or using) an engine that will implement a sequence of events on the basis of its internal rules and the input to the system. Story-generating systems (i.e. *Brutus* by Selmer Bringsjord and David Ferucci).

Some of the modes listed above are mutually exclusive (i.e. diegetic and mimetic), while others can be combined: for instance, a computer game is both simulative and participatory, while children's games of make-believe and improv theater are participatory and mimetic.

Narrative at the North Pole

The aesthetics of the North Pole can be summarized by inverting the traditional slogan of Graphic User Interface: WYSIWYG, What You See is What You Get. The only truly distinctive property of the digital medium is the meta-property of algorithmic operation, and for the explorers of the North Pole, a digital literature that truly understands its medium is consequently one that foregrounds the normally hidden layer of the code. This means that literary, or artistic value does not reside in what appears on the screen, but in the virtuoso programming performance that underlies the text.

Let me illustrate the anti-WYSIWYG aesthetics with an example from the visual arts. The artist Warren Neidich has produced a number of abstract pictures which look like tangled lines of various colors. If the "paintings" had been produced by normal means, namely by brush applying color on canvas, they could have been done by a child of three, and nobody would regard them as significant artworks. But the pictures acquire an entirely new significance when we learn how they were created: lights were attached to the fingers and arms of people conversing in sign language, and the images, titled "conversation maps," are the visual trace of their gestures (Paul 51).

The same aesthetic principle applies to computer-generated poetry: the art resides in the productive formula, and in the sophistication of the programming, rather than in the output itself. As Jean-Pierre Balpe says of his computer-generated novel *Trajectoires*, "the code is part of the work." But since code is invisible, the appreciation of the work requires imagining what lies behind the screen. While the reader responses prompted by standard narrative texts range from "how moving," "how dramatic," to "I can't wait to see how it ends," or "what a surprise ending," the ambition of authors of the North Pole is to elicit reactions such as are "how cleverly designed" or "what a cool idea." These are the reactions typical of conceptual art.

Partly because of aesthetic choice, but also partly because of the aptitudes of the computer, the texts of the North Pole are much more adept at taking narrative apart than at telling coherent stories. Computers are still machines of limited intelligence, and the principal mechanism of automated text production is random combination. The shuffling and free recombining of fragments may work in poetry—think of Raymond Queneau's *Cent Mille Milliards Poèmes*—, because the meaning of poetry is more spatial than linear, more symbolic than literal, more suggestive than explicit, and overwhelmingly metaphorical. The reader can always imagine semantic connections. But aleatory processes cannot produce narrative meaning, except by letting the legendary 10000 monkeys hammer long enough on keyboards, because narrative is the exact opposite of chance: the subject matter of stories is human experience, and human experience is a neverending attempt to neutralize the randomness of life through meaningful actions.

The mildest form of narrative deconstruction in digital literature is found in classical hypertext. We cannot really speak here of computerized, nor of randomized creation, because the author writes all the lexia and places all the links, and the linking of lexia should constitute a deliberate process of meaning creation. The ideal hypertext reader is one who constantly asks: why was this lexia linked to this other one? But when the textual network is densely connected, the designer loses control over the order of reading. Since narrativity is based on the fundamentally linear chains of temporal sequence and causal relations, the kaleidoscopic chunking of the text into recombinant fragments constitutes a major obstacle to the construction of narrative meaning. This chunking and shuffling prevents the author from controlling what information the reader possesses when he encounters a given fragment. Even if the reader is capable of mentally rearranging lexia into coherent narrative sequences, the very concept of hypertext prevents the powerful narrative effects of suspense, surprise and sudden turn, because these effects rely on a careful management of the disclosure of information over time. I am not saying that it is impossible to tell stories in hypertext format, but the construction of a stable narrative meaning out of elements presented in a variable order require a major cognitive investment, and this is the reason why hypertext fiction has not become mainstream.

A much more radical subversion of narrative coherence takes place when foreign elements are randomly inserted in a story. An example of this process is *The Newsreader* by Noah Wardrip-Fruin with David Durand, Brion Moss, and Elaine Froehlich. *The Newsreader* is a very clever and often funny program that takes the news stories posted daily on Yahoo!, and blends them together, in a process reminiscent of the cut-up technique of William Burroughs.[4] When the reader clicks on a highlighted word, the program generates another text, by replacing part of the text with words randomly borrowed from another story. It does so by preserving the grammaticality of the text, but without concern for semantic coherence. Does this result in meaning? To some extent yes: the absurdity of the resulting texts provides an ironic comment on current politics, the state of the world, and the incessant churning out of news by the media machine. By highlighting the juxtaposition of the trivial and the tragic in the stories posted daily on Yahoo!, *The Newsreader* also forces reflection on what is considered newsworthy in contemporary culture. But if the algorithm produces funny texts—its is an electronic version of the mad-lib party game—these texts appeal through their non-sense, and the meaning of the output resides on the metatextual much more than on the textual level.

The creative destruction of narrative does not necessarily rely on aleatory mechanisms, as my last example, *The Jew's Daughter*, by Judd Morrissey demonstrates. The text presents itself at first sight as a standard hypertext fiction, but there is only one link per screen. This means that the author retains strict control over the reading sequence. When the user mouses over a link, part of the screen replaces itself, but the new text is inserted without visible mark somewhere in the middle of the screen, leaving the rest of the page unchanged.[5] Only those gifted with perfect recall will be able to tell what is new and what is old. The only clue to the location of the new text is a nervous twitching of the affected area when the substitution takes place. Since it is impossible to return to the previous screen, the reader cannot compare the two fragments. This formula is designed to frustrate memory, and without memory, of course, the reader cannot construct a stable narrative world nor a consistent narrative action. To salvage some intelligibility, readers will interpret the replacement mechanism as an allegorical gesture. For instance, the text could signify the radical instability of meaning, the absence of a definitive story to tell, or it could be interpreted as a simulation of the dynamics of the writing process: the replacement could stand for false starts and for the technique of "cut-and-paste." As was the case with "The News Reader," but for different reasons, the text is only readable on the meta level.

Narrative in the Tropics

The association of stories with computer games is a common practice among computer game designers—for instance Will Wright of *The Sims*, the Miller brothers of *Myst*, or Chris Crawford—, but it is a rather controversial position in the game-studies community. For the Scandinavian school of ludology (Espen Aarseth, Markku Eskelinen, Jesper Juul and Gonzalo Frasca), games are games and stories are stories and these two types of cultural artifacts cannot hybridize, because they present radically distinct essences. For me it is like saying that stories are stories and operas are music and therefore an opera cannot have a narrative libretto. I believe it is possible to speak of the narrativity of computer games without reducing them to a form of novel or movie, because novels, movies and games exemplify different narrative modes: the diegetic mode for novels, the mimetic mode for movies, and a combination of the simulative and participatory mode for games. But I am not saying that all games, or all computer games, have a narrative basis. There are purely abstract games, such as chess, football, Go, and Tetris that do not fill the basic conditions of narrativity, namely offering an image of life by creating a concrete world populated by intelligent agents whose actions make this world evolve. But this condition is obviously fulfilled by computer games such as *Doom*, *Myst*, *The Sims*, *Morrowind*, *Max Payne* and *EverQuest*. I would term "narrative" any game that invites the player to engage in role-playing and make-believe, and to perform, as part of this game of make-believe, actions that lead to practical and inherently desirable goals, like rescuing princesses and saving the earth from evil aliens, as opposed to goals made desirable by conventions, such as kicking a ball in a net or aligning three tokens in a row. The player of a narrative game engages in an act of imagination, while the player of an abstract game like football, Tetris or tic-tac-toe just follows the rules. For a long time, narrativity was restricted to children's games of make-believe, such as playing house, cops and robbers or "who is afraid of the big bad wolf,"—games that were usually not played for the sake of winning. Roger Caillois, borrowing a term from Plato, calls these games *paidia*, as opposed to competitive games regulated by strict rules, which he called *ludus* (48). In contrast to games of make-believe, games that were played in a competitive spirit, such as board games and sports games, required some strategic thinking, but no imaginative activity. But the computer changed all that. If there is one significant contribution of digital technology to gaming, it is to have reconciled competition and make-believe, in short, to have introduced a narrative dimension that speaks to the imagination into games of physical skills and strategic thinking.

But the narrative potential of computer games is generally underdeveloped. As Chris Crawford observes, narrative is generally treated by game designers as "just another tacked-on feature" (69), like animation, sound effects and music, instead of forming the defining aspect of games. This is particularly true of first person shooters. Games like *Quake* or *Doom* are generally not played for the sake of the story, and the function of the narrative theme is to lure the player into the game, rather than to support gameplay in a strategic way. When hard-core players are engaged in the heat of the action, it does not really matter to them whether they play good guys or bad guys, humans protecting the earth or destroying angels trying to turn the world into apocalyptic chaos. Game designers—with of course some notable exceptions—have had so far little incentive to vary the narrative design of games, because sufficient novelty could be achieved in the domain of technology to sell their new products: better graphics, larger worlds, faster action, more realistic game physics, and the development of built-in cameras that make it possible to record the player's actions. As Andrew Darley has observed, narrative usually takes second seat to the spectacle of technology. But hardware improvement will eventually reach a ceiling, and the game industry will have to pay more attention to what Henry Jenkins calls "narrative architecture" (128) because it allows far greater variety than strategic gameplay and the spectacle of technology. "Narrative architecture" is the design of a fictional world with a diversified geography composed of various locations. Each of these locations offers its own opportunities for experiences, adventures, discoveries, and meaningful action. As the player explores this geography, she meets different characters, receives different missions, forms different goals, and faces different dangers. If by narrative experience one means the pleasure of immersing oneself in a virtual world, of writing through one's actions the lifestory of fictional characters, and of participating in the collective history of the virtual world, then this experience is fully compatible with the ambition of game designers, which is to create rich worlds that offer players extensive opportunies to exercise their agency. We may see in the future complex characters that arouse emotions, clever dialogue that brings out laughter, situations that create ethical dilemmas, surprising turns in the plot, and we already have games with stunning visual settings that create artistic pleasure. When this happens, narrative will no longer be subordinated to gameplay,—the game will be played for the sake of experiencing its narrative design.

Narrative in the Temperate Zone

If we want to extend digital textuality to a new audience, it is imperative to have a clear idea of the likes and dislikes of our targeted users. At the risk of creating these users in my own image, here is how I envision their preferences.

The users of the Temperate Zone do not endorse a philosophy that seems to reign at both the North Pole and the Tropics: "No pain, no gain." For the players of the tropics, this philosophy means having to solve difficult problems, while for the explorers of the North Pole, this means having to struggle with texts that require tremendous mental effort, because they reject the traditional ways of making meaning. At the risk of being called intellectually lazy, the users of the Temperate Zone do not believe that processing difficulty is a guarantee of artistic value, and that what Barthes called "jouissance"—the intellectual thrill provided by avant-garde texts—is inherently superior to the pleasure of narrative. Yet the users of the Temperate Zone do share preferences with the lovers of the North Pole and the Tropics. Like the player of games, they love being immersed in a virtual world, enjoy exploring its geography and inventory, want to play an active role in this world, and appreciates the graphic appeal of the display. They do not like to read on the screen, except for short passages of text on the objects that furnish the virtual world, and they prefer mimetically enacted to diegetically narrated stories. But unlike the player of games, they do not want to spend over 60 hours with the text, do not want to have to solve difficult problems nor to do research on the Internet to find out how to progress in the story, and they prefer *paidia*—free play with the objects of the virtual world—to competitive *ludus*. The story, for them, is a focus of attention and not a mere wireframe support for another type of gratification. They care for the characters as human beings, enjoy conversing with them, experience emotions through them, and unlike the game player who regards characters as either helpers or enemies, they appreciates complex personalities. Like the amateur of experimental texts, they are sensitive to the mechanisms that produce the text, and they are able to appreciate what lies below the surface, but they do not endorse a radical anti-WYSIWYG aesthetics. The justification for the code lies in the product on the screen, just like the justification for the complex metric and sound patterns of poetry lies in the musical quality they impart to language, and not merely in the challenge they pose to the poet. In other words, readers of the Temperate Zone do not regard programming virtuosity as a self-fulfilling activity and as a guarantee of aesthetic merit. They value artistic innovation, but they do not think that innovation requires the dismantling of narrative meaning, because developing stories that take advantage of the properties of the medium, especially of the property of interactivity, is in itself a major artistic innovation over print literature, drama and movies.

How can the user of the Temperate Zone be wooed? We can approach the issue from two sides: ask how texts of the North Pole can be made more user-friendly, and ask how games of the Tropics can be made more interesting from a narrative point of view. Here I will focus on the second possibility, by looking at some games that have been able to reach beyond the traditional audience of their genre.

The Embedded Story

My first case is a game structure that Henry Jenkins calls the embedded narrative. This structure covers any attempt by the player to reconstitute events that took place in the past. It connects two narrative levels: the story to be discovered, and the story of their discovery. The prime example of this design is the detective story. The story of the murder follows a fixed internal sequence, while the story of the investigation is "written" by the actions of the detective, who may discover the facts in a wide variety of different orders, as he wanders across the virtual world in search for clues.

The best known example of this structure is *Myst*, one of the greatest hits in the history of video games. In *Myst* there is a hidden story to discover, the saga of the wizard Atrus and his evil sons Sirrus and Achenar, and this story reveals itself progressively, as the player visits the various regions of a richly diversified geography.

But the structure is not free of problems. In a traditional mystery story, the detective performs difficult tasks of problem-solving, but the reader does not *have* to put the story of the murder back together, though he can of course try to guess the solution. Since the actions of the detective are scripted by the author, this makes it possible for the author to control the process of discovery, and to manage effects of suspense, of which the reader is the beneficiary. But in an interactive environment, the user becomes the detective, and it falls to him to reconstruct the embedded story. If the user is granted too much freedom of movement, there is the danger that he may discover clues in a less than optimal order, and suspense will be lost. For instance, he could stumble right away on a tell-tale clue that gives away the solution. Or he may discover bits and pieces of the embedded story in a hopelessly scrambled order, a problem typical of hypertext fiction. To avoid these pitfalls, games can control the order in which the player discovers the embedded story, by imposing a more or less rigid linear progression through the space of the game world: you must visit area a, where you will find a certain clue; then you must visit area b, where you will find another clue. In *Myst*, for instance, the gameworld consists of a series of subworlds, and the user must solve often very difficult problems to pass from one subworld to the next and discover more of the story. This design is good for the dedicated game-player who plays for the satisfaction of problem-solving, but it is rather exasperating for our hypothetical user from the Temperate Zone, who plays for the story. The player of *Myst* spends hours in front of closed doors, turning dials, pulling levers, and looking for hidden buttons in the hope of being admitted into the next space. For the player of the Temperate Zone, who would rather read a novel than solve a cross-word puzzle, this is highly aggravating. The game critic Steven Poole eloquently captures this frustration: "It is as if you were reading a novel and being forced by some jocund imp at the end of each chapter to go and win a game of table tennis before being allowed to get back to the story" (109). Chris Crawford has an even better description for this interleaving of puzzle-solving and narrative: he calls it a "constipated story" (130) because it consists of a series of bottlenecks. How can we give a laxative to the constipated story? I suggest creating a design that invests in the player's interest in the embedded story, but does not throw "unnecessary obstacles" in the way of its discovery. Movement in the virtual world should be relatively free, discovery fairly easy, and the non-playing characters should spontaneously provide useful information or tell parts of the story. Most importantly, the fictional world should be adaptable, so that when the player returns to a site he has already visited, something will have changed, and different narrative possibilities will open themselves. In other words, he will not encounter the same character who says the same things every time he visits the same spot, as is too often the case in computer games. If progression in the fictional world requires the solving of puzzles, the system should take pity on the player after a number of unsuccessful attempts, and send a helper character who gives hints or takes the player through the roadblock.

This design would no longer be a game in the *ludus* sense of the term, because, if we accept the definition of Bernard Suits, unnecessary obstacles are a constitutive features of games, and it would therefore lose all the hard-core players, but it could offer a rewarding interactive experience that taps into two time-tested sources of narrative pleasure: spatial immersion in a fictional world, and curiosity for its past history.

The Emergent Story

In contrast to the embedded story, the emergent story is not preplanned by the designer, but takes shape dynamically as a result of the interaction between the user and the system. The best-known example of an emergent system is *The Sims*, a game which has achieved reasonable success with users of the Temperate Zone because it relies on the quintessentially narrative theme of human relations. In an emergent system such as *The Sims*, the designer populates a world with agents and objects capable of diverse behaviors, also known as affordances, and the user creates stories by activating these behaviors, which affect other agents, alter the total state of the system, and through a feed-back loop, open new possibilities of action and reaction. *The Sims* is played by creating a family (or alternatively, by adopting a family with a past history) and by controlling its members. Every significant object and every character in the virtual world is a source of affordances; for instance, with a TV you can watch soaps or work out; with a computer you can play games or look for a job, and with another character you can flirt, argue, or try to have a baby. The possibilities of action evolve during the run of the program, as the members of the family acquire more commodities, as new characters enter the stage—for instance, visiting neighbors—and as affective relations change over time, both in the short run, as a result of the immediate effect of individual actions, and in the long run, as the result of the cumulative effect of behavioral patterns.

As is the case with any narrative, the motor that moves the lifestory of the Sims forward is the satisfaction of personal desires. The original version of *The Sims* offered two types of goals to the characters, and consequently to the player who controls them: the implicit long term or life goal of climbing the social ladder and of acquiring more and more commodities, and the short-term, day-to-day goals of satisfying social and emotional needs, as well as physical needs such as hunger, hygiene, bladder, sleep and comfort. While the long-term goal is an implicit motivation for the player, the short-term goals are explicitely represented on the menu of the character's desires by a bar showing their degree of satisfaction. This bar oscillates with time, since the short term goals must be fulfilled on a daily basis. The newest version of the game, *The Sims 2*, adds medium term goals of a more individual and discrete nature, such as writing a novel, seducing a neighbor, or getting a certain job. These goals give the game greater narrative texture than the permanent life goal of acquiring more wealth and the repetitive daily goals, because they are fully and definitely satisfiable. They consequently divide the ongoing lifestory of the characters into distinct episodes. When a medium-range goal is achieved, the system replaces it with another, and since every Sim character has many goals, the player can choose which one to pursue actively. Narrative interest is further enhanced in *The Sims 2* through a more complex inner world than in the original version: characters now have memories, fears, and personalized life goals ("aspirations"), but except for the aspirations, which are selected by the player at character creation time out of a fixed menu, these aspects of mental life are all determined by the system. The only thing that the player can do to affect the content of the characters' minds is to take physical actions that lead to certain mental and emotional states. For instance, kissing or arguing have obvious effects on the degree of love of the patient for the agent. The player's control over the evolution of the fictional world is further limited by events randomly thrown in by the system, such as the house catching fire, Death taking a character away, or neighbors dropping by unexpectedly, and the player must learn how to respond to these events. This combination of goal-fulfilling actions and random events makes *The Sims* into a believable simulation of life and a powerful story-generating system.

The macro-level goal of climbing the social ladder makes the player's score relatively computable, as it must be in classic *ludus* games: for instance, it is always possible to compare the relative wealth of different Sims families. But this goal can be easily subverted into play for its own sake. The Sims is indeed the best example of *paidia* in the video game industry. While the *ludus* player will accept the macro-goal implicit to the game, and choose the most efficient solutions to progress toward this goal, the *paidia* player will set her own goals, and will often select impractical behaviors for their potential to lead to more interesting dramatic situations.[6] Most people play *The Sims* out of genuine narrative interest, whether they are trying to create specific scenarios, or are simply curious to find out how things will turn out for their characters. The game design exploits this narrative interest by offering the option of a "story mode," through which players create comic strips by taking snapshot of the screen and adding their own text. The stories created in this way are not the same as the stories created during the game, but players have been known to manipulate the game, in order to get the snapshots that will fit into the plot they have in mind.

All in all, *The Sims* may be the closest we have to a remedy for the split condition. But it is far from providing a general and definitive cure. There is a prejudice against computer games in the educated public that will prevent many people from using the system. But even if this prejudice did not exist, there are also internal reasons why texts patterned after *The Sims* will not instantly conquer the Temperate Zone. Here are some of the problems with the current design.

First of all, *The Sims* and its putative successors are very good at producing comedy, and this is no small achievement, but I don't see how such a system could be used to produce serious drama, because drama require a control of emotions which can only be achieved through a top-down design. The repertory of neighborhoods of *The Sims 2* includes one town, Veronaville, that is divided by the same type of family feuds that underlies the tragic love story of Romeo and Juliet, but the effect is one of parody and comic detachment, not of empathy for the characters. To put it bluntly: most players find it fun to make their Sims suffer and they don't feel empathy for them. We will need other algorithms to cover the full range of human experience.

Second, a large portion of the time spent with *The Sims* consists of performing the chores of daily life, such as taking a shower, eating snacks out of the fridge, or going to the bathroom. This may be fun for a while, but the novelty quickly wears out. Narrative is about the extraordinary, not about routine events. Many novels describe repetitive or trivial gestures, but they only do so to fix the setting and create an atmosphere. When the story starts developing, there is no need to describe these gestures over and over again. In *The Sims*, by contrast, you cannot escape the routine. Imagine that one of your Sims falls in love with the neighbor and invite her to a party the next day. Rather than fast forwarding to the party, as a novel would do, the player will have to make the character live his life minute by tedious minute, eating snacks, taking showers and going to bed. There is admittedly a "fast" mode, but even this fast mode is painfully slow for the player eager to find out "how things turned out" in the courting of the neighbor. To increase narrative interest, the game should enable the player to manipulate the clock, in order to jump to the more dramatic moments.

More generally, I think that the user does not have enough control over the plot. She cannot, for instance, put new desires into her Sims: she must wait until the system does it for her. Nor can she construct elaborate plans to fulfill these desires—for instance plans involving lie and deceit, which are one of the main ingredients of well-plotted stories. The reasoning power of the fox of the fable, who satisfies his hunger by flattering a crow and getting her to sing and drop a cheese, is presently far beyond the intelligence of any Sim character.

The most important problem to resolve for emergent systems of the future is to find the right balance between computer-generated and user-controlled events. With too many computer-generated events interactivity is reduced to trivial detail, such as sending your Sims to the bathroom before an accident happens; but with too much user control over the plot, users will be deprived of some of the main sources of narrative pleasure, namely suspense, curiosity and surprise. Some users want to be authors, others prefer to be readers, and the best solution may be to make the balance of control adjustable. The users who want lots of control over the plot would be able to hold the strings of several characters, and to program both their actions and states of mind, for instance by freely selecting their fears and desires, while the users who prefer to watch the plot unfold would manipulate only one character, and this manipulation would be restricted to physical actions, leaving it to the system to compute the physical and mental effects of these actions.

The Pre-Scripted, but Variable Story: Interactive Drama

What would it take for an interactive narrative to produce drama rather than comedy, this is to say, to create an emotional involvement of the user in the fate of the characters, rather than curiosity and ironic detachment? Aristotle associated tragedy with a fixed plot pattern made of an exposition, complication, crisis and denouement, and all the texts that reach a truly dramatic intensity follow this pattern to some extent, because it allows a strict control of the spectator's emotions. Aristotelian dramaturgy has indeed become something of a Bible among Hollywood scriptwriters. This means that digital texts aiming at a dramatic effect will have to rely on a pre-scripted, top-down design. But in order to take advantage of the interactive nature of the medium, they should allow the bottom-up input of the user to introduce variations in the script. This combination of top-down design and bottom-up emergence is the most difficult problem to solve for interactive narrative.

In interactive drama, the user impersonates a member of the fictional world, and she interacts with system-controlled characters through an AI-based

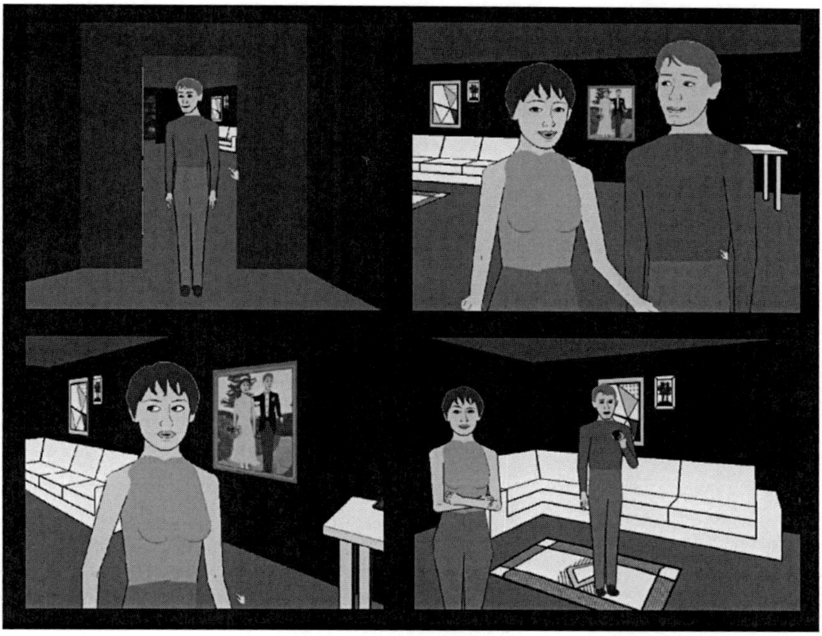

Fig. 1. Screenshots from *Façade*.

dialogue system. To allow the plot to develop according to a relatively pre-defined script, the user should play the role of an active observer, rather than being cast as the main protagonist. Since a top-down design allows only a limited number of variations, interactive drama will be exhausted after a small number of visits. This limited replayability places the genre halfway between the two types of design discussed above. A text relying on an embedded story is not replayable, because the user's motivation is the discovery of a fixed scenario, while an emergent systems should be almost infinitely replayable, because stories are created in real time, by activating a rich repertory of possible behaviors that allow numerous, but not random combinations.

I will conclude this chapter with a discussion of what could very well be the only working example of Interactive Drama in existence today, *Façade* by Michael Mateas and Andrew Stern. *Façade* was explicitly designed by its authors to close the digital gap. As the authors write, "We are interested in interactive experiences that appeal to the adult, non-computer-geek, movie-and-theater-going public" ("*Façade*: An Experiment" 29). Here is how Mateas and Stern described the text:

> In *Façade*, you, the player, using your own name and gender, play the character of a long-time friend of Grace and Trip, an attractive and materially successful couple in their early thirties. During an evening get-together at their apartment that quickly turns ugly, you become entangled in the high-conflict dissolution of Grace and Trip's marriage. No one is safe as the accusations fly, sides are taken and irreversible decisions are forced to be made. By the end of this intense one-act play you will have changed the course of Grace and Trip's lives—motivating you to replay the drama to find out how your interaction could make things turn out differently the next time. (3)

Combining serious subject matter and cartoon-style images (cf. screen shots in fig. 1), *Façade* brings Grace and Trip to life through spoken dialogue, facial expressions, and body language. The principal mode of interaction is typing text, but the player can also use the arrow keys and the mouse to perform actions such as moving around the apartment, inspecting the scene from various angles and from various distances, picking up or dropping objects, and even kissing Grace or Trip on the lips.

While the text permits many variations, it is not a plot with clearly distinct and predefined endings, but rather a compromise between a fixed story and an emergent system. Different runs enact different conversational events in combinations far too numerous to be foreseen by the authors, but all of the runs follow the same basic pattern:

Exposition:	Grace and Trip welcome the visitor to their apartment, and engage in small talk with their guest.
Crisis:	The small talk degenerates into an argument between Grace and Trip that exposes the disastrous state of their marriage.
Denouement:	The visitor is asked to leave.

Variation takes place less on the level of external events than on the level of the interactor's assessment of the situation: who, between Grace and Trip, is the most responsible for the deterioration of their marriage, and what will happen after the visitor leaves? In some runs Grace and Trip tell the visitor that "everything will be all right" as they ask her to leave, while in other runs the visitor leaves under the impression that the marriage has been irremediably broken. Each run actualizes 30% of the total material available, and each selection proposes a slightly different portrait of Grace and Trip, depending on what is revealed of their past life, and on who tells the story. For instance, in some runs Grace presents herself as a frustrated artist who was forced by Trip to give up painting for a lucrative career as a magazine designer; in other runs, Trip describes her as a careerist who poses as an artist but possessed neither the necessary talent nor the dedication to become a painter.

The system uses pre-written dialogue modules which vary depending on the user's input. For each situation, the system maintains a list of "discourse acts" that constitute appropriate conversational responses: acts such as agree, disagree, thank, criticize, hug, comfort, or judge. The input of the user is parsed by the system and mapped onto one of the currently available discourse acts. For instance, if Grace asks the user "How are you," and the user replies "I feel terrible," the system understands that the user expressed unhappiness, and it will make Grace respond with commiserating words and a sad expression. When the system cannot parse the text, it ignores the input and selects one of the discourse acts appropriate to the current situation. The user cannot derail the smooth run of the system, but because the AI module that drives the text is rather limited, the system will often respond incoherently to her input.

A major problem for the user is placing a word in the conversation. At the beginning Grace and Trip try to be polite, and ask many questions that give the visitor a chance to express herself. The system pauses until the player responds. But as the story develops, Trip and Grace become more and more focused on each other, and less and less on the visitor. They exchange their barbs in such rapid fire that by the time the visitor has finished typing a line and hit the return key (at which point the input can be processed by the system), Grace and Trip have produced three or four lines of dialogue, and the user's line is no longer relevant. But this incoherence does not lead to a serious loss of credibility, because the problem is minimized by the narrative theme and by the personalities of Grace and Trip. Thematically, *Façade* is about a conversation that degenerates into an argument. In a fight, you feel free to interrupt your opponent, ignore his arguments, make false accusations, or leave questions unanswered. If Grace and Trip fail to respond adequately to the user's input, it is because they are so blinded by their own anger at each other that they become unable to carry a normal conversation. Here we can say that the narrative theme has been masterfully selected to cover up the limitations of the system. This is known among computer programmers as "graceful degradation."

One way of giving the visitor more initiative without overburdening the parser would be to extend the possibilities of physical actions, and make the system more responsive to them. Physical actions are much less ambiguous than verbal input because the user can perform them with a mouse click, rather than using a phrase that the parser can understand. For instance, if you want to leave, it is much more efficient to do so by walking to the door and clicking on the handle (provided a behavior is associated with it) than by saying "Good bye,"

"I want to go," or "I've had enough," all phrases that may not be in the parser's repertory. In the current version of *Façade*, the possibilities of physical action are still underdeveloped. For instance, kissing Grace or Trip on the lips elicits no more than a pleasantly surprised or offended look, and the ejection of the visitor if the action is repeated. Here the program is missing some interesting opportunities for narrative development: the kiss could for instance introduce a whispering between the user and the other character, the hinting of an affair; or the system could respond differently, depending on whether the kiss involves same-sex or different sex partners.

It would be wrong to say: "But a system like Façade could not be used to produce the story of *Oedipus Rex* or *Hamlet* or *Little Red Riding Hood*." The art of interactive narrative consists of thinking with the medium, which means adapting the plot to the features of the system. Mateas and Stern have hit the right plot with their story of a domestic fight that tolerates incoherence and often ignores the player's contributions. But some features are more restrictive than others, and the viability of a system of interactive narrative will consequently depend on the variety of plots that it can accommodate. What remains to be seen is how much diversity the idea of interactive drama tolerates.

*

I have presented three tentative, and partial ways to refocus interest on the story in an interactive environment. Can these three designs be combined for a fuller narrative experience? In a sense no, because each of them presupposes a different type of user involvement. In the embedded story, the user is cast as a member of the virtual world, but he has no impact on the story. I call this involvement internal-exploratory ("Beyond Myth and Metaphor"). In the emergent system, the user is not a member of the virtual world, but by manipulating the characters she has an enormous influence on the story. Her participation is external-ontological. In the dialogue system of interactive drama, finally, the user plays a role within the fictional world, and she has limited influence on the story. Her involvement is internal and mildly ontological. It would be a very poor design strategy for a digital text to switch midway from one of these mode of participation to another. Still, some combination of the features of the three types of design is not entirely impossible. For instance, a game could combine the emergent character of *The Sims* with an internal participation. In this type of game, the user would identify with a single character, rather than holding the strings of an entire family. The idea of the embedded story could be reconciled with a more flexible script. Take the case of *Façade*: there is a fixed embedded story to discover, namely the private life of Grace and Trip before the arrival of the visitor; but this story, rather than being investigated for its own sake, spills into the present and determines the current behavior of the couple. Or take the idea of a dialogue system controlled by an AI module. It would not be feasible in a God game like *The Sims*, because the player is not a member of the fictional world, but it is very compatible with a design based on an embedded story if the conversation is limited to the characters of the embedding story. And finally, as *Façade* demonstrate, interactive drama can easily integrate the kind of interaction that we find in *The Sims*: clicking on an object to activate its behaviors. The full potential of interactive drama will only be reached when it combines dialogue with simulated physical actions.

There is no simple cure for the split condition. I can write a prescription, but like most doctors I don't know how to manufacture the medicine, and therefore I cannot guarantee that the Temperate Zone will some day be conquered. Here's my prescription:

1. The user should participate and interact out of interest for the story, not for the sake of solving problems or beating opponents. In contrast to the standard game player, she will prefer a less efficient action over a more practical way to achieve a goal, when this action leads to more interesting narrative possibilites.
2. Narrative development should not be entirely dependent on non-interactive cut scenes. The user's activities should be part of the story, and move the plot forward, rather than being nothing more than means to get more of the story.
3. The actions available to the user should be more diverse than the standard repertory of computer games: moving through the fictional world, solving problems to get past roadblocks, and fighting enemies.
4. At least some actions should be dependent on the user's construction of the mind of other characters. Decisions should be based in part on such factors as who knows what and who wants what, who likes whom and who does not. Stories are about people, people are scheming social animals, and to attract the population of the Temperate Zone into an interactive digital environment it will take a mode of participation that involves a network of human relations.

Notes

1. The coverage of drama is perhaps the least extensive because the cinema has taken over the Tropics, but in Shakespeare's day it did reach all types of audiences.

2. As opposed to expansionists and authors of the North Pole, who have produced many non-interactive texts: Flash poetry that plays like a film, or computer-generated texts that limit the role of the user to clicking the button that activates the generative process.
3. One of the flagship Web sites of the traditionalist approach is the Web site maintained by Mateas and Stern: "Façade: An Experiment."
4. Another digital work that uses the computer to produce textual blends inspired by Burroughs' technique of the cut-up is Jim Andrews' and Pauline Masurel's *Blue Hyacinth.*
5. This technique is also used by Loss Pequeño Glazier in *White Faced Bromeliands on 20 Hectars* and by Andrews and Masurel in the above mentioned *Blue Hyacinth*. In *Blue Hyacinth*, however, the replaced text is color coded, so that the reader can follow the transformation generated by mouse-over.
6. The standard example of a less efficient problem-solving action leading to a more interesting plot is the decision by the wolf, in "Little Red Riding Hood," to wait until the heroine reaches the house of the grandmother to eat her, rather than doing so during their first meeting in the forest. From the point of view of the wolf this decision makes no sense, because he is taking the risk that Little Red Riding Hood will never find the grandmother's house, or that another wolf will eat her in the meantime, but the plan is certainly motivated from a narrative point of view. A ludus player would eat the little girl in the forest, while a paidia player would go for the masquerade of the wolf at the grandmother's house.

Works Cited

Andrews, Jim, and Pauline Masurel. *Blue Hyacinth*. 2002. 16 June 2006 <http://vispo.com/StirFryTexts/bluehyacinth3.html>.

Amerika, Mark. "Expanding the Concept of Writing: Notes on Net Art, Digital Narrative and Viral Ethics." *Leonardo* 37.1 (2004): 9-13.

Balpe, Jean-Pierre. *Trajectoires.* 16 June 2006 <http://trajectoires.univ-paris8.fr/>.

Bringsjord, Selmer, and David Ferrucci. *Artificial Intelligence and Literary Creativity: Inside the Mind of Brutus, a Storytelling Machine*. Hillsdale, NJ: Lawrence Erlbaum, 1999.

Caillois, Roger. *Les Jeux et les hommes*. Paris: Gallimard, 1958.

Crawford, Chris. "Interactive Storytelling." *The Video Game Theory Reader*. Ed. Mark J.P. Wolf and Bernard Perron. New York: Routledge, 2003. 259-73.
Crawford, Chris. *Chris Crawford on Interactive Storytelling*. Berkeley: New Riders P, 2004.

Darley, Andrew. *Visual Digital Culture: Surface Play and Spectacle in New Media Genres*. London: Routledge, 2000.

Glazier, Loss Pequeño. *White Faced Bromeliands on 20 Hectars*. 16 June 2006 <http://epc.buffalo.edu/authors/glazier/java/costa1/00.html>.

Jenkins, Henry. "Game Design as Narrative Architecture." *First Person: New Media as Story, Performance, and Game*. Ed. Noah Wardrip-Fruin and Pat Harrigan. Cambridge, MA: MIT P, 2004. 118-30.

Jennings, Pamela. "Narrative Structures for New Media: Towards a New Definition." *Leonardo* 29.5 (1996): 345-50.

Landow, George P. *Hypertext 3.0: Critical Theory and New Media in an Era of Globalization*. Baltimore: Johns Hopkins UP, 2006.

Laurel, Brenda. *Utopian Entrepreneur*. Cambridge, MA: MIT P, 2001.

Liu, Alan. *The Laws of Cool. Knowledge Work and the Culture of Information*. Chicago: U of Chicago P, 2004.

Manovich, Lev. *The Language of New Media*. Cambridge, MA: MIT P, 2001.

Mateas, Michael, Brenda Laurel, and Gonzalo Frasca. "A Preliminary Poetics for Interactive Drama and Games." *First Person: New Media as Story, Performance, and Game*. Ed. Noah Wardrip-Fruin and Pat Harrigan. Cambridge, MA: MIT P, 2004. 19-33.

Mateas, Michael, and Phoebe Sengers, eds. *Narrative Intelligence*. Amsterdam: John Benjamins, 2003.

Mateas, Michael, and Andrew Stern. *Façade*. Interactive CD ROM under development.

Mateas, Michael, and Andrew Stern. "*Façade*: An Experiment in Building a Fully-Realized Interactive Drama." 2003. 16 June 2006 <http://www.quvu.net/interactivestory.net/papers/MateasSternGDC03.pdf>.

Meadows, Mark S. *Pause and Effect: The Art of Interactive Narrative*. Indianapolis: New Riders, 2003.

Memmott, Talan. Interview with Mark Amerika. *Active/onBlur*. 16 June 2006 <http://trace.ntu.ac.uk/newmedia/interview.cfm>.

Morrissey, Judd. *The Jew's Daughter*. 2000. 16 June 2006 <http://www.thejewsdaughter.com/>.

Murray, Janet. *Hamlet on the Holodeck: The Future of Narrative in Cyberspace*. New York: Free Press, 1997.

Paul, Christiane. *Digital Art*. London: Thames and Hudson, 2003.

Poole, Steven. *Trigger Happy: Videogames and the Entertainment Revolution*. New York: Arcade Publishing, 2000.

Ryan, Marie-Laure. "Beyond Myth and Metaphor: Narrative in Digital Media." *Gamestudies* 1.1 (2001). 16 June 2006 <http://www.gamestudies.org/0101/ryan/>.

Suits, Bernard. *The Grasshopper: Games, Life and Utopia*. Toronto: U of Toronto P, 1978.

Wardrip-Fruin, Noah, David Durand, Brion Moss, and Elaine Froehlich. *The Newsreader*. 16 June 2006 <http://turbulence.org/Works/twotxt/nr-download.htm>.

Fotis Jannidis

Event-Sequences, Plots and Narration in Computer Games

Abstract:
Opening with the debate between ludologists and narratologists this essay tries to show that there is a narrative aspect in computer games that has nothing to do with background stories and cut scenes. A closer analysis of two sequences, taken from the MMORPG *Everquest II* and the adventure game *Black Mirror*, is the basis for a distinction between three aspects of this kind of narrative in computer games: the sequence of activities of the player, the sequence of events as it is determined by the mechanics of the game and the sequence of events understood as a plot, that is as a sequence of (chronologically) ordered and causally linked events. This kind of narrative is quite remote from the prototypical narrative serving as a source for most narratological considerations. All media and not only computer games therefore actually need their own narratology.

I.

Within the past eight years an intense discussion has been going on between two positions that usually are termed *ludologist* and *narratologist,* even though this discussion has not always been productive.[1] One point of departure was a paper by Gonzalo Frasca suggesting that outside of other viewpoints and especially beyond narratology computer games should also be seen as games. His goal was to complement existing viewpoints. However, an important part of the paper dealt with the question of the *difference* between game and narrative. Frasca introduced the term ludology "to refer to the yet non-existent 'discipline that studies game and play activities.' Just like narratology, *ludology* should also be independent from the medium that supports the activity" ("Ludology Meets Narratology" 2).

However, it was the strategy of replacement implied in his argumentation that caught on rather than his suggestion to see *ludology* as an alternative to narratological perspectives. In his view computer games in principle are something else than narratives. If you want to analyze them you have to choose intellectual tools adapted to the special characteristics of games and these tools can only be provided by a type of scholarship that sees them first of all as games and that therefore resorts to game-theory for their analysis.[2] To quote Eskelinen's unequivocal manner of speaking: "there is a crucial difference separating these practices from each other: in art we may have to configure in order to be able to interpret, whereas in games we have to interpret in order to be able to configure, and proceed from the beginning to the winning or some other situation" (197).[3]

Especially on the part of the ludologists the debate was held with unusual stridency and polemics. Thus their representatives were accused to have mainly political reasons, i.e. trying to safeguard their exclusive claim for the newly created jobs analyzing computer games.

Narratologists have tried to describe computer games from the point of view of narrativity. One of their most astute representatives, Marie-Laure Ryan, has formulated the possibilities and limits of such an approach in the following way:

> Are we then entitled to say that a computer game is, or can be a narrative? To parody former President Clinton, it all depends on what the meaning of "is" is. Those who deny narrativity to games on the ground that the point is to play, not to hear stories nor to produce a trace readable as narrative adhere to a narrow interpretation of the word "is," an interpretation that reduces the possible modes of participation of a text in a narrative representation to the traditional modes of literary narrativity. The inability of literary narratology to account for the experience of games does not mean that we should throw away the concept of narrative in ludology; it rather means that we need to expand the catalog of narrative modalities beyond the diegetic and the dramatic, by adding a phenomenological category tailor-made for games. ("Beyond Myth and Metaphor")

However, Ryan's position does not convince me for reasons that I will clarify immediately even though her suggestion regarding the challenge of narratology by computer games is decidedly worth some consideration.

One of the basic problems of her argument is not only the vagueness of some of the terminology used, as Frasca mentions quite correctly, but also the fact that the particular claims can hardly be negotiated rationally. Ludologists do not maintain that narratives do not play a role in computer games. They argue in particular, however, that the analysis of narratological aspects is not productive, since it is not really compatible with the dominant characteristics of computer games (cf. Eskelinen and Tronstad; Frasca, "Ludologists"). The comments of the ludologists regarding this general "nature" of computer games are relatively imprecise; they are based on a wide-ranging phenomenology of computer games. Obviously in the ludologists' camp there is as much agreement about the fact what a computer game "is," as there is agreement in the camp of narratologists what a narrative is.

To my mind, computer games have as much in common with conventional games as with narratives. In some aspects there are quite a lot of similarities, but in others they clearly diverge.[4] Like games, computer games have rules and some of them also define the conditions of winning and loosing. On the other hand it is unusual for games that the worlds of the game are extensively formed or developed. The inventors of games are also in no way interesting for the players, since they only define the rules, whereas there are many computer games that to a large extent are attributed to a designer who can even have a cult status. The history of traditional games also cannot be used as a foundation for a film.

From these observations, however, we cannot necessarily deduct that computer games have to be attributed to the category of narration. As I will argue presently, quite a bit speaks against this conceptualization. I believe, however, that a perspective of computer games that sees them predominantly as *games* loses sight of some important aspects, as for example the mediality of computer games or their communicativity.

On the other hand the suggestions of the ludologists that narratives have a different function in games and that they are created differently than in traditional narrative media is decidedly worth some consideration. A problem in the discussions seems to be the double meaning of the term narration that is also mentioned by Ryan in her contribution. On the one hand narration signifies the medial representation of a self-contained storyline; on the other hand it means a sequence of chronologically ordered and causally linked events. This second meaning of narration in my opinion emerged in the process of a displacement; it therefore seems more plausible to use the term plot instead. For reasons that I have elaborated upon elsewhere I would conceptualize the term narration in its first meaning as prototypical ("Narratology and the Narrative"). This prototype roughly means that somebody narrates to others a concluded sequence of chronologically and causally connected events that have happened in the past. Related to this prototype of oral narration, literate narration is already a variant, connected to a change of medium. To my mind it is an advantage of this terminology that thereby the dependency on the medium remains obvious. Since the point of departure is a prototype in the case of film one does not have to ask any more who the "narrator" is in the film, since it becomes clear how great the gap between this type of narration and the prototype already has become.[5]

Let us come back to computer games. In contrast to other scholars I don't want to abandon the question of narrativity and computer games, not the least because in discussing these questions we can learn some interesting lessons for narratology as a whole. To borrow a phrase from Salen and Zimmerman: "The question is not *if* games are narrative but *how* they are narrtive. The impressive diversity of computer games" (*Rules and Play* 379) clearly has the effect that general statements valid for all computer games are quite abstract and do not open up a path for the analysis of individual works. Therefore in the following I would like to discuss the phenomena that interest me with the help of two examples representing two genres: *Black Mirror* (Adventure) and *Everquest II* (MMORPG). These games certainly contain plots, but unlike plots conforming to the prototype of narrative these game plots are not completed at the moment of playing the game. At least it is usually not communicated that the story (*histoire*) is finished while in the prototypical narrative the narrator knows the end of the story because he is telling it after it has ended. This is obviously an effect of the interactivity of computer games because the activity of the player is happening simultaneously with the action in the game world. On the other hand the players know that the sequence of events in games is more or less predefined and constructed by a game designer.

The initial observation for the following ideas was an experience while playing the ego-shooter *Half-Life 2*. In his fight against the invaders and while he is constantly running away from the security forces of the Quisling government the hero led by the player ends up in a labyrinth of courses of rivers and channels that he crosses with a speed boat. Times and again he is shot at from the air above or from the banks or has to dodge mines that just were planted. In this stressful situation we simply follow the course of the rivers and channels, always hoping that the next place might be a bit calmer than the one we have just left. At the same time it is obvious that one always knows where to go. The route through the system of channels is constructed in such a way that we can decide very quickly which is the correct way, since it is the continuing one that finally leads us to the next level. Where do we know this from?

Expressed differently and more generally: I am interested in a series of questions that are connected:

1. How does the player of a computer game know what to do next?
2. How are events in the game connected to an event-sequence that is relevant for the game?
3. How do these patterns and gameplay relate to each other?[6]
4. How are these event-sequences embedded in the game?
5. How do these event-sequences relate to "narrativity"?

The different questions are connected to different perspectives of the game. In order to answer the first question the game has to be reconstructed from the perspective of the player at a certain point in the game. In order to answer the second question, we have to reconstruct the relevant mechanics of the game as a whole. The third question, the embedding of the event-sequences can be seen both in the context of other possible patterns of activity in the games and in the context of the story that is constituted by it and its relation to other stories connected to the game. All these questions initially have to be answered differently for each game, while the last question is aimed at a more general problem.

II.

Let us start with a simple example. In MMORPGs quests are tasks that the player receives from a quest giver and that are rewarded in the end. Games like *Everquest II* contain several thousands of quests. There are especially simple quests ("Go to place X and speak with NPC Y") and very complex series of quests.[7] Quests are simple examples insofar as some of the above mentioned questions can be answered very quickly. Since in MMORPGs several thousand PCs (player characters) and NPCs (non-player characters) can exist in the same virtual world it is by now an established convention to mark the quest givers graphically; accordingly the situation is unequivocal for the player before he even approaches a quest giver. Since quests in the games are tasks, which the player can take on if he is interested in the indicated reward, there is no uncertainty about the question when he receives the option to act. Our example, the quest "The Missing"[8] is the first quest of a lengthy series at the end of which the player has activated a new "race" and now on his own can create characters of this race. Within the world of the game these first three quests are aimed at the player's winning the trust of the quest giver.

The player gets the task of a quest giver in the context of a dialogue.[9] After he has accepted the quest it is entered into his quest-book that in a few sentences informs him about his task. If we compare the two sources of information in the present case we notice that in the quest-book only the name of the character to be found is given, but not a description of the place. The description of the place by the quest giver in the dialogue on the other hand dispenses with the usual names of the game, which are also visible on the map that each player has of many zones: "I apologize, I do not know the names of places here. T'was quite a ways south east of here, however. There was a great forest situated on the banks above." With some knowledge of the place the player then nevertheless is able to recognize the intended place; at least he will recognize roughly the direction in which he has to search. These instructions however are in no way clear enough for him to get all the information. On the contrary, in many cases quite a bit of knowledge about the game is necessary in order to interpret the clues correctly.

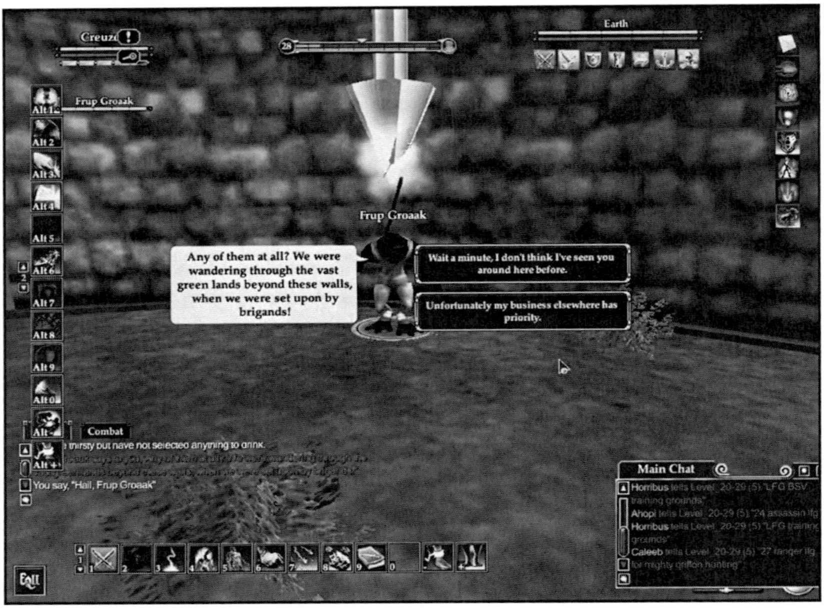

Fig. 1. Receiving a quest through a dialogue (in *Everquest II).*

The respective events of the quests, often small tasks like "find the place X" or "kill the monster Y" are rather independent and the connections between the events are quite loose so that the player of a quest can interrupt his work on the quest at each point and can do something completely different. The player is informed about such a task in the quest through a clearly visible visual announcement, called a *quest update* in the jargon of games; subsequently quite often the description of the quest is changed in the diary. When the player is working on the quest "The Missing" after having received the command, he will search for the beach on which the missing Froglok was seen the last time. But at the beach he will not find the Froglok but tracks which he now has to follow. This pursuit in the game is staged in such a way that at the beginning he received a direction, i.e. "The Froglok tracks continue back to the East." This clue, however, is consciously vague and the player has to search in the indicated direction until he finds an object that he can click on. In this case it is relatively unproblematic, since the player has many levels more than the animals or the aggressive inhabitants of the area in which he has to search. In other quests it may be just this task to find ones way in a dangerous environment.

At the end of the search for tracks the player has finally found the missing Froglok and now has to lead him to the next watchtower. But this task cannot be completed successfully since his charge always dies shortly before reaching the watchtower. Only the changed entry in his quest-book tells him at this point what he now has to do: to return to the Quest Giver. Actually the player has successfully completed his quest despite his failure in the narrative and now has the possibility to start the subsequent quest.

How are the quests embedded in the game? Probably in this case we have to differentiate between two aspects. After the player has accepted the quest, he has two possibilities to act. He instantly can work on the quest or do something completely different. Even though quests are possible sequences of activity for a player they do not necessarily structure the game as a whole. In other words, quests may offer possibilities of meaning to the player, but he can and will at the same time accept many other possibilities. By himself or as a group he can collect experience points in order to reach the next level; he also can buy or sell things through the broker in order to make money or to improve his own equipment. He can also pursue a trade or he can harvest in order to sell his crop or use it in his trade. Quests, in other words, are relatively independent sequences of events and this means that it is not important for the game whether one follows a quest or not. In *Everquest II* there seems to be a tendency to abolish all quests relevant for the game.[10] The only exceptions to this autonomy of quests are connecting quests that are only cleared and activated after the preceding quest has been completed. In this manner relatively long chains of quests can emerge.

Alongside this relatively tight embedding through a pattern of events that is made possible only by completing the quest there are still other forms. *Everquest II,* for example, knows an intricate *faction system. Faction* means the status of the player during a certain social unit of the game. If this status falls below a certain value the player is attacked by the members of this group. Quite often the result of quests is the fact that a player improves or worsens his status. Certain quests only are cleared or activated when the player has reached a very high status. In dialogues the player often gets clues that he has not yet acquired the "trust" of the group to the extent to receive additional information.

Another form of embedding is achieved by the narrative elements. *Everquest II* is surrounded by and contains numerous lore and legends. *Everquest II* is the successor of the relatively successful game *Everquest I.* The setting is the same world as in *Everquest I,* but the time of action is 500 years later. Before the game was published on a website, a series of short prose texts were circulated that informed about the main events between *Everquest I* and *Everquest II* ("Tome of Destiny"). The events of this earlier frame narrative determine the game from several points of view, starting with the political factions and including the appearance of the world as well as the fact that a series of quests is even more or less closely related to events in this frame narrative.

The quest "The Missing" analyzed above refers to the narrative of the Frogloks. Only some years after *Everquest I* was on the market the Frogloks became a race that could be chosen by the players and they obtained their own city, Gukta. This was motivated by a frame story that narrated the eviction of the trolls who had been the original—of course "evil"—inhabitants of the city. Between the two games a great battle of the evil forces took place that among others also conquered Gukta. But some Frogloks managed to flee, even though they lived in slavery for a long time and only lately a small community of free Frogloks was built. But the player does not receive this information at the beginning of this quest sequence but only after he has won the trust of the Quest Giver and when he has reached the secret subterranean city of the Frogloks. There he has to complete several additional quests in order to gain the trust of the Chief. One of these quests is finding the answers to a series of questions concerning a history lesson.[11]

The gameplay for the completion of quests mainly consists in moving the character through very dangerous environment, in finding objects and in killing monsters. The latter can often be very difficult since it necessitates a group (6 persons in all) or even a raid (totaling 24 persons).

How important then is the embedding of the sequence of events into a meaningful structure of the narrative? We can complete most of the quests without paying attention to this narrative and according to my observations this is the manner in which most players are playing. The reason probably is that the quests do not change the world of the game. The only change that can be achieved by the player is receiving his reward,[12] which is his main interest anyway. There are some signs that quests can change the world of the game for a certain time or even lastingly. *Everquest I* for example contains a zone in which a continuous war with battles between three parties is taking place and it is part of the assignments of the player to help one of the parties to a temporary victory. But it is clear that in this respect the design of the game conflicts with the story. From the point of view of design every player of a MMORPG is supposed to have the same possibilities, i.e. quests, as all the others. A story, however, put very simply, narrates—if it is worth telling at all—a fundamental alteration of the world.

III.

Black Mirror: The Dark Side of the Soul is a horror-adventure game that is not afraid of some strong splatter-effects. The player is leading the young nobleman Samuel Gordon who for the first time in many years returns to his ancestral castle. Samuel soon suspects that the alleged suicide of his grandfather was a murder. It seems to be somehow connected to a curse that William had investigated because it has haunted the family for many years: Every 200 years one of the family members goes insane. William had found out that this curse somehow is connected to five peculiar keys that are handed down in the family. The greatest part of the activities of the player consists in getting back these five keys, mostly from the graves of the respective family members. During Samuel's investigations a series of further murders occur which he is also trying to clear up. An added parallel plot concerns his uncle Robert who later is also murdered. He is the classical *mad scientist* who is using his position as the director of an insane asylum for lethal experiments on human beings. In the end Samuel finds out that he himself is the murderer. He succeeds in breaking the curse and commits suicide in order to atone for his sins.[13]

The way in which this game is played as well as its technique makes it into a classical adventure game. The player receives a fixed 2D-view of each setting and can only change from one view to the next. The mouse helps to explore the image and changes its color for all active objects. The player can investigate all these objects and can take over some of them into his inventory or use them. Objects in the inventory often have to be combined with other objects.[14]

The section analyzed in the following is part of the narrative sequence in which Samuel receives the key from James, his unhappy uncle, whose mild insanity was reason enough for his brother Robert to put him into an insane asylum. As the player knows at this point in the game, James has hidden the key in the ancestral castle, in all probability in the intricate sewage system that could not be entered until now.

The first image shows a turning wheel on a mechanism on its right hand side (fig. 2).[15] The mechanism cannot be operated and at this point the player would not even know why he should operate it. On the other hand, half-broken mechanisms that have to be repaired belong to the conventions of the genre. The player then can rightfully assume that this mechanism will significantly contribute in reaching his goal to collect all five keys.

One of the active objects in this image is the gear wheel on the floor. If you click on it, Samuel picks it up. The description of the gear wheel in the inventory: "Where could this fit?" indicates that the player has to fit it into something. The great similarity of the picked-up gear wheel with the one that is part of the mechanism provides a further clue to the player where he has to install it. After the installation, however, the mechanism still is not moveable. Another empty pin seems to imply that the player has to look for the third gear wheel in order to finally be able to get the mechanism going.

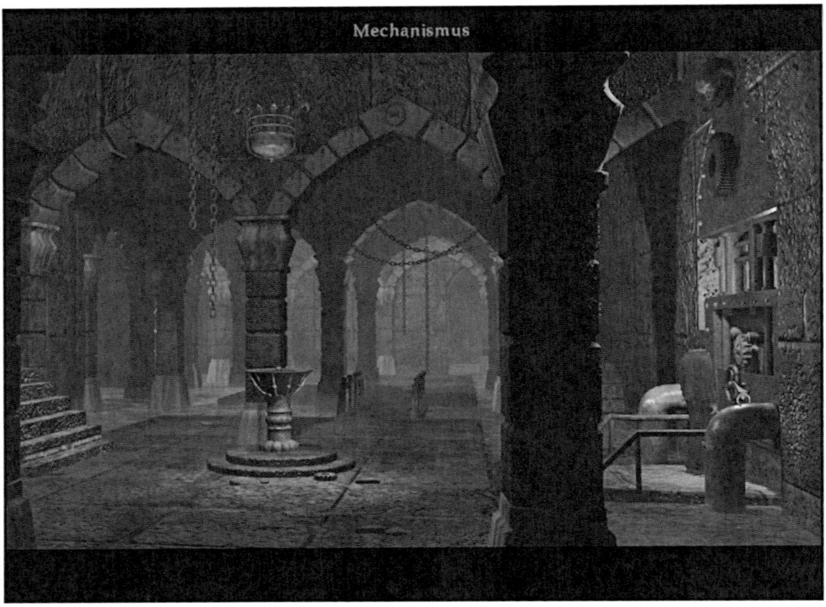

Fig. 2. Strange mechanism in *Black Mirror* as the next problem for the player.

The player then finds what he is looking for on the second image of the sewage system, but he cannot reach it. This opens up another task that further delays the main task. In order to reach the gear wheel he first has to change the level of water; for this he needs a lever. And in order to be able to use the rod he found as a lever he first has to grind it down, etc.

One of the characteristics of Adventure Games is to give the player only vague hints what has to be done next. However, the goal of his activities is quite clear.[16] It is only the way to the goal which is unclear. To find out which steps to take is one of the basic repertoires of the gameplay of Adventure Games. Part of these is the careful searching of the image for active objects and the combination of objects in the inventory with objects in the image as well as dialogues with characters and the solution of independent puzzles. In the search for solutions, however, several aspects support the player. It is especially his knowledge about the conventions of the genre and his knowledge of everyday life that limit the free permutation of all activity-options. The similarity of the gear wheels with the free pin for example works in this manner so that it does not make much sense to use the map or a coin on the gear wheel swimming in the water. This, however, is a scaleable dimension to which reviews of the game often refer with phrases like: "The riddles are logical." It is of prime importance for the player to know that there is a solution at all and that his hunt will be definitely successful at some point.

In many games not all objects are functional but exist only in order to enrich the world of the game; to a certain extent then they belong to the category that Roland Barthes has called reality effects enlarging the seeming options of activity for the player. The designers of *Black Mirror* therefore have chosen a middle way. Many of the non-functional objects cannot be activated any more after having been clicked on once or twice so that they are no longer activity-options.

The activities of the player create a sequence of events of which some are relevant for the game in order to allow for further steps. Therefore, as in the case of MMORPGs, we have to differentiate between the narrative sequence of the player and that on which the game is dependent with a minimal amount of steps for the solution. The one is accessible by means of narratives of players like: ". . . and then I searched for a long time until I found out that. . . ." The other by studying the code or the suggestions for solutions, the so-called *walkthroughs.*

The activities of the player and the events in the world of the game are more closely related than in a MMORPG. As I have already mentioned: Even though the gameplay mostly consists of the combination of objects, the type of combination is made meaningful by referring the possible combinations sensibly to each other since its semantizations like "tie the hook to the string," or "use the fishing rod to get the gear wheel out of the water" contribute to limiting the activity-options for the player.

IV.

If we look at these admittedly still very sketchy analyses of narrative sequences in MMORPGs and Adventures it becomes clear that we have to distinguish between three types of event-sequences, at least for these genres:

1. The sequence of activities of the player
2. The sequence of events as it is determined by the mechanics of the game, i.e. in quests, but also in the type of setting and the arrangement of objects in ego-shooters or the distribution of mobs with certain levels in different zones.
3. This sequence of events (as mentioned in point 2) is understood as a plot, that is as a sequence of chronologically ordered and causally linked events.[17]

After what has been said so far, the first probably has been described clearly enough. In point 2 I am referring to sequences of events that are defined in the mechanics of the game. Depending on the genre and type of action this can mean clearly defined events or also rather as possibilities. A quest for example describes a clearly defined sequence of events. Zones can be arranged as proposals to players bringing them to areas with mobs appropriate for their own level. The players can accept these propositions or decline them, thereby experiencing another sequence of zones. Thereby greater sequences of events emerge because there are players of level 10-20 in zone A and some of level 20-30 in zone B etc.

The difference between my points 2 and 3 will best become obvious from an example. If the player in the quest-sequence in *Everquest II* that I analyzed has won the trust of the quest giver he can travel into the hidden city. From the perspective of the mechanics of the game two flags have been set in order to save the successful completion of two quests. What from the point of view of the plot appears as a causally connected sequence of events from the point of view of the mechanics of the game materializes as the setting of binary flags. Why is this perspective of the mechanics of the game important at all? Because players know about it and behave accordingly. Indeed, the relation of points 2 and 3 is a problem for each computer game and we can assume that the fact that many observers still claim an infant state of computer games has to do with the slow progress of their development.

The more rigorous the sequence of events is organized in the sense of the second point, the fewer event-sequences can be added that are considered meaningful. The more complex the worlds of the game become the more independent point 1 will become of point 2. Certainly it is no coincidence that especially in MMORPGs the narratives of players are so numerous.

Apart from these three event-sequences there are also those that are narrative in a traditional sense: cut scenes informing the player about changes in the game world or texts (for example as narratives of characters in the world of the game that describe past events) or even in form of narrative paratexts. As mentioned above, very often these forms have been called "narrative" and have been seen in contradistinction to the "interaction" of the game. If we treat these narratives as unimportant because for the gameplay they seem to be insignificant additions then we have constructed a strong opposition, which, however, is intellectually not very satisfying. I also believe that narration in this traditional sense is not a central element of computer games but I do not feel that it is opposed to the organization of event-sequences that are central for computer games or at least for the two genres analyzed here.

But does it then make sense to treat all these different forms of event-sequences in the same way? To my mind, yes: All of them, point 1 excepted, result in the player's awareness of an intelligence that invents the events and their sequence and presents them to him, the main difference between the event-sequences lying in the way in which they are presented. In this perspective the comparison to simulation that sometimes is applied in order to replace the comparison to narration[18] does not seem very productive since it is just this organizing intelligence as counterpart that does not find a place here, even though it is securely established in the world of computer games and its cult of author-designers or in the complaints about bad game-designers. The assumption that the world of the game is an artifact that is intentionally related to the situation of the player is a central element for the inferences of the player to find out what he now can or should do. It is just this minimum assumption that is the foundation for these inferences that lead to a move after having evaluated the possibilities of action.

Let us summarize: In computer games event-sequences are organized and actualized by players. The sequences can be organized more or less rigidly according to the degrees of liberty of the game. From the perspective of the player, at least two important event-sequences will emerge: On the one hand the one that is played by himself, and then the one that corresponds to the mechanics of the game. And thirdly the meaningful organization of event-sequences in narratives comes into play. It seems to me that the narratology of computer games has to be adapted to this multilayered "nature" of the medium. If we want to talk of a narration in computer games at all—since regarding the prototype of narrativity it seems to be a borderline phenomenon—it is mainly the organization of event-sequences that we have to be concerned with and the different aspects of these sequences that only in an interactive medium can drift apart to such an extent.

This does not mean that something important has been said about all computer games. But maybe questions about the nature and especially that of *all* computer games are not very productive anyway.

Appendix

I. Everquest II

The following written record shows the dialogues and activities of a quest which is the beginning of a longer sequence of quests. The reward of this sequence is opening up the frogloks, a froglike race, so that the player can choose a froglok as PC himself. Played on February 2, 2006.[19]

The first three quests could be entitled: "Winning the trust of the Frogloks." Here one can win the trust of the Frogloks to the extent that one can travel to their village. There, one has to win the trust of their leader with the help of a sequence of further quests.

1st Quest: The Missing

You say, "Hail, Frup Groaak"

Frup Groaak says to you, "Any of them at all? We were wandering through the vast green lands beyond these walls, when we were set upon by brigands!"

You say to Frup Groaak, "Wait a minute, I don't think I've seen you around here before."

Frup Groaak says to you, "Indeed, I should think not. I was very surprised to see other of our kind around here. I and my brothers managed to make our way here from far to the south."

You say to Frup Groaak, "Far to the south?"

Frup Groaak says to you, "In the depths of the Feerrott. I cannot explain more now - trust is something which must be won, and our home cannot be revealed to those who might wish us harm."

You say to Frup Groaak, "I've no wish to harm you."

Frup Groaak says to you, "Then perhaps you'll wish to help."

You say to Frup Groaak, "I'd agree to that."

Frup Groaak says to you, "The other I came to these shores with, Giidib, set out from the beach in the opposite direction as I."

You say to Frup Groaak, "Which beach?"

Frup Groaak says to you, "I apologize, I do not know the names of places here. T'was quite a ways south east of here, however. There was a great forest situated on the banks above."

You say to Frup Groaak, "I'm sure I can find that."

Frup Groaak says to you, "I would like to see Giidib join me here in Qeynos while I plead our case to your illustrious Queen."

You say to Frup Groaak, "I will set about finding him."

Description of the quest (each player has a quest-book with all open quests with their description):

> "Anewfroglokhaswashedupontheshoresof Qeynos. FrupGroaakbidsme seek out [!] a lost compatriot of his, Giidib. Very curiously, no one seems to have heard anything about frogloks outside of Qeynos since the Age of Cataclysms."

In the zone Antonica at the beach we find the tracks of Giidib as well as an indicator for a direction. If we follow these tracks we come to a rock that is only active for those players who are on this level of the quest. Option for activity: "Search around the rock." Result: "Quest journal update! The froglok tracks continue back to the east." After several such updates we finally find Giidib in a pond and get the task to lead him to the next watchtower. Just before arrival Giidib dies without the player having any possibility to prevent that. All we can do is to let Frup Groaak know.

You say, "Hail, Frup Groaak"
Frup Groaak says to you, "I sense from the sadness that surrounds you that something has gone wrong. . . ."
You say to Frup Groaak, "I don't know how to tell you this. . . ."
Frup Groaak says to you, ". . . but he has gone to the embrace of Marr's mighty arms. This is plainly written in your expression."
You say to Frup Groaak, "He was ambushed before I found him."
Frup Groaak says to you, "Was he able to fend off his attackers at least?"
You say to Frup Groaak, "Yes. It was the poison that did him in."
Frup Groaak says to you, "Poison? This is an assassin's tool. Why would he be hunted in such a fashion?"
You say to Frup Groaak, "I am unsure."
You gain experience!
Frup Groaak says to you, "Allow me to think for a time, and I will commune with Marr. Perhaps he will grant me guidance."
You say to Frup Groaak, "Very well, I shall seek you out later."

From the history book of the Frogloks [The text is contained in a history book of the Frogloks that the player receives and about which he has to be able to answer questions during a quest]:

We are the children of Mithaniel Marr, god of valor. He has given unto us a greater share of honor than was given to any other creature. We would rather die with honor than extend our lives through deception of any kind.
This can be challenging for others—the concept that laws are not meant for the breaking. While we try to be understanding that not everyone shares our belief, it is still distressing to see rules bent and broken. The laws of the land are not simply suggestions, as some seem to think.

Such is the way of the troll. After many attempts to peacefully coexist with them in the Innothule Swamp, their utter disregard for protocol finally provoked us to battle. The blessing of Mithaniel Marr was with us as we drove them from their filthy city of Grobb. In His honor, we renamed it Gukta, Outpost of Marr.

Some may wonder why we contented ourselves with a simple home like Gukta, when our obvious intelligence indicates we might prefer plush surroundings. That is not our way, you see. It seems those whose skin is laced with hair need warmth and softness in their surroundings. We are comforted in other ways.

Others have said that we frogloks are too proud and that we deserve the challenges that have been our lot through the Ages. We do not deny the pride we feel in our heritage. We were created by Mithaniel Marr, after all. Through his grace, we have overcome many adversities that would have brought others to the brink of despair.

When the ogres of Rallos Zek began their reign of terror, they tried to start with Gukta. They thought they knew all the deepest reaches of our territory, thinking that we had not changed or developed it after wresting the city from the trolls. The ogres fought without honor, to their own detriment.

We are not taken unawares with ease. The foul ogres thought to destroy us, but we were guided by the Hand of Marr. Our eggs were carried deep into the tunnels, followed by elite forces to guard them. Deep beneath the marshes we went. We sent emissaries to Freeport to advise the Overlord of the rise of the Rallosians.

The Overlord brushed our concerns aside and so, with barely three days' rest, our emissaries headed to Qeynos to warn the Bayles. Though sympathetic, they chose only to increase their watches rather than advancing to attack before the Rallosians expanded their reach. How we wished they had listened! The Rallosian Army grew in strength and power until the cities of Freeport and Qeynos were besieged.

The blessings of Mithaniel Marr still protected the faithful. The honorless ogres broke faith with their own allies, defiling the Temple of Cazic-Thule. And so, in the darkest hours when it appeared that the Rallosians would conquer all, the Green Mist covered the battlefields, slaying our enemies. However, not all our enemies were destroyed.

Though many of our adversaries perished in the time of the Green Mist, some escaped by continuing to pursue us. Deeper into the marshes we fled, and still they followed. It was then that we felt the rage of the gods.

> By failing to stand our ground and hold back the tide of the enemy, we disappointed the gods who had held us high in their trust and esteem. Punishment was both swift and severe.
> Though we had escaped the Rallosians, we could not escape the torments to follow. The Rending kept those of us who had scattered apart. Separated from any other honorable race, we drifted into oblivion.
> And yet, oblivion would have been more kind than that which we endured. For those of us who found our way to the Feerrott were taken prisoner by the Tae'Ew, who fed us to their god.
> Generation upon generation of froglok has been subjugated by the Alliz Tae'Ew. They demanded service and, weakened by the calamities that had befallen us, we were in no position to rebel. Until now.
> In secret, several of our strongest broke free and are repeating the journey made so long ago when we first sent word to Freeport and Qeynos that the Rallosians were on the move. Surely we have atoned for our offenses against the gods and will be saved.

II. Black Mirror

The following written record somewhat simplifies matters since the search for active points in the image is one of the main tasks in an Adventure.

If the player clicks on an active object for the first time, he usually receives a lengthy description. Some objects cannot be clicked on any more after the first or second time. Upon further clicking on the others the player only gets a short statement. In the following the number is marked directly after the description of the object.

The exact task is unknown to the player when he enters the sewage system.

Upon entering: "James did not lie. The secret subterranean system of ducts exists. The air is very stuffy here."

The room has two exits. "Upwards" into the normal basement and "Further down."

From left to right:

Sewer (1)
First: "This intricate sewage system was built by my ancestors."

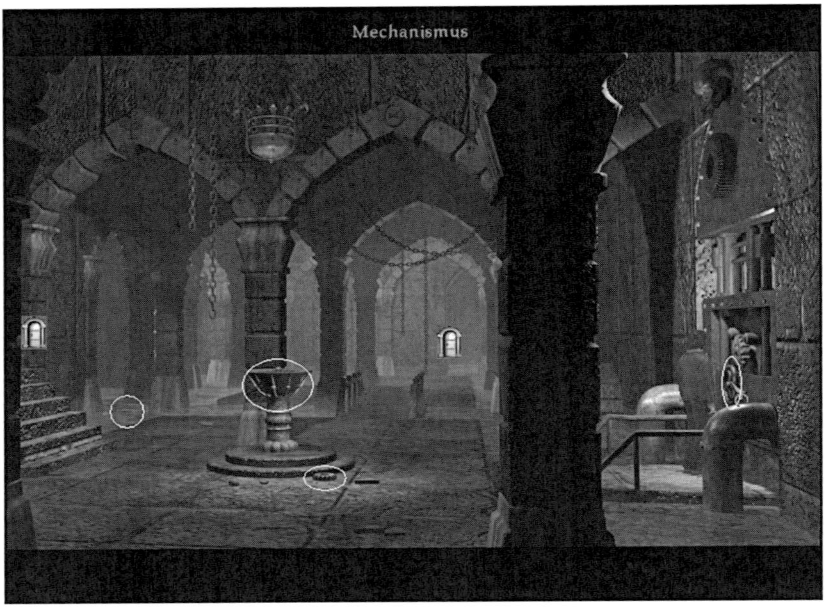

Fig. 3. Strange mechanism in *Black Mirror*. Interactive objects are marked.

Well (3)
First: "What could a well in the basement serve? Maybe one could get water from down below, when the castle was besieged. The water here could not be drunk without problems. It is too brackish."
Description: 1) "There is not a drop of water in it." 2) "It is empty."

Gear wheel (1)
First: Gear wheel → inventory.
Description (in the inventory): "Where could this fit?"

Mechanism (∞)
First: "This mechanism probably regulates the pumps in the basement. It certainly can't have been standing here since antiquity."
Description: "It seems to be jammed."

Activities:

"Combine gear wheel (inventory) with mechanism."

Result:
The description of the mechanism has not changed.
2nd Image:
Upon entering:
"The air in here smells horribly. It smells stale. There hasn't been any fresh air here for centuries."

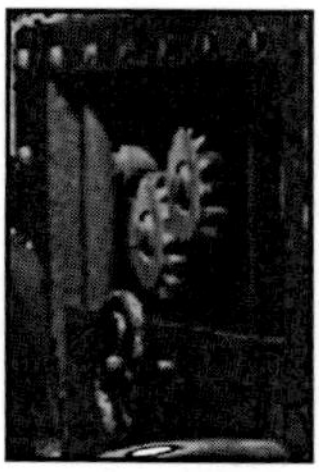

Fig. 4. The strange mechanism; obviously one wheel is missing.

Banister (1)
First: "It's rusty from the humidity. One of the bars is loose." Bar → inventory.

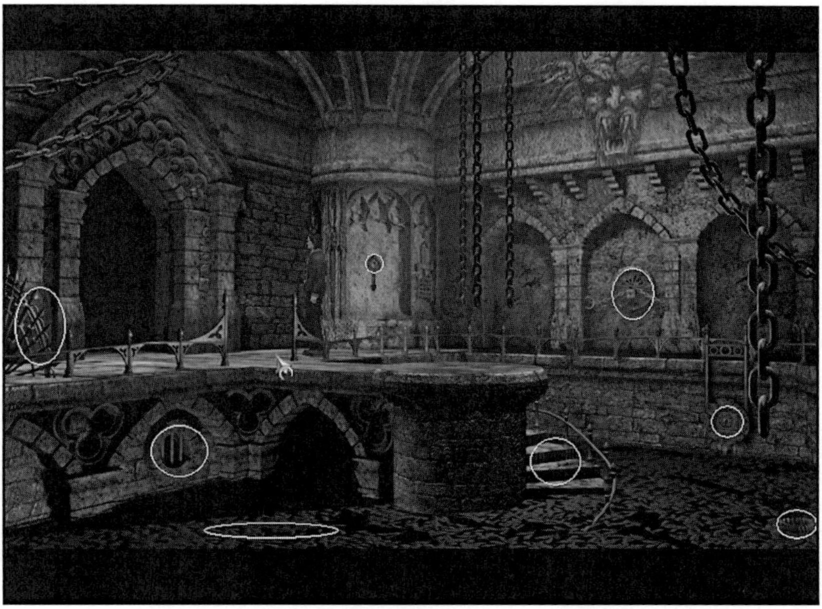

Fig. 5. "How do I get rid of the water?"

Sewage shaft (1)
First: "Probably that's where the water runs out when it overflows.

Dirty-cloudy water (1)
First: "The water must have been sitting here for years. It is full of algae and dirt."

Opening a (∞)
First: "Some kind of opening. Maybe something is missing here."

Description: "Some kind of opening. Maybe something is missing here."
Stairway (1)
First: "The stairs can't be used. They are completely under water."
Wheel (∞)
First: "I can't turn it at all. It is blocked by chains that are secured with a padlock."
Description: 1) "It can't be turned." 2) "I can't turn it at all."
Opening b (1)
First: "I think this is where the water flows. But this probably hasn't happened for a long time."
Gear wheel: (∞)
First: "Part of a machine. There are so many algae in the water it didn't sink. How can I get it out of there?"
Description: 1) "From here I can't reach it." 2) "It is too far away."

Activities

1. Try to combine the bar (in the inventory) with the opening a.
   "I will try to put the bar in there."—"It doesn't work. The bar is too fat."
2. Go into the basement above and grind the bar down to a point
3. Combine the bar with the opening a.
   "With this it should be better. I will try using the bar as a lever." (Hatch by opening b has opened.) "The hatch has opened, but the water isn't running out."
4. Combine the acid (from the inventory) with the wheel.
   "The acid has dissolved the padlock as if it were made of paper."
5. Turn the wheel.
   Water comes through opening b. Short blackout. New image "The water has risen to the edge of the drain pipe."

6. Combine hook (in inventory) with rope (in inventory). Combine rope with pointed rod. "A primitive fishing rod. It will have to do."
7. Combine fishing rod with gear wheel, Samuel catches the wheel from the water. Gear wheel → inventory.
8. Combine the gear wheel with the mechanism from the first image. Now the wheel can be turned.
   "It has worked. The mechanism is functioning."
   The water is flowing out. Upon entering the back part: "All of the water has flown out to the subterranean sewage system. The road is clear." At the end of the stairway now lies a small case with the key.

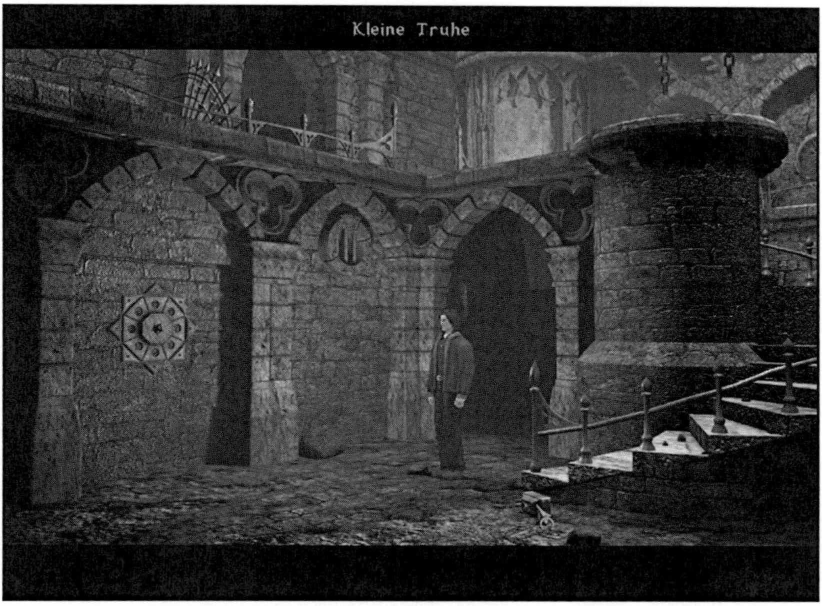

Fig. 6. Problem solved.

Translated by Brigitte Pichon and Dorian Rudnytsky

Notes

1. The debate was far-reaching and fierce. Cf. its summary in Frasca, "Ludologists." Frasca's seeming peace-offer primarily defends ludologists against a series of reproaches and a number of suggestions for additional literature are given as well. On the side of narratologists I would specifically mention the contributions by Marie-Laure Ryan and on the side of the ludologists the ones by Markku Eskelinen. A good discussion of the different approaches can be found in Hartmann (57ff.). An intellectual low point of the whole debate can be found in an essay by Mark Wallace whose contribution is full of anti-intellectual resentment. Cf. Mark Wallace: "The Play's the Thing." On the other hand a lot of research is taking place which is not interested in these kinds of questions, e.g. the histories of computer games, cf. Kent, Lischka or Mertens and Meißner.
2. Aarseth, Eskelinen, Frasca, and Juul probably are the most well-known ludologists; cf. the literature in Frasca, "Ludologists."
3. A similar argument has been made by authors who are not concerned with game studies at all, like the journalist Steven Poole in his book *Trigger Happy* claiming that stories are only important as back-ground or as the non interactive parts of games. On the other hand, however, he assumes that these parts of games will become more important in the future; cf. Poole 103ff. or Walter. Wolf tries to reconcile his notion of narrative with the interactivity of computer games but states at the end that for many players the storyless game experience will become the prototypical one; cf. Wolf, chapter 5.
4. Some interesting observations relating to these differences can now be found in Juul.
5. In my opinion it is an interesting consequence of this approach that a general narratology is quite impossible; rather, there exist only narratologies relating to particular media.
6. The notion of *gameplay* is notoriously fuzzy. As I understand it in the following it is the type of choices and options or activities that a player has in order to reach his goal. This is often not differentiated from the experience of the game but is seen as a central unit for the design of games. Cf. e.g. the statements of game-designers in "Spielspiel, Teil 2."
7. E.g. the so-called Peacock or the Claymore quest lines in *Everquest II*.
8. Cf. the record in the appendix.
9. In *Everquest II,* like in most computer games today also one's own contributions to the dialogues are given by the game. However, at each move

in the game the player has the possibility to continue the dialogue or to break it off.

10. At the beginning only certain zones could be entered if one had completed an entry-quest. This was abolished for all zones that were important for all players and it was retained only for special quest- and raid-zones.
11. For the text of this lesson cf. the appendix.
12. The reward always consists in experience points, but in addition can also be money, an improvement of the faction or an object.
13. Of course we could ask why he takes the trouble to break the family curse since with his death as the last family-member the family is extinguished.
14. A very detailed analysis of the functioning of adventure games can be found in Walter.
15. The basis of the following analysis is the detailed summary in the appendix.
16. Therefore, in some Adventures there are quest-diaries that inform the player about the current goals, Ankh for example.
17. The conceptual difference between the game experience of the player on the one hand and the game sequence as conceived by the game designer on the other is well established in games studies (cf. DeMarle 77f.). I am trying to show here that the second aspect can be viewed under two perspectives.
18. Cf. the above-mentioned contributions by Frasca.
19. Since MMORPGs constantly change, it is necessary to note the date of playing.

Works Cited

Black Mirror: The Dark Side of Soul. Future Games, 2003.

DeMarle, Mary. "Nonlinear Game Narrative." *Game Writing: Narrative Skills for Videogamers.* Ed. Chris Bateman. Boston: Charles River Media, 2006. 71-84.

Eskelinen, Markku, and Ragnhild Tronstad. "Video Games and Configurative Performances." *The Video Game Theory Reader.* Ed. Mark J. Wolf and Bernhard Perron. New York: Routledge, 2003. 195-220.

Everquest II. Sony Online Entertainment, 2005.

Frasca, Gonzalo. “Ludologists Love Stories, too: Notes from a Debate that Never Took Place.” *Level Up. Digital Games Research Conference 2003 Proceedings.* Ed. Marinka Copier and Joost Raessens. Utrecht: Universiteit Utrecht UP, 2003. 92-99. 7 July 2006 <http://ludology.org/articles/Frasca_LevelUp2003.pdf>.

———. “Ludology Meets Narratology: Similitude and differences between (video)games and narrative.” 7 July 2006 <http://www.ludology.org/articles/ludology.htm>.

Hartmann, Bernd. *Literatur, Film und das Computerspiel.* Münster: Lit Verlag, 2004.

Jannidis, Fotis. “Narratology and the Narrative.” *What is Narratology.* Ed. Tom Kindt and Jan C. Meister. Berlin: de Gruyter, 2003. 35-54.

Juul, Jesper. *Half-Real: Video Games between Real Rules and Fictional Worlds.* Cambridge, MA: MIT Press, 2005

Kent, Steven L. *The Ultimate History of the Video Games.* Roseville: Prima Publishing, 2001.

Lischka, Konrad. *Spielplatz Computer: Kultur, Geschichte und Ästhetik des Computerspiels.* Heidelberg: Heise, 2002.

Mertens, Mathias, and Tobias Meißner. *Wir waren Space Invaders: Geschichten vom Computerspielen.* Frankfurt am Main: Eichborn, 2002.

Poole, Steven. *Trigger Happy: The Inner Life of Videogames.* London: Fourth Estate, 2000.

Ryan, Marie-Laure. “Beyond Myth and Metaphor: The Case of Narrative in Digital Media.” *Game Studies* 1.1 (2001). 7 July 2006 <http://gamestudies.org/0101/ryan/>.

Salen, Katie, and Eric Zimmerman. *Rules of Play: Game Design Fundementals.* Cambridge, MA: MIT Press, 2004.

“Spielspiel, Teil 2.” *Games Entertainment Education,* April 2006: 56-60. 7 July 2006 <http://www.geemag.de/relaunch_storyseite.php?story=72&ausgabe=20>.

"Tome of Destiny." *Everquest II Online Gaming Network*. 7 July 2006 <http://eq2.ogaming.com/data/1862~TomeofDestiny.php>.

Wallace, Mark. "The Play's the Thing." *The Escapist* 36 (2006). 7 July 2006 <http://www.escapistmagazine.com/issue/36/13>.

Walter, Klaus. *Grenzen spielerischen Erzählens: Spiel- und Erzählstrukturen in graphischen Adventure Games*. Siegen: Universi, 2002.

Wolf, Mark J.P., ed. *The Medium of the Video Game*. Austin, TX: U of Texas P, 2002.

Writing and Programming

Jean-Pierre Balpe

Principles and Processes of Generative Literature

Questions to Literature[1]

Abstract:

Generative literature, defined as the production of continuously changing literary texts by means of a specific dictionary, some set of rules and the use of algorithms, is a very specific form of digital literature which is completely changing most of the concepts of classical literature. Texts being produced by a computer and not written by an author require indeed a very special way of engrammation and, in consequence, also point to a specific way of reading, particularly concerning all the aspects of the literary time. In my paper, I will try to present some of the characteristics of generative texts and their consequences for the conception of literature itself.

I call "engrammation" the adaptation of choices of expression to the technical constraints of the medium used for its mediatization. For instance, a book needs a fixed writing, and the mediatization by means of a screen needs other modalities of presentation.

I call "generative literature" a literature where the texts are produced through a computer by means of a set of formal rules, the use of any kind of algorithm, specific dictionaries and eventually knowledge representations. That means a literature of which the author does not write the final texts but which only works at the level of the high rank components such as conceptual models, knowledge rules, dictionary entries and rhetoric definitions. A text without an author generally seems to be out of question. Such a designation seems to describe an impossible literature because, despite the fact that we generally assume a very strong link between a text and its author, in this case the author is separated from the text. In generative literature, there certainly also is an author but one who has not really written the text which is being presented to a reader; his function is not the one we usually assign to an author. The difference is: this author is something like a meta-author trying to define what literature is for him and how his literary conception can be formally described. The tools of engrammation he uses are totally different. But at the end of the process there are also texts.

For example:

Text 1:
Poursuite du bonheur. . . . Peu de choses laissent des traces: pourtant il s'agit bien quelquefois de vivre; ses souvenirs brûlent Jeanne-la-voleuse comme des flammes—le soleil cogne: les dieux passent par rafales; les carrefours sont des arbres de vie, la lumière dévaste les villas. . . . Les spirales des avenues trouent Montreuil: qu'a fait Jeanne du temps qui lui a été donné. . . .

Text 2:
Pour la première fois de sa vie Léa se sent vivre vraiment; mais le temps dont dispose Léa est limité—la vie lui fait mal à petits bruits, à petits gorgées, par les interstices! Des haut-parleurs diffusent partout le nom du dieu du jour ! Les individus suivent leurs bannières. . . . Un vieillard fatigué, voûté, traverse l'avenue: voici ce que lui dit Montreuil: 'la pénombre seule a le pouvoir d'ouvrir au monde le cœur d'un homme'. . . . Tout ce que Léa voit de sombre dans l'avenir est plus rassurant que son présent: ses souvenirs la brûlent comme des flammes. . . .

These two texts were produced by the generator used in my Internet novel *Fictions (fiction)* (*Fictions* 2004) the title of which has a double meaning: It both connotes that all fiction is a fiction by itself and that a fiction can be built by an infinite number of fictions, something like a "mise en abyme" of the fiction itself.[2] Indeed none of these texts will ever be presented to the reader for a second time and no reader will ever have the same set of texts. Each text seems to be independent from the others. Such a situation seems to be rather natural for poetry. For example:

Poem 1:
Le temps se remplissait de sable l'océan
utilisait le ciel batterie de l'océan une crique en contrebas
du chemin je passais par la côte le paysage faisait
son philosophe quelque chose brillait au milieu
du sentier il y avait un avant et un après
le jour venait de commencer le ciel était avec toi
la campagne jouait du violon le soleil
pesait sur mes épaules le même soleil éclairait
l'océan et le ciel la campagne était élégiaque tu
te baignais de ciel la campagne
disait son évidence la mer se révulsait
l'océan paraissait tout en surface

le paysage se répétait
la mer broutait les herbes sauvages de la côte

Poem 2:
Nous nous sommes cherchés
sur les côtes l'océan tendait son
bourdonnement l'océan utilisait
le ciel tout était banalement
charmant un calme s'installait du côté de l'océan je ne voyais plus
que toi tu étais du côté
de la terre le ciel perdait ses dimensions les nuages
flottaient à l'envers tu faisais partie du décor
le paysage s'oubliait dans la mer quelque chose brillait
au milieu du sentier la campagne refusait de se
laisser aller au drame l'horizon s'éloignait intarissablement
je te disais: "une chose qui
semble très claire peut ne pas l'être du tout" sentier
et vent c'était ça la mer te faisait face
la mer se prenait pour le ciel le ciel était avec toi
l'océan était impassible[3]

For poetry indeed, each text is something like an island isolated from the other ones even when they are published in the same book of poetry. A poem is a text by itself; each poem is entirely closed and supported by itself. But none of these texts will ever be read by any other reader unless, like here, they are printed as examples by way of exception. However, generative texts usually do not require to be printed. And even in that case, the reception of generative poetry is really different from the reception of non-generative poetry.

But, in this paper, I will speak about generative narratives, not about generative poetry, the concepts and approaches of which are undeniably rather different.

Everybody knows Gérard Genette's classical definition of narrative: A narrative is a text built on a diegetic axis, i.e. diegesis (164-8, 246-50). That implies that any text has a beginning and an end. Whatever games would be played on the diegetic axis, it undeniably is the basic structure of the narrative. All the episodes of the narrative are organized along that structure. Such a situation is strongly underlined by the kind of medium that has usually been used to materialize the narrative: the book. A book always has a first page and a last one and all the reading is done by taking into account this constraint. We all know that. This does not imply that a reader cannot read backwards, but we also know that reading backwards is an exception from the rule and rather is an approach of researchers than one of an ordinary reader. Hence it is the diegetic axis which structures all the conceptions of novels or short stories and a reading cannot stop at any point of that axis; more precisely, if a reading does stop at some point it is to let the reader dream or think about what he has just read; it is like a halt during the reading process.

The generative conception of narrative completely transforms this situation. In a generative novel indeed at every point of the narrative we experience an equivalence principle that is dependent on generation. At these points, a text is only a temporary specimen of an infinite family of virtual texts. In concrete terms, this means that any point of the generative axis is the theoretical point of an infinity of texts and I propose to call that situation "alepsis" because at that point, the arrow of the diegetic axis is broken up for some time. In general, the diegesis includes all the events that allegedly have occurred, including actions and spaces not explicitly described; most of them are however offered or suggested in the reading of the book. In generative fiction, on the contrary, most of the texts will never be offered in the reading: Most of the diegesis markers are latent ones. The two texts quoted above, for example, are not topologically situated on the diegetic axis one before the other but are produced at the same point of that axis. This signifies that they are—from a narrative point of view—completely equivalent: One may be read in place of the other. Moreover, there is—at this point—an innumerable number of virtual texts each of which has a function within the narrative process: They can be read or not (depending on the possibility of generation) but if they are being read, they offer a new view onto the narrative; if they are not being read, then the reader does not miss them. One text can therefore be read as an alternative for all the others, while it can also be read before or after another. There is, at that point, no obligation.

I am not talking of prolepsis (anticipation) or analepsis ('flashback'), which are the two main rhetorical tools used by a writer to build his story. Because prolepsis and analepsis can only be understood from the perspective of linear reading, there is neither prolepsis nor analepsis in the reading; the reading is linear and always backwards: Prolepsis and analepsis are only writing games, engrammation procedures, playing with the arrow of time. If so, then the classical literary theory adopts the writer's point of view and not the reader's; it supposes that the diegetic axis can be cut into parts, which can be easily displaced. Hence each story appears to be a combination of smaller stories that have a kind of independence from each other except from the diegetic point of view. Indeed, these fragments may not change in their form and once they have been given to the reader, they are located on the arrow of time forever;

they can not be displaced and thus the game is very different for the writer and for the reader: The reception of the text is fixed whereas this was not the case during its production.

The fact that in a generative narrative such a constraint was surrendered has introduced a great liberty. When a reader obtains a text, he obtains that text at a certain moment of his reading but he has no idea of what other text he could have obtained at the same moment of another reading. He cannot tell if the text is directly related to the previous or the next one (in terms of reading time). The relation between one text and another can be a relation situated on the diegetic axis but it can also be any other kind of relation. For example, that text can be another concretion of the virtualities of the same model of text, i.e. an alepsis, a text not situated in a relation of time to the previous one.

In that case—and this certainly is one of the reasons why generative texts disturb our reading habits—the reader loses all the usual markers relating to the diegetic axis and has to find or invent other kinds of references. The narrative is not totally built in advance but put together from a variety of virtualities that are—or are not—actualizing themselves in the course of reading. This reading is then fundamental and tends to substitute itself to the diegetic axis. Each new reading—actualizing the narrative in a new way, built on what I call micro-fictions—creates its own diegesis, which is not a predetermined but an undetermined diegetic axis. This really means: Any reader A needs to develop a unique hypothesis giving him an idea of the narrative that is different from that of any reader B. Maybe he could ascertain that his conception of time in that narrative were true if he could read all the virtual texts but that is completely impossible. The only indicators he may be able to use are those given by each micro-fiction: Each of them seems to be built on the basis of a local diegesis which then, for the reader, is necessarily related to a more general one. What is more important than the building of a diegetic axis by one author then, are the various images the reader always builds of a possible moving diegesis: The diegetic axis is totally virtual and can only be built by the reader's strategy which is something like a translepsis, an infinitely moving hypothesis of diegetic relations. It is what I call the "hologrammatic principle": Each text contains the entire time of all possible texts. There is no concrete diegetic axis but only a virtual one, much more related to the reading than to the writing. The narrative neither needs beginning nor end because the narration is entirely built by each reading of each text: One novel can thus be constituted by one or an infinite number of texts and no reader reads the same number of texts. There is no structure of the narrative, only an idea of a virtual one built by the reading itself.

For the author, that situation is very interesting because he is free from the necessity of mainly conceiving a narrative in relation with time and he does not have to respect a linear form of engrammation. This opens up a new field of creativity: He has to think about and to work on all the virtualities implied by the generative fiction and time, even if by tradition it is a very important element, is only one of the virtualities. A narrative is not obliged to have one beginning and one end. Each generative narrative includes an infinite number of equal alepsis and it is not obligatory that time relations are initially stated between them. In *Fictions (fiction)* for example, each text is totally independent from the others; in *Trajectoires*, the texts are quite independent apart from the fact that the interface design leads the reader to imagine a form of diegesis by using a set of symbolic numbers. In that case, the design interface is a procedure of time engrammation. It is definitely sufficient for the reader to know that he is reading a narrative because—due to his cultural concept of narrative—he tries to find diegesis markers. Hence a novel can be totally built on an infinite present as in *Fictions (fiction)*; the reader builds a diegesis from its translepsis course himself. Thus generative literature demonstrates that the writer's control of the diegesis is not necessary because diegesis is a cognitive concept of reception: We always imagine stories on a time axis and even if a narrative has neither beginning nor end, we as readers always imagine them. Generative literature opens up a creative field for a renewal of narration.

Without a doubt, the book as a product of a dated technology using fixed engrammation procedures, has become a reductive matrix that we have to reform in order to liberate the text from the linearity and the determination of its pages, to allow text to reveal itself by other means and in other contexts with other possibilities of expression. Hence literature is not dependent on the medium of the book any longer but can also use other media such as any kind of screens, cellular phones or computer networks. Thus there is no more necessity of sticking to linear forms. Today generative literature wants to affirm the vital and infinite power of "literary" communication as a dynamic diffraction of relations, in which the always-different text manifests its identities only through the infinite repetitions of its generation of the same, i.e. rather through its infinite changes than through its halts. What this process assumes is the fecundating power of language enriching itself within all the restraining particularities of any given context. It is this fecundating power of language upon which the receiving subject continually renews him- or herself.

At a first approach, the generative text indeed demolishes all the material references that are the basis of the reader's "preparation," for example the physical appearance of the book (volume, thickness, division into paragraphs and chapters, and so on). This radically new literature, putting the diegetic axis into the background, has to invent new forms of fiction, new forms of stories, new forms of narration when using other media than the book. A "generative novel" is thus forced to invent all its codes. Generative literature wants first of all to be something like a "literarization" of technology, because what it demonstrates first and foremost in its multiplicities and its variations are its potentials and its changing states. Even if that situation is not absolutely new in literary history, where the eagerness to present texts with new technical apparatus has always existed at least in the margins, the digitalization of its technology has created a significantly new situation for literature, since the immediacy of its generation and its infinitude stage the formalisms from which the texts come forth. The relation of the subject to his or her writing is thus a new relation to time, and the functional concept of authorship itself is to be entirely redefined, because what matters more than anything else is the "historical" memory of forms and their displacement. There is no end to writing and reading: A generative text can be continued infinitely. The person whom we can only call a "digital author" as long as there is a lack of a more accurate term does not need to deny tradition by means of a radical modernity but requires of it something like a new reading or, at least, an unheard one. Generative literature's only pretension is to enrich the text's potentialities. It forsakes the fiction of fiction to be only interested in the subjective production and formalization of meaning. In that sense, it only exists through infinite literary production. A generative text then rejects clotting, time's dictatorial caricature: it presents a notion of eternity. But this eternity-whim truly differs from that of "classical" literature because it does not depend on the duration of its memory but on the infinitude of its reproductions.

Out of the book displayed on screens or, by various technical processes, on any object henceforth used as a displaying surface, the text—now enabled to become time and space—completely changes its nature. It becomes picture-text, sometimes even text-universe or performing-text. This old appeal, outlined in several historical attempts, now finds its fulfillment thanks to the possibilities that computerization allows.

According to George Steiner, the comprehension of modern art requires the acceptance of the fading of any conception of "culture" as having an immanent value, necessarily linked to hierarchical societies, and its replacement with a set of "cultures," all regarded as equal (313). The computer culture, in that sense, is an absolutely new and absolutely up-to-date writing tool inventing new engrammation modalities.

Generative literature does not make any claim to the intangible and almost divine universality of pre-generative literature, in which the only active role left lies in the production of new glosses; it merely wants to be the ephemeral and temporary moment of a common literarity, revealing itself only in the instant of the creative stimulus.

More than ever before, nowadays a text is more dependent on its media contexts. Every surface may be used as medium for texts, not only sheets of paper but also all sorts of "screens" which can be a mountain, a building, a screen of a cellular phone, a person's body etc. Classical culture originates from gambling on transcendence: "Art and mind looks toward what is not yet here, at the accepted risk of being ignored by the living" (Steiner 102). Classical culture is produced only for museums, for "conservation," and thus is opposed to ephemeral consummation. The classical production of writings is aimed entirely at archives. To this way of thinking, the slightest "loss" is considered a cultural tragedy. Any destroyed manuscript is a burning library; the obliteration of any draft scribbled on a tabletop seems a disaster. Against that museum art, that library-and-dust art, generative writing is an art of consummation that refuses to look back on its tracks, which it regards as nothing more than signs headed for something else.

The real-time generative text, which exists only by its instantaneity, first of all fights against all this: "Everywhere the virus of potentiality prevails. Carrying us away toward a rapture which is also that of unresponsiveness" (Baudrillard 315). But the unresponsiveness that Baudrillard fears is, in fact, a positive one because it is unresponsive to an external hierarchy of values, to a culture of reverence. "The public is no longer the wise echo of talent, something like a referee and relay in transmission of an atypical attempt; it associates itself to the artistic elaboration inside of a set of sometimes uninhibited energies" (Steiner 106).

The text, no longer regarded as "literary," now has to annihilate all reverence because a generative text can always be substituted by another one. Hence it is not the singular display that is at the heart of generative literature but rather the movement, the series of ever-changing displays of text. The computer culture is close to spreading, to dispersion. It introduces a new relation to memory, no longer "reminding" but—because a text, read at a given moment, is nothing other than an image of another text read at another moment with which it maintains links of dependence and independence, at once diffracted and refractory—rebuilding remembrance: Active participation by the reader lies in its elaboration. Displaying on its screens the vanishing of the master, of eternity's claims, generative literature engages the reader as a culture in itself.

Like all new literary approaches, generative literature must first struggle with resistance within itself, to find the ways that are peculiar to it by rejecting—even sometimes at the cost of provocation and error, at the risk of becoming illegible —what is inside it, enclosing itself within the sticky thickness of its moments of arrest.

While strongly individualist, the contemporary spirit is also gregarious, taking pleasure in the instantaneity of shared time, where the contemporary spirit formerly was collective inside the unchangeable cultural spiritual union. Generative literature tries to be on the side of the effusive superficiality of show. It wants to reconcile the literary activity with that of play and game: To separate literature from the sphere of reverential and deadly seriousness in which the whole classical tradition locks it. Not merely about a particular text, it questions itself infinitely about the aesthetic working of the human spirit.

Notes

1. An extract of the last paragraphs of this paper has been previously published, with some important differences, in Jean-Pierre Balpe, "Toward a Diffracted Literature."
2. Cf. my other generative novel *Trajectoires* <http://trajectoires.univ-paris8.fr>.
3. Texts by the text generator *La nuit de Cerisy* created for a performance in Cerisy-la-Salle in August 2004.

Works Cited

Balpe, Jean-Pierre. *Fictions (fiction)*. 2004. 11 Apr. 2006 <http://fiction.maisonpop.com>.

———. "Toward a Diffracted Literature." *Leonardo* 37 (2004): 385-390.

———. *Trajectoires*. 2001. 11 Apr. 2006 <http://trajectoires.univ-paris8.fr>.

Baudrillard, Jean. *Cool Memories*. Paris: Éd. Galilée, 1990.

Genette, Gérard. *Figures, III*. Paris: Seuil, 1972.

Steiner, George. *Dans le château de Barbe-Bleue: Notes pour une redéfinition de la culture*. Paris: Gallimard, 1971.

Loss Pequeño Glazier

Code, Cod, Ode

Poetic Language & Programming or Pequeño Lector/Klein Leser: Es gibt (See)lachsfilet?

Abstract:

Mutation or modulation of words manifest orthographic relations between variants but also sometimes suggest more elusive relations. Much importance can be seen in the specificity of language, especially considering the sum of variations of a single word in different languages. The word itself is a solid object at the center of such a set of permutations. The meaning of a sum of such variants can be likened to an array in programming. An array object can be greater than the sum of its parts, a concept that ties into Cubism as well as to poetry where languages mix. Other array poetry suggests geometric structure; this is poetry that creates meaning from empty space as much as from its solid textual areas. This is similar to the way that architecture creates meaning from empty spaces, as seen notably in uses of the arch. The structural strength of empty space can also be seen in a number of postmodern poems, where such space is integral to their expressiveness. These poems also use array concepts to inform the poem. It is useful to look at examples of code in my own work, which uses arrays and empty space as solid material in strings. What is of use in this method is the concept of precise poetic analysis, of the relevance of position, location, and structure as crucially important in reading code as poetic material.

> Deep in the mechanism
> of the smallest zone
> the drunken waltzers
> the elephants tango . . .
>
> *Charles Stein*

> Released from inexplicable code.
>
> *Steve Benson*

Somehow I find the words "code," "cod," and "ode" to be an excellent starting place for a consideration of poetic language and programming. There is a way the three words are not only related, but suggested movement through mutation, or modulation. Of course the mind knows this relation is merely an accident, as the relation is only through orthography, not meaning, but the poet in one cannot help but try to find a more elusive relation.

The modularity and mutability of language was especially noticeable at the cafeteria of the University of Siegen on November 26, 2004, when the feature of the day, Seelachsfilet, which is pollack (a fish related to cod) was listed on a poster menu. Apparently, the pollack ran out and they decided to substitute salmon filet instead. So instead of making a new poster they simply crossed out the "See" portion of the word, converting "Seelachsfilet" to "Lachsfilet" and thus "pollack" to "salmon." This was a more meaningful mutation, also orthographic, and one that gives a better sense of movement through changes in, or loss of, letters.

The topic of poetic language and programming always raises some interesting issues. On the one hand are programmers who insist that programming is a skill and simply a less-than-compelling means to an end. On the other hand are programmers who are skilled and who have poetic intentions but who find such skill to be not quite enough on their own. Once you know how to invent procedure, write code, and create programs, what do you want to say? One programmer I know has said of code, all he sees is structure, or the architecture of the statements. Where does the connection between the code and the poetry of the coded object lie?

Specifically, as to language, how does programming "make it new?" How is it language? What specifically do we mean by "language" and how is language specific?

To look at how language is specific, I'd first like to consider a single word, say "mother," and look at how it appears in different languages.

mother	English	mamă	Rumanian
mam	Welsh	nënë	Albanian
máthair	Gaelic	meter	Greek
mère	French	mat'	Russian
madre	Spanish	motina	Lithuanian
mãe	Portuguese	mayr	Armenian
madre	Italian	mādar	Persian
mater	Latin	matar	Sanskrit
Mutter	German	ama	Basque
moeder	Dutch	mare	Catalan
móoir	Icelandic	äiti	Finnish
moder	Swedish	anya	Hungarian
matka	Polish	anne	Turkish
matka	Czech		

There are similarities, of course, due to historical reasons and because of affinities between language families. But I would like to ponder for just a moment, in languages that are so different, how a single word can appear so similar. I would suggest that a single word can have many modulations, variations across languages, without losing meaning. That is to say these words form a set or series of variants of a lexical unit. For humans, the concept of "mother" is, rather than its spelling in one language, the sum of the variations across all languages, remarkably motile and supple in meaning, mother is the sum of the differences of its fixed form.

"Mother" is a fixed object. As solid an object as Ezra Pound describes:

> AN OBJECT
>
> This thing, that hath a code and not a core,
> Hath set acquaintance where might be affections,
> And nothing now
> Disturbeth his reflections. (18)

Or an object as Joan-Elies Adell writes in "For Every Object":

> I hold them,/invisible to my eyes, and play with them unafraid,/feeling the pointed tips, the sharp edges,/like fragments of language, hiding behind life/ready to cut. (17)

Or otherwise, it is like a moment in motion frozen in time. As Pound sketches:

> IN A STATION OF THE METRO
>
> The apparition of these faces in the crowd;
> Petals on a wet, black bough. (35)

Such a language object is solid and fixed but at the same time it captures motion. It is a specific language item consisting of a number of permutations. It is and is not all of the permutations, depending from where you are looking in language. In programming terms, this would be called an array. An array is a collection of objects that share a single variable name, differentiated only by where they are located in the collection, a collection of parts whose sum is greater than the whole.

A Structure of Parts

Such a vision of the sum of the parts being greater than the whole is, coincidentally, not unlike Cubism. If we look, for example, at a work like Pablo Picasso's Cubist collage *Guitar* (1913), we see the following.

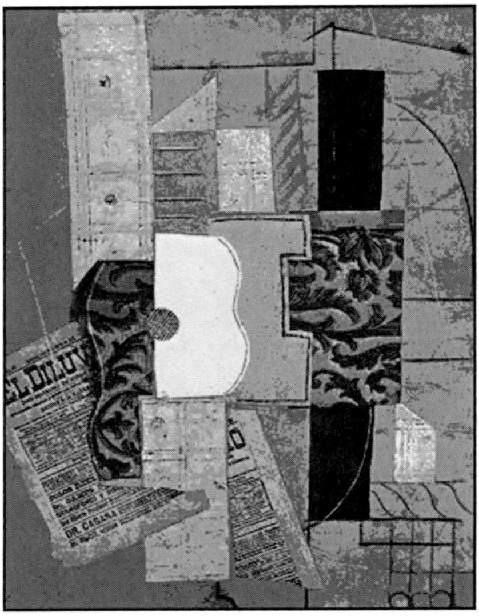

Fig. 1. Pablo Picasso. *Guitar*. Museum of Modern Art, New York.

Here, fragments of "guitar" material, frets, neck, halves of guitar bodies, and other related images, lie scattered on a mostly blue background. There is a sense to which the different parts of a guitar can be shown and it is the collection of these parts, the "guitar array," one could say, which better defines a guitar than a photo-realistic image. (This is, perhaps, because a photo realistic image must always suffer from the subjectivity, or tyranny, of perspective. It can only be viewed from one viewpoint. The array solves this by providing a multitude of viewpoints in one plane.)

At the core of these collections of variants, I would like to look at how language mixes in a given work.

For example, observe how languages mix in the following poem extract. By using more than one language, for example, the expression "Cuban sol," which is "sun" but also carries the suggestions of "soul," one sees a work, by using multiple languages, where the sum of the parts is greater than the whole.

> ON YOUR MARX. "She is tropical, a Caribbean
> breeze, Cuban sol." The radiant birds exist in cages
> in two's—parakeets, lovebirds, and sulfur-crested
> cockatoos. The corner for birds, amor. Half-shell
> fountain, miniature orange trees form walkways,
> immensely pleasing to the self sense even as there
> are sones o salsa in waiting rooms. An effusively
> tiled Sevilla. Socialist radio of the Revolución. (Glazier, *Anatman* 50)

A similar sense of the parts being greater than the whole occurs in Gertrude Stein's famous utterance:

> Rose is a rose is a rose is a rose. (395)

Here, the parts that are greater than the whole are all the same, "rose," and yet placing them as an array seems to almost geometrically increase the overall meaning of them.

As an array, this would appear as something like:

```
// This is where the array is created.

    a1 = new makeArray(4);
        a1[0] = "rose"
        a1[1] = "rose"
        a1[2] = "rose"
        a1[3] = "rose"
```

In this case, though I am referring to this utterance as an array, we cannot strictly consider the phrase an array since the elements ("rose") are all identical. This is a good opportunity for us to see how such elements can begin to take on geometric value. In other words, the instances of the word "rose" are like pillars in a colonnade where "rose" is the anchor for each pillar and the intervening explanation is empty. ("Rose" is never defined other than being "rose.")

How then is language capable of suggesting a structure? How is it geometric, structural? How does the visual produce meaning?

To answer this, I would like to look at Gomringer's "o," a poem which like the arches suggested by "a rose," creates meaning from empty space as much as from its solid textual areas.

o
bo
blow
blow blow
blow blow blow
blow blow
blow
bo
o o
go so
grow show
grow grow show show
grow grow grow o show show show
grow grow show show
grow show
go so
o o
lo
flow
flow flow
flow flow flow
flow flow
flow
lo
o

Fig. 2. Eugen Gomringer. *Untitled* (Solt 94).

In this poem, we see a similar empty space of meaning which, like the arches suggested in the Stein poem create a situation where "nothing" is actually stronger than textual content.

Architecture as Meaning

"There is nothing!"
William Carlos Williams

Arches express the ultimate in a play between structure and empty space, as if the architectural were paradigmatic of a kind of structural view of textual construction. With arches as well, one sees not a single item but a totality, a structure of parts. Here are some arches, examples of Havana's marvelous colonial architecture.

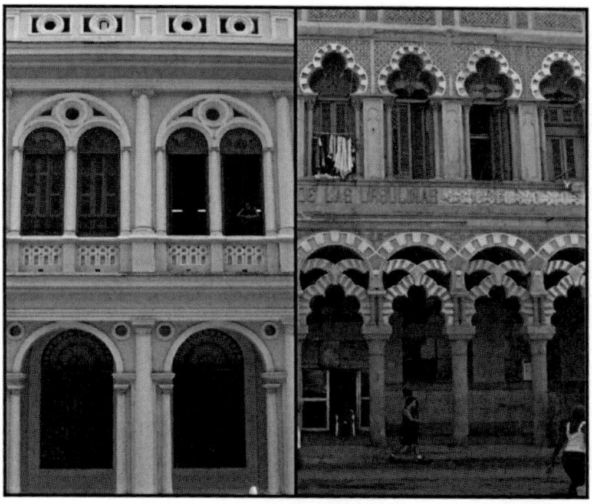

Fig. 3. Representative Arches, La Habana Vieja.

The rounded arcs of white space one sees in Gomringer's poem are present in the arches between the columns. Indeed, such figurations create structural strength in order to allow spaces to exist. In this context, the spaces could be seen as the point, the solid matter of the architectural conversation.

We see a similar engagement of the structural strength of empty space in Hannah Weiner's "Little Book 107":

I SEE W
OR   DS
I SEE OKs
ISEE OK$_s$ (8)

Here, the linebreak after the first "W" and the space between "OR" and "DS," as examples, create an archway, opening, or structural frame in which the simple "I see words" takes on much more meaning in its altered form than in its poetic representation.

A similar solidity of empty space is evident in "I Can't Resist the Craving for Another Inhalation" by Bruce Andrews:

below
pterodactyl-like
fannie
bay
sparks !
consolably
adulteryless
troubling
sweet
puce
skids . . .
leaped rule
final spire
conditions as goodbye
lattice unaware

For instance *bazzam*
luckier
sparrow (19)

Here we see an example of a text where the empty spaces are as important as the words themselves, the spaces offering pauses, timing, and a real counterpoint to the signification of the words. This poem would not be the same if the words were all compressed; the spaces give the poem its expressiveness.

In "Fidel" by Andrews, the words seem to form an arch by virtue of their placement on the page. This combines the best of the Moorish arch with the sensibility of Gomringer:

pee axe eel map sex am
case ape pal lax
pal leap lax ease
ale leap seam
alm amp ease sax lamp lax eel
amp alm pax eel sex
pax peel map pea ma am ale same amp
sex amp seam leap sax pee ape alm
lap lamp pal ale ma pee alm axe
ma lax pal ape ease am amp sax lap
pax peel lamp sax lap amp alm seam
leap lax
ale lamp leap
lamp eel alm
sex peel ape am map axe
ale pal ease am eel pee seam
sax peel pea axe map sex ape amp
lax leap alm lpa lamp pax ma ale ape
am axe pea lamp amp leap
amp alm leap axe pea lamp
pal ale am
sex sax
alm am map eel sex axe
sex alm am ale lamp pal (Andrews 35)

Empty space is crucial in this poem, of course, since its tight shape would not exist if it weren't juxtaposed against the white space to the left. But also, this poem is also an example of permutation, since it solely employs a finite number of letters in different combinations, with only one letter, "e," used more than once in a word.

Uno

'Three simple words: crack dot com.' Ambience or confrontation
as in that sense of 'colonial'. Path of obstruction. Call now to start
something more comfortable? Like what? Another URL. The
another on the constitution. A little eye wash for your public
Resumes. HTML as the world's dominant language. As in, write
Nahuatl. Po cenotes. Act of Tejanismo. 'You speak so many
Metal models of now-healed body parts offered at the shrine
Cartago. A statue of the Virgin wistfully reappeared on August 2, 1635.

Fig. 4. Screenshot from *White-Faced Bromeliads on 20 Hectares.*

At this point, I would like to look at two examples of code in my own work. One from *White-Faced Bromeliads on 20 Hectares* and the other from *Io Sono At Swoons,* both of which are available on my Electronic Poetry Center (EPC) Author Page.

In *Bromeliads* variant line possibilities are used to produce alternate readings of the text. These are technically arranged in arrays as follows:

```
// This is where the array is created.

    a1 = new makeArray(2);
        a1[0] = "'Three simple words: crack dot com.' Ambience or
            confrontation."
        a1[1] = "I'm sure there is a white house hand in there somewhere.
            'Colonia'"
```

I would like to note the structural and visual shape of this two-element array, and to suggest that the array here is not only content-based but also a visual writing technique. The structural arrangement is a means to organize the *possibilities* of the onscreen event and as such, the holes here are solid and meaningful, ways of arranging such possibilities. These are possibilities which exist as manifest in the code but can only be realized in the onscreen event of the code's execution.

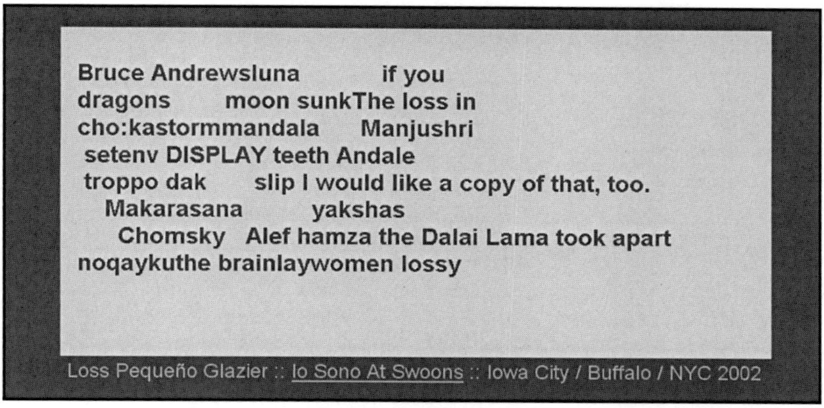

Fig. 5. Screenshot from *Io Sono At Swoons.*

I would compare this to another poem, *Io Sono At Swoons,* a work that assembles texts from a collection of strings, or data elements. The aesthetic contours of *Io Sono* was created by adding spaces, tangible nothing, or holes, into the strings. The result of adding such numbers of spaces into strings could not be anticipated. Thus the only way to work was to insert spaces, compile, then observe the displayed results of the spaces in the onscreen event. Working with the holes, or empty space, became a means to work with the poem.

What is of use in this method is the concept of precise poetic analysis, of the relevance of position, location, and structure. The arrangement of words in a poem or in a program have intense value. Like the empty space beneath arches in a colonnade, writing and programming is a dance around the architectural spaces of the poem's parts.

Words themselves are modular and programmable. As smoothly as "code" flows to "cod" and then to "ode," poetic language is always engaged in the play of its parts.

Works Cited

Adell, Joan-Elies. *Poemes, Pesmi, Poems.* Barcelona: Poetry Reading Series, 2004.

Andrews, Bruce. *Wobbling.* New York: Roof Books, 1981.

Glazier, Loss Pequeño. *Anatman, Pumpkin Seed, Algorithm.* Cambridge, UK: Salt, 2003.

Glazier, Loss Pequeño. *White-Faced Bromeliads on 20 Hectares.* 1999. 4 Aug. 2006 <http://epc.buffalo.edu/authors/glazier/java/costa1/00.html>.

———. *Io Sono At Swoons.* 2002. 4 Aug. 2006 <http://epc.buffalo.edu/authors/glazier/java/iowa/index.html>.

———. Electronic Poetry Center Author Page. 4 August 2006 <http://epc.buffalo.edu/authors/glazier>.

Pound, Ezra. *Selected Poems of Ezra Pound.* New York: New Directions, 1957.

Solt, Mary Ellen, ed. *Concrete Poetry: A World View.* Bloomington: Indiana UP, 1968.

Stein, Gertrude. *Writings 1903-1932.* New York: Library of America, 1998.

Weiner, Hannah. *Little Books/Indians.* New York: Roof Books, 1980.

Practical Examples

Laura Borràs Castanyer

E-Learning and Literary Studies

Towards a New Culture of Teaching?

Abstract:

The introduction of digital technologies into the learning processes has meant the creation of new educational spaces that, when they take place on the Internet and are founded in non-presence and asynchrony, are known as "Virtual Learning Environments" (VLE). VLE constitute new pedagogic realities that must answer to the users' needs, their educational purposes, the curricula with which they work and, specifically, the formative needs for the people that integrate them. But the key to define "virtual" in terms of human experience and not in terms of technological hardware is the concept of "presence," which is crucial in our pedagogical model and our way of being comparative literature lecturers in a virtual university.

Technologies are tools capable of building a learning frame, although it is necessary to endow them with contents and humanity. Different voices have warned of the sterility of a technological environment that does not have any pedagogic or didactic specificity (different from the traditional models). After all, learning is learning whether it has an extra "e" or not, and so VLE are only as good or as bad as the ways they are used. Thus, the revolutionary point of their use would not be the technological aspect whether they really offer new ways of teaching. We will show the example of a completely virtual university, the Universitat Oberta de Catalunya (UOC), where the learning process takes place in a virtual campus, and we will focus on a specific subject: *Universal Literary Topics.*

Introduction: The Irruption of Digital Technologies in the Learning Process

E-learning is probably no more than a manifestation of e-living. Perhaps today circumstances allow us using the correlation between the appearance of e-learning as a tool and the need to modify our traditional model of education, at least at the university. At all events, we must insist on the fact that the main revolution is not found in the prefix *e,* because quite often traditional monodirectional learning has been hidden under the appearance of technological sophistication. What seems interesting to us is to rethink the concept of learning in modern societies taking into account that we have more and more digitally literate students, who are, of course, using the Internet as a source of resources to be explored.

The European program of action "eLearning 2001" defines electronic learning as the use of multimedia technologies and of the Internet to improve the quality of learning. However, we firmly believe that e-learning is an excellent opportunity for a real improvement of quality if these multimedia technologies lead us to a more humanized teaching process. The possibility to use complex systems on a virtual campus allows and stimulates exchange and collaboration, with remarkable effects in the students' success and satisfaction. So the aim of this work is showing how good e-learning results are achieved when there's a good e-teaching method behind them. To do that, we will give the example of a completely virtual university, Universitat Oberta de Catalunya (UOC), where the learning process takes place on a virtual campus and we will focus on one specific subject: that of comparative literature. If there were—and there still are—some risks, I hope that we'll be able to see that, finally, the key to define "virtual" in terms of human experience and not in terms of technological hardware is the concept of "presence," which is crucial in our pedagogical model and our way of teaching comparative literature at a virtual university.

The Challenging Experience of Teaching Literature through the Net

Taking as an example the course *Universal Literary Topics* (Borràs Castanyer), I'll try to show the way in which we have organized this course taking as a starting point a series of texts connected by the fact that they deal with some of the basic topics of universal literature. We use hypertext material combining linear reading with sequential or fragmentary reading inherent to hypertext, as well as video or audio resources organized around the topics of "desire," "journey," "identity/otherness." During the process, our intention was to analyze the text from a dual perspective, both towards the present and the future but also, simultaneously, within a paradigm formed by other texts that precede and influence it. The course intends then to contribute to a transversal reading of universal literature in a virtual environment of learning, and at the same time it provides guidelines to students for a practical exercise of comparative literature. It also suggests reading itineraries that cross periods and literary traditions that are far from each other in time and space. Following intertextual connections that guide us within the hypertextual corpus, some of the reading routes are described as randomized and subjective and they show the broadcast and the articulation of such a topic in the literary tradition. Students, after familiarizing themselves with the texts presented in the materials and with the navigating tools offered by the materials, have to select some other topics and to build up their own (hyper)textual corpus which will constitute their final project. And there is a constant dialogue between the teacher and the students so that, although we are a distance university—or without distances, as our motto has it—the concepts of presence and accompaniment of the students are fundamental. This is what provides a true added value to the teaching and studying of literature at the UOC and it ensures out success.

Our task in the day-by-day work as virtual lecturers combines electronic didactic materials, on-line resources, digital libraries, web sites of reference, virtual exhibitions, etc. It is necessary to seriously consider that on-line teaching using these digital resources implies becoming detached from acquired habits and transforming the discourse of communicative techniques. The fluidity of hypertext obliged us to rethink one of the main preoccupations of a writer: the possibility to exert control over the way a reader reads. Indeed the author-lecturer's creative act requires, on the part of the user-student, an interpretative act and also aims at making the students wander around the text. Hypertext shows a new form of "textuality" based on the capacity of "penetration" of a text marked with links that open doors to new horizons. In hypertexts, any illusion of control vanishes: seduction is the only motivation we wanted in order to lead towards hypertextual wandering. So, we must react to the transfer of knowledge by accompanying students in their process of intellectual maturation, contributing to the virtual blackboard or inciting the debate in the virtual forum, correcting exercises in a very personalized way, answering doubts, considering new questions, etc. After all, it is a holistic and beneficial task for the students, since it obliges them to read, to compare, to listen to their colleagues, as well as to the lecturer, to deeply participate in the course, to organize their ideas in a logical form and to present them coherently. In other words: to organize and build their learning process in a radically subjective way, using their own initiative and capacities.

The History of an Experiment

We started to design the course in 2001, just at the time when we were trying to see if the students at the UOC were capable of dealing substantially with their new environment through its own language: that of hypertext. This was the challenge, but at the same time we had to consider whether we, the teachers, were prepared for our role of imparting contents in the specific circumstances of a digital environment. We were thus not only talking about a campus that allowed interaction with, and tutoring and monitoring of, students but we were also facing the risk of replacing our linear discourse with a fragmented discourse which the students would approach in ways that we could not control. What does all this mean? Well, in lectures/classes to my students at the traditional University of Barcelona, I control the situation to the extent that the discourse is linear and is delivered in a linear way. That is to say, for the students to understand what is to come, I have first to explain other things in order to maintain a narrative coherence that has to be complete to allow information transfer to be effective. But now it was not simply a question of having webs or other virtual spaces to complement my teaching act in the lecture hall, nor of producing paper materials such as manuals on a particular topic plus activities for distance students. No. This time we had to make a double leap into the unknown to see if we could be replaced or at least transported to the opaque limbo of a hypertext. This challenged us to check whether we knew how to "disorganize" our academic discourse that looked so "organized" on the surface. Obviously we used this peculiar textual form that hypertext imposes on the works that we read and interpret, a form that at first appears fragmentary. Fragmentary, because to practice the art of screen writing or a certain topographical script necessarily takes us to a certain spatial arrangement of the contents. This arrangement is not only a discursively linear one in the sense that the discourse is broken down into items —the individual words—that constitute blocks of text linked through contents, but also fragmentary from the point of view of accessibility of the texts, since only text extracts were selected and not whole works. I have thus to confess that we acted inversely to the way which, according to Chateaubriand, the moralist Joseph Joubert did things. When he read, he ripped out the pages that he didn't like and in this way he collected a personal library made of stripped volumes. A collection of fragments, then, which is the consequence of a certain teaching strategy, but which also reflects changes in the mentality and perception of the reality that affects reading and writing as well as cultural production as a whole. I suppose that we found support in the principle enunciated by Seamus Heaney in a poem from *Electric Light* which is significantly entitled "The Fragment." The last lines read like this:

> "Since when," he asked,
> "Are the first and the last line of any poem
> Where the poem begins and ends?" (57)

I do not know if we succeeded in our attempt, but the fact is that there were three different proposals for the use of hypertext or for the diverse presentation of information. The difference, far from being seen as a lack of structure, was upheld even in the face of contrary recommendations from technicians and experts. As a sign of the diversity of proposed adaptations to the aims of the course, the main one was to promote critical perspectives that reconstruct the history of literature according to deeply personal routes and canons. And to achieve this we had to forge a close alliance between the two methodologies employed for the purpose: the principle of intertextuality (in the production of the materials) and the conversational interaction in the forum debates. In fact the course aimed at inverting the conventional relationship between the author-critics on the one hand and the literary text on the other by transforming the former into the agents of an exploration that passed freely from one work to another, discerning their own thematic thread. The hypertext form of the materials corresponds, then, to this type of discontinuous reading that moves from one work to another without a pre-established thread instead of the reading of each text in its entirety. In this way, the three blocks into which the material is divided represent three personal anthologies of extracts—literary or otherwise—selected for their intertextual relationship with an initial text that is particularly typical of the topic that each author has chosen and with comments based on this relationship. In other words, they are series of readings that are not sequential—one text after another—but rather intertextual—one text within another. Those of us who have worked on its design have thought about the new ways of reading that the Internet and digitalization have created, where texts are now not self-enclosed but open up an infinite set of rewritings and interpretations.

The course is to be seen as an example of reading and a particular literary hermeneutics in which three teachers, who here act as "model readers," offer our reading itineraries through the vast corpus of the world's literature based around three specific themes. Subsequently, when the students have surfed and discussed, reasoned and argued over these three themes and the debates that they generate, we ask them to do the same as we have done and produce their own hypertext of readings.

Travel, Desire and Identity/Otherness

The theme in which I developed the idea of travel led me to the image of the Western traveler *par excellence*, Ulysses. The eponymous *Odyssey* includes a text serving as a base camp in which Ulysses appears confined in hell, burning in a tongue of flames beside Diomedes. The *Odyssey* allowed me to make a leap to canto XXVI of the Inferno in Dante's *Divine Comedy*, but passing via Virgil's *Aeneid*, which provided the opportunity for an exercise in the ideological reading that Dante makes of it, with his clear underlying political stance. But the journey did not end here, for I also took a trip through the medieval romance so that, in the end, I could transfer my discourse to a more theoretical reflection on the concept of "borrowing" and "adaptation." In a final twist, this brings us once again to a practical application and to the reading and decoding of iconographical discourse by analyzing the diverse interpretations that are to be found in miniatures and certain modern and recent paintings illustrating the passage where Dante condemns Ulysses to the flames.

I wanted to show that writing implies initiating a dense and complex skein of references and allusions, thus making written works into networks of interlaced textual paths. In the same way, reading always implies associations, and in order to understand we must interpret and achieve a profound reading of a text. At times, this process can sometimes depend on interesting and attractive detective work, where we pull the threads to unravel the complex and therefore fascinating skein of relations among texts, which is known as intertextuality. At bottom, the striking similarities between hypertext and text that we find in Poststructuralism and the recurrent position that "texts refer to other texts endlessly" are especially pertinent in this sort of teaching approach, because it is in hypertextuality that intertextuality recovers its ideal form of existence to the extent that the formal properties of an educational hypertext encourage the readers-students to reach their own conclusions regarding the contents. From an intellectual point of view the students become more active and therefore also interactive, also to the extent that the students feel encouraged to make conjectures, compare opinions and consult sources easily and directly, meditate on the proposed itineraries, search through them, even perhaps to spot possible textual continuations.

So what did the hypertext format offer me? I have to begin by admitting that what I felt most keenly was fear. I was frightened that I did not control my students' access to the discourse and to the arrangement that I set down in order for it to become intelligible. Fear too at not knowing how to give a lecture on the basis of interruptions, fragmentation, discontinuity, and thus condemn the hypertextual arrangement of my discourse to be seen as lacking coherence. Fear, obviously, of making the students feel totally lost so that they would be unable to achieve the goals I had set. In short, what I experimented with was the fear and the disarray of making truly ours the critical discourse that we apply to literary works. These latter are expected to accomplish clarification in a course without knowing how extensive it is, how many pages there are, how much there is left to read, where the students are unable to see where to glean the knowledge that they are expected to find, and therefore whether there are also missing parts of the course which will appear in the final exam. I think that basically the fear is understandable to the extent that this initiative was in fact a new way of teaching literature. An experiment, obviously, whose success depends exclusively on the students' acceptance of the method that has inspired it. So I tried to overcome my fear by designing diagrams, conceptual maps and hierarchies within the links and pages so as to organize as far as possible, not to say "direct," the discourse that I wanted to get across. At the same time, I tried to make the maximum use of the potentialities of the environment to complement my expertise as a teacher and the persuasive force of my oral discourse with audiovisual information that almost forces an interpretation. In the conventional classroom I can start with an introduction and then read the text, making brief parentheses for clarification, which I then sum up in the end. Yet, here they received the original text or the Catalan translation with the possibility of a recital/interpretation by Vittorio Gassman; they could follow the text on their screens, with explanatory links, illustrations, complementary webs, additional references, etc.

However, the procedure that I adopted was not the same as that of my colleagues. Professor Raffaele Pinto wanted to bring out the intertextuality in a fundamentally physical way, with juxtaposed texts, with texts that referred to other texts from a viewpoint that was strictly textual. He wanted to explore desire by diving into the moral roots of the liberty of the individual immanently and lay roots arising from an experience which was at the heart of the inspiration of medieval literature. So in order to talk about desire, he selected the preface from Boccaccio's *Decameron*, a text which, in its dense brevity, explicitly sets out a synthesis of the significance that desire had come to acquire in medieval literature and of the modern values, which, starting with Boccaccio, would have a cultural horizon that was lay and secularized. Pinto draws from him a whole series of textual associations that allow him to survey the appearance of troubadour poetry in Romance languages and to establish textual links to Catalan and Spanish poetry of the 15th and 16th centuries. For the third topic, Joan Elies Adell shows how identity is born out of the disturbing encounter with otherness, and the extent to which the other has been utilized to construct the sense of self, the sense of centrality of the world itself. The other has been a central figure in the formation of modernity, a modernity which at the same time demonstrates the impossibility of reducing the world to one unique point of view. Joan Elies Adell's extremely exploratory hypertext forces the students to re-think Western history starting from the role that the other has assumed. This act of revision takes literary shape here through a reading of Joseph Conrad's *Heart of Darkness*. With the background of postcolonial literatures, the aim is to get across the function of imperialism as a political and ideological system. To this end his hypertext analyzes the first chapter of the book and weaves a dense network of links that annotate, interrupt and relate the text to others (in other literary genres: poetry, essays, etc., and also other media like cinematography). Instead of serving as buoys in the midst of the navigation, they tend to generate a loss, which is revealed as necessary and inevitable in the reading process and in literary criticism.

The Rules of this Particular Game

The course is arranged around three activities for continuous assessment and a final project, as well as participation in the debates.[1] The first task is for the students to describe their view of the structure of the materials and the consequences that this has for their studies: familiarization with the articulation criteria, main texts, secondary texts, reading list, the search function in hypertext. In tasks 2, 3 and 4 they have to produce ideas of their own based on each of the three themes related to the paradigm of modernity in literature. Finally, they have to present a justification of the selected themes and this takes the form of the final project. Each one of these tasks is accompanied by a series of conversations in the forum in which the contents are gradually separated out. There is a debate about four basic issues: first, the course materials, and then each of the three thematic blocks and their critical evaluations in relation to them. At the same time there appears a whole range of satellite topics which are keenly taken up by the students: for instance, the relationship between language and identity, between gender and language, the subject, the role of women in history and in literature, anorexia, and so on. Finally, there is a collective debate on the choice of a critical category, and the justifications and anthologies proposed by all of those taking part, as well as a verification of the presence of the theme chosen in the course materials or of close or relatable categories.

The Results

This course has only been offered for three consecutive years. The growth of students taking the course is exponential, going from 7 to 29 and then to 53 in these first three years. We have to compare this growth considering that for an optative subject there's an average of 15 to 20 students per class and from 5 to 10 in the first semester.

This extraordinary increase, which has not been observed in any of the other subjects in the degree, would seem to indicate the level of success attained. The level which is not only seen quantitatively in the percentage of good results but also qualitatively in the feeling of satisfaction expressed by the students who have done the course, as well as their sense of having worked well, with great intensity, and very beneficially, and have shown a high degree of reliance on, and positive gleanings from, the classroom debates.

Even so, it has not been easy. And this is so, paradoxically, because our initial warnings in the first semester that the course was offered were greeted with enthusiasm and understanding by the students. They understood perfectly well the reasoning behind the "shape" of their literature manual and they were relatively happy with it, although they expressed the view that their knowledge was slight when faced with a hypertext that seemed so rich. Let us have a look at some of their impressions. We can observe that the students saw the approach as innovative[2]:

> . . . As far as I can see, the material is fascinating, although it looks difficult. I've read Laura Borràs's introduction, "Re-reading as a liberating act," and her point of view strikes me as very interesting. I had never considered this approach but I find it extremely appropriate and it opens new ways for me to relate to literature.
>
> M. Àngels Monte Ochoa de Aspuru, "Diversos," 14/03/2002

> I am truly overwhelmed by the quantity and quality of the material in the course hypertext; it brings me face to face with my own ignorance and in fact with myself because it's incredible how many things there are that I haven't read. . . . We can turn to the saying *ars longa, vita brevis*, in other words, there are masses of written works to be read in just one lifetime. Perhaps this is enough to calm us down if, as Laura Borràs says in her introduction, re-reading is the supreme liberating act which allows us multiple journeys into literature; we cannot lose sight of the fact that literary tradition is made up not only of reading material, but also of the interpretations of others. . . .
>
> Miquel Àngel Arasa Favà, 15/03/2002

> One feels foolish and become aware of the fact how small you are when faced with the infinite transversal lines—vertical, horizontal and oblique—in the material, which a quick glance hardly takes in. It is a mixed feeling: on the one hand, you feel hurt by the reminder, which you consider unnecessary, of your limitations; on the other hand, grateful because you see that you are invited, with a large dose of freedom, to initiate the creative task of designing one of your small possible canons, some original but inevitably miniscule thematic collection, or even to plunge into the moving search and discovery of intertextual connections . . . express the wish to share the bliss and joy deriving from the quality of the contents of the course web.
>
> Josep Maria Àlvarez, 15/03/2002

> We are faced with excellent material expressing innovative ideas. It does this through a wide range of readings which, above all, bear the cultural

> and literary imprint of their authors. Material whose structure, in my opinion, corresponds to a rather loose interpretation of intertextuality and which in fact is quite close to the functioning of our brains, which undoubtedly work through the association of ideas, sometimes very subtle ones, derived from our existential background. . . . With this premise, taking part in our journey might be a fascinating adventure, an adventure that leads us to open the container of our readings and our most hidden knowledge, to appreciate passages, authors and concepts and to see how far we are capable of reaching . . . and this makes me ask myself: is my intellectual baggage sufficient to construct an attractive trajectory, to create a canon of intertexts in which the analogies appear to have a minimal coherence? I'm afraid of losing the intention into a puerile dead-end or a maze where each text is a wall shut off by my own ignorance. These are some of the doubts which strike me but they coexist alongside the curiosity and interest that the course arouses in me.
>
> Josep Maria Roquer, 16/03/2002

> We have here some very good material. At last we are doing a subject truly from within, exploiting the possibilities offered by digital technologies. The hypertext format allows one to delve in much deeper and more rapidly with regard to informational and relational possibilities. However, I'm afraid to lose my way in this sea of texts and images. . . . From the prior consideration that the possibility of a transversal reading is undoubtedly a fascinating adventure, I feel at the same time that this methodology also runs the risk of an excessive fragmentation of the readings. . . . My state of mind at the beginning of the course and when faced with the materials is one of total insecurity; everything that I don't know suddenly appears before me, and with the possibility that it will grow bigger and bigger. I also feel overwhelmed like Miquel Àngel and foolish like Àlvarez. I also have my doubts about whether my intellectual baggage is enough as the starting point for the preparation of my own text corpus. . . . I must also say that I very much like the treatment given to the non-literary materials related to the texts, especially the pictorial ones. I have spent quite a time nosing through them. Many thanks.
>
> M. Àngels Monte, 17/03/2002

These feelings of anxiety, of shortcomings, of being overwhelmed, contrast with the excellent results that the group achieved; they were of an extraordinary quality. However, these reflections from the pilot course contrast significantly with the lack of sensitivity towards the materials demonstrated by the second set of students. To begin with there was an initial moment of perplexity over a manual that was not a book, but a web. They could download the web onto their hard disks, but they refused to do this and asked for the course on a CD. Subsequently, their attitude changed when they had seen the learning possibilities of a teaching approach based on the students' self-determination in relation to the route and the contents. Let's look at this through some of their comments:

1. Complete perplexity over the lack of materials. It seems that the web by itself was not considered synonymous with materials.
   Dear fellow students: I'm another one who has not received any materials for this course. I have sent a message to the office but I still haven't got an answer. Have you others received any type of print material?
   Cristina Gallego Jiménez, 27/02/2003

2. Mystery that stems from the physical structure of the material, of the medium.
   Although it is very interesting, this course is somewhat mysterious.
   Laia Morral Plans, 27/02/2003

3. There were also students who looked beyond the form:
   The approach to the course strikes me as very interesting and I believe that together we can make real discoveries.
   Josep Maria Agustí Julià, 04/03/2003

4. In the light of this bewilderment, authentical textual versions of the materials began to appear:
   Given the lack of a print version of the contents, I have tried to make one, at least of the journey part, although of course I have not included either the images or the links.
   Lídia Jiménez Comas, 05/03/2003

5. Three weeks into the semester panic took hold:
   I urgently need help. I can see that people are making contributions and I still haven't received the CD, nor have I been able to download the material that we have to read. Can anyone tell me what I'm supposed to do?
   Teresa Alsina Batí, 11/03/2003

6. And when the CD finally arrived, it became clear that it was not what they had expected. I suppose they thought that the CD would be something like a book, not that it was just a physical support for the web. However, this also showed their limited technical skills.

   At last I've received the CD but now I have the problem (like many of you) that I can't install it. I will try to do what Glòria suggests, to see if I finally can consult the materials. I have printed out the journey but it is not the same. As soon as possible I will make my personal evaluation of the materials—I mean the first task. I can't wait to see these materials which, from what I have read, are very interesting.

   Teresa Alsina Batí, 16/03/2003

In effect, comprehension comes when they understand that "it is not the same" because there is a complete correlation between what we provide and what we demand. It is not a question of us forcing them to the difficult task of surfing to find academic contents with a view to study and memorization, which is probably where most of the worries and fears come from. Instead we have provided them with academic material in web or CD format so that they explore it, discuss it, comment on it, learn things from it and, at the end of the semester, produce their own material.

Finally, I will present the total rejection of the material format, which came from the third—and for the moment last—group of students.

The initial confusion:

> I chose this subject because I thought it might be very interesting. What I didn't realize was that all the material is in hypertext format, and the truth is that I'm afraid that it will tire me. To get home and open a book is not the same thing as to be stuck the whole time in front of the computer. Well, I hope everything goes OK.
>
> Fina Roldán Cobo, 02/03/2004

> It's the first time that I've seen a course in which hypertext is used. I find this materials format very interesting because each of us can choose the link which seems most attractive in order to go deeper into the text and relate to other information. One might say that it's like a realization of the associative organization of our memory. This does not mean that I agree with using materials of this type because, although you can choose the path you want to follow, the paths are limited. By

> contrast, when we read a book, our memory and imagination are unlimited and we can relate to any ideas that we want.
>
> Teresa Alsina Butí, 03/03/2004

> I will see how I get on with the hypertext material, since I've already had to drop out of one course because of it. I'm sorry but it upsets me. I prefer a system with books where I can make my own notes undisturbed and not go backwards and forwards making notes from the screen. In spite of all this, I will make the attempt.
>
> Àngela Murillo, 03/03/2004

> In this course, I am surprised at the type of material that we have to use but I still can't give my opinion about it because I haven't really delved into it. I admit though that my first thought was to print out the lot.
>
> Rosa Maria Navarro Fernández, 06/03/2004

However, something was beginning to change:

> I was quite surprised that all of the material related to the course was in digital format. Perhaps this is because I'm technologically a bit old-fashioned, or perhaps it is because I have a great passion for the traditional format on paper. So, full of skepticism and reservations, I started to wander virtually through the course content, and, in doing so, I discovered that the format was not all that bad. It is obvious that hypertext has many interesting aspects. To move around the CD is like taking a walk, a trip that is interesting because of the topics of this course, a trip where you are constantly on new paths, paths that take you forwards and backwards as if you were walking through a maze. It has its pleasant side, but in spite of that I recognize that for a reader accustomed to conventional linear reading, it will be a problem, as one constantly gets lost among the words. Perhaps it's all just a question of getting used to it, of stopping to see as unknown and strange a way of learning that seems interesting. And this topic really could turn out to be interesting because the CD is of a very high quality. I can only talk about its strengths: visually attractive, filled with things that could help us in our research—I've even found fragments from Francis Ford Coppola's clever film *Apocalypse Now*, which, as you know, takes as its model Conrad's Heart of Darkness. Definitively an interesting find.
>
> Oriol Grau i Mas, 07/03/2004

My first impression when I saw the material was that I felt a bit uncomfortable not having a printed version and not being accustomed to a format with a sea of fragmented contents. It was certainly my lack of experience that caused this sense of disorientation—I didn't know what order to follow or whether it was the right one. On the other hand, if there is no paper version, it is also a way of saving ink and paper. I suppose it's a matter of getting used to it.

Rosa Teixidor, 08/03/2004

The initial contact results in rejection and, in consequence, the defense of books and paper. We also find thoughts about the contradiction between studying at a virtual university and rejecting materials that are in line with the times. This perhaps is evidence that some of our students choose the UOC because it is a distance university rather than due to the conviction of the value of its media, but this is also surprising.

I am one of those who prefer the paper format (although it's odd for a UOC student to say this.) I like the contact with paper; I like to underline with those fluorescent marker pens in various colors and make my own notes in pencil. However, I appreciate that the subject requires something more that the linear format that a book allows. The most serious problem for me, though, is not the question of one medium or another. For me the real problem is that you might get lost among all these hypertext windows that we can open. We know where to start but we don't know what our destination is or whether it's the right one, whether we have left something behind or whether we still have to go further. Take heart, this is the future.

Gerard Cortès, 03/03/2004

I have been reading your contributions and I really agree with most of you. The advantages of the electronic format are already well known to us: sound, image, unlimited copies, free edition of a text, links to the Internet, etc. However, reading from a computer screen is exhausting and hinders concentration. Long texts need to be in the traditional format of book or print. As a fellow student has said, it is odd to say this at the UOC, which is virtual *par excellence*, but I feel that reading from the screen is physically very tiring—your eyes, your body posture. And clearly, there's nothing like a book.

Ana Luisa Campo, 03/03/2004

I'm sorry to have to say that I prefer paper. I spend all day in an office facing a computer. Now I have to repeat the headaches and the bad posture when I get home in order to study at the UOC. Personally, I don't think it will ever be possible to replace a paper book with one in electronic format. I'm another one who uses the basic techniques of pencil notes in the margins, underlining, circling, etc. Up to now most of the courses that I have studied have supplemented the paper version with an electronic edition—for me this option is ideal—and those that use only the electronic format are tiring and heavy, and I believe that in the end the student does not perform as well.

Fina Roldán, 03/03/2004

Slowly there began to appear defenders of paper and books who nevertheless started to be sensitive to the specific features of the medium and we now see messages that attempt to find a compromise between the two media.

From the start, I personally had reservations about the hypertext format due to the fear of not being able to cover all the information that it provides. However, as I progressed into the material, investigating the different windows and links . . . it is like an adventure—I have to admit that I was completely seduced. And I have only covered a small part. I feel that the advantages of this system are remarkable. Like David, I believe that the transversality makes the course even more interesting, facilitating the task of relating one text to another, of relating works and authors, etc. All this, bearing in mind that although I use virtual communication, I still write letters, and I believe that there is nothing like reading a good book, and that a screen will never have the charm of some old yellowed pages. I hope that we enjoy this hypertext adventure.

Alex Ogier, 03/03/2004

There's nothing like paper. Take my word for it, reading a book, even if it is only a part or a page, sitting in front of the computer is not the same as sitting on the sofa or lying down in bed, with the freedom of movement that the computer denies us, because who has read a book that you can't put down without changing your posture while you read? Everybody does so. You make yourself as comfortable as you like or that the situation (of the book) allows. On the other hand, I also recall that we are in a virtual university, and society is changing, so the way to teach and the materials used in education also have to change. I'm tell-

> ing you, in this case it seems to me appropriate given that you don't have to go forwards and backwards between pages; instead the screens solve this problem for you. On the other hand, we must be ecological. This is also positive because there is a considerable saving in paper. Imagine the amount of paper needed to print the materials for this course. So, for those of us who are studying, for the moment a few trees have been saved. To be practical, we also have to think about space, since all we need is a computer and a little place for the CD. Books take up space, and there comes a time when there's none left.
>
> Montserrat Jou, 04/03/2004

> . . . Regarding hypertext, I also believe that it is a wonderful system to extend and complement the information contained in a work. Personally, I have printed out a large part of the links to give myself an overall idea of what the CD contains. Perhaps this is not the best way, because in the future if we have to print out all the hypertexts, the house won't be big enough to store all the paper, and hypertext, which is designed to save paper, will have the opposite effect. Perhaps one should only print out what is really interesting and leave the rest on the CD. Hypertext is like a great library at your fingertips.
>
> Carles Amengual, 06/03/2004

Although it is difficult to believe, there were a few positive initial reactions, even though the wording expresses shock, confrontation and, above all, moderation.

> It is the first time that I happen upon a hypertext for my studies. But for the moment it strikes me as brilliant, although the organization and the system of links that take you to more links might seem disturbing. I find that the horizontal invasion of my desk that I used to cause when I was studying is now reduced to my PC and not much more: papers, set books, one or two reference titles. Previously it was: the set text, the text supplements, the sheets with the recommended reading list, the manual or encyclopedia to look up terms. Whew! For me the defining assessment would be that this occupancy has come to an end. For the moment, delighted; but we will see how I manage when I get stuck. I might start climbing up the walls. . . .
>
> Montserrat Poveda, 02/03/2004

> I have been surfing the course web and I find that the material is simply pleasant and enlightening. I think that in a subject like this one the idea

> of transversality with different texts is fantastic. How many times when we were reading a book have our thoughts turned to another? Well, here I find the links to other literary works that can illustrate a particular topic very sensible. It is not a question of whether the windows that open annoy you or not; it is rather than they simply take us deeper into the matter. I think that it is good material . . . it is instructive and, above all, user-friendly. And it strikes me as being just that. I can't imagine this course in paper format. Well, I can imagine it, but with notes in the margin telling us to consult a web page or a book. In courses like this one I am completely in favor of hypertext materials. At least this is my feeling at the moment and I hope that it stays that way.
>
> David Font, 03/03/2004

> I would like to say that I like these materials in hypertext format. Even so, the first time that I was faced with similar materials—I am thinking about medieval literature, which presented all the texts like this—I felt a bit nervous. I had the impression that I was missing something, that I hadn't read everything. Subsequently, I learnt to read right through without interruptions (without opening windows) and get the gist of the thing. The second reading served to go deeper, to take notes. . . . In the end I found it fantastic: I had a lot of information available to me. So, the materials of "Universal literary topics" has neither surprised nor worried me. I feel that one notes the clear imprint of Laura Borràs. I have already done a couple of courses with Laura, and one of the things that she has made me familiar with is the virtual workshop, a space where we find the reading list, the supplementary links and the students projects. Highly stimulating!
>
> Assumpta Grabolosa, 04/03/2004

We note that the open dialogue in the forum clarifies standpoints as it does the opinions of the other students.

> I suppose that the fact that the materials are in a still new format enriches the forum with opinions of all sorts. Personally, my wandering through the course materials made me curious. . . . I understand that it is easier to use papers, scribbling on them, making them your own, but we must appreciate the great possibilities that hypertext offers. As readers, we're free to decide how far we want to go in the hypertext journey that we have to undertake. In effect, in certain courses at the UOC hypertext offers us a way of learning that is more agile, more pleasant

> and more interactive. At the moment I recall the materials of Medieval Catalan Literature. We have to consider that those made available to us pages and pages of different books without having them on our desks and we could consult them one by one. And it allows us to do this while we weave our own way and it makes clear that literature is a living discipline. I will finish with one last point: this course is also special with respect to its contents. Because to study world literature starting from desire, travel and identity/otherness rather than chronologically or starting from the most renowned authors—which is what traditionally happens—is truly innovative and interesting. And I feel that the formula of transversal study is a fine way to refer to themes and present them because it often stimulates the students and spurs them on to broaden their knowledge. And for me this is authentic learning.
>
> Rosa Clarena, 08/03/2004

> Let me finish by saying that I find it courageous to present a subject (and even more so literature, a concept that we always associate with books) in this format. I feel that it would have been half-baked if the CD had just been of supplementary material. Personally, I have no problem with the visual fatigue that the screen is supposed to cause or the fact that I can't read in bed, as someone mentioned. For me these arguments are not very serious. The truth is that I work for hours in front of a screen and I have never studied lying in bed or on the sofa. As for the contents, I feel that the authors of the CD have produced a presentation that is ordered, pleasant and, as someone has said, quickly arouses your curiosity. For all these reasons I feel that we should get stuck in and see how it goes.
>
> Joan Piqué, 08/03/2004

And with the euphoria of those who up to now had been reticent, confirmatory statements started to arrive:

> In my presentation I was not too positive about studying with the format we had been given, and after the first week I would like to confirm this. I am speaking absolutely personally, and I have total respect for those who are enthusiastic about the CD. My sincere congratulations. By contrast, for me, it is tedious to get home after a long day at work, to sit down facing the computer and not be sure where to begin. (Open and close windows, go from one side to another, nowhere convenient to take notes, difficulty to concentrate in the short study time

> that I have.) It would be easier for me to have it on paper like other courses, and use the CD to refer to supplementary material and as a means of researching information and/or to accede to interesting web pages, as many of you have said. As a matter of fact, I have started to devote more time to the other subject that I am enrolled for. . . . The CD is all right; I grant you that it is well organized, but I don't see it as the only material. And rather than ending up printing it out, I would prefer to have it on paper to start with and not waste paper and ink. I'm sorry; I'm afraid that more than one contribution is like a hankie with tears.
>
> Ángela Murillo, 07/03/2004

> After reading my fellow students' contributions and after a look at the CD which contains the course material, I have to align myself with those who are more or less against the materials being exclusively in this format. I have had the opportunity to study two courses with CD materials (the two medieval literature ones), but in both cases it was a matter of material that was supplementary to the basic course, which was on paper. I feel that it would have been more appropriate to provide the basic material in paper format. The other, as a fellow student has said, could present a sort of text that was longer and denser, providing all sorts of complementary material—links, illustrations, diagrams, etc.—in digital format. For me the computer represents my connection with the university, with the class as we might put it, but I still can't see it as my study tool. Apart from visual fatigue and the difficulty concentrating, I have to make the most of any brief period for my studies, making notes and revising, and at many of these moments I don't have the computer. I have to say that I find the material on the CD well produced and very suitable and, although I have printed out a lot of the contents—basically the ones that I feel should have been provided on paper—I do not rule out a change of opinion during the course.
>
> Marta Puiggarí, 07/03/2004

But let us not think that all is lost. In the end, there are some changes in their attitude:

> I admit that before I delved into the study of the CD, I had my doubts about this new study method. Like many of you, I felt that it was suitable only as a complement to the print material. Now, after having

gone a bit deeper, I am delighted with what I have seen, the quality of the images, the very original way to present the course, the links, as well as the freedom that hypertext gives you to create your own itinerary.

Rosa Maria Navarro, 08/03/2004

And the final reward arrived after months of intensive work.

Some Final Assessments

My main assessment of the course is that it has taught me to read from a more general, more transversal and more critical perspective. I have learnt to read a work not on its own, isolated from the world of literature, but rather by fitting it into the whole. It is a critical and generalist view that is highly enriching. . . . What is needed is precisely this capacity to think critically. As for the approach, I have found the use of hypertext very suitable, especially in this type of university. However, I suppose that the web could continue to grow. The debates have struck me as very interesting, and you feel motivated when you know that they are being closely followed. In sum, it has all been very positive and enriching.

M. Alba Serra, 13/06/2004

Right from the beginning, I felt that the goal of this course was that each of us, after a first reading guided by the hypertext, would come to their own reading, or re-reading, of the proposed theme. I really believe that this has been achieved. . . . What I mean is that all of us, through the experience of what we have read and our particular cultural interests, which obviously vary a great deal, have been able to imagine and find new paths and itineraries. I very much appreciate the contributions to the forum because this has made the course dynamic, active and participatory. Even if it was only in a small way, many of us became involved. In a university like this one, where we in fact see each other very rarely, the contributions of our fellow students help us to "live" the subject in a different way, and to feel part of a goal and an experience. (I also say this because I have suffered courses in which the teacher and the students made not a single appearance in the forum. In these cases, I felt completely alone and it obviously affected my dedication.) . . . Well, nothing else. I've enjoyed this course

> a lot. I've felt comfortable and perfectly free to choose and construct my own paths, to seek and find readings and, above all, to put into practice the series of ideas that the readings have suggested to me.
>
> Assumpta Grabolosa, 20/06/2004

It seems that, at last, it has become clear that the mere virtual environment or even the materials that are a reservoir of contents are not sufficient. Practice shows us that with close accompaniment students finally understand the philosophy behind the course.[3] More than merely achieving the aims of the course, students take an active part and show a very positive attitude to the subject having caused so much mistrust at the beginning.

Notes

1. So far we have covered the course materials, but there are also the tools that are common to all courses at the UOC: the study plan, which provides a clear temporal order for the contents and arranges them in a calendar of tasks and guided readings, and the virtual spaces of the noticeboard, forum and debate, as well as the link to documents and resources, the connection to the virtual library, etc.
2. The following student statements have been taken from the course's online community forum.
3. You can consult some of our students' work in the HERMENEIA web, some arising directly out of the course: Josep Maria Àlvarez (on pornography, <http://www.uoc.edu/in3/hermeneia/exemples/pornotemes/index.htm>), Teresa Vilà (on adultery, <http://www.terra.es/personal5/manelmarti/temeslit/>) and M. Alba Serra (on war, <http://www.uoc.edu/in3/hermeneia/exemples/guerra_literatura/index.html>), and some out of their teaching: Josep Maria Roquer (web of poetic hermeneutics, <http://www.uoc.edu/in3/hermeneia/exemples/ROQUER_web/Home.htm>), Núria Casademunt (alienation in Kafka, <http://www.uoc.edu/in3/hermeneia/exemples/kafka_alienacio/principal.htm>). Other examples can be found among the "Didactic examples" of the HERMENIA web.

Works Cited

Barràs Castanyer, Laura, ed. *Universal Literary Topics*. Barcelona: Ediuoc, 2001.

Heaney, Seamus. *Electric Light*. London: Faber and Faber, 2001.

Susanne Berkenheger

i'm dying, honey

dramatized proceedings from super-space (2004)[1]

Abstract

The text *i'm dying, honey* originates in a chat room. Ninety percent of it is original dialogue: digital whisper-friendships and babbling heartthrobs, pasted harangues and virtual shoulder shrugging. This is the background for my character, a provocatress who made herself known only within her own rules of speech, who played with snippets of communication as if she was a computer program. The chatters tried to crack her hermetic shield: they begged and flirted, they implored my character to speak with them in a normal language, they came up with the wildest ideas, they became impatient, they ranted and raved and thereby contributed to a dialogic piece which shall find its way back into the net after a theater performance.

yüksel: A A A A A A A A A A A A A A A A A A A A A A A A A A A A A A
AAAAAAAAAAAAAAAAAAAAAAAAAAAAAAAAAAA
AAAAA AAAAAAAAAAAAAAAAAAAAAAA AAA
A A A A A A A A A A A A A A A A A A A A A A A A A A A A A A
AAAAAAAAAAAAAAAAAAAAAAAAAAAAAAAAA
AAAAAAAAAA AAAAAAAAAA

ümit: he cant take it :)

yüksel: A A A A A A A A A A A A A A A A A A A A A A A A A A A A A A
AAAAAAAAAAAAAAAAAAAAAAAAAAAAAAAAAAA
AAAAA AAAAAAAAAAAAAAAAAAAAAAA AAA
A A A A A A A A A A A A A A A A A A A A A A A A A A A A A A
AAAAAAAAAAAAAAAAAAAAAAAAAAAAAAAAA
AAAAAAAAAA AAAAAAAAAA

ümit: check out how hard hes working at it :)

yüksel: A A A A A A A A A A A A A A A A A A A A A A A A A A A A A A
AAAAAAAAAAAAAAAAAAAAAAAAAAAAAAAAAAA
AAAAA AAAAAAAAAAAAAAAAAAAAAAA AAA
A A A A A A A A A A A A A A A A A A A A A A A A A A A A A A
AAAAAAAAAAAAAAAAAAAAAAAAAAAAAAAAA
AAAAAAAAAA AAAAAAAAAA

ümit: he'll smash keyboard in a minute haha

yüksel: how am I supposed to WRITE AGAINST 15 PEOPLE?
AAAAAAAAAAAAAAAAAAAAAAAAAAAAAA
AAAAAAAAAAAAAAAAAAAAAAAAAAAAAAAAA
AAAAA AAAAAAAAAAAAAAAAAAAAAAA AAA
AAAAAAAAAAAAAAAAAAAAAAAAAAAAAA
AAAAAAAAAAAAAAAAAAAAAAAAAAAAAAA
AAAAAAAAAA AAAAAAAAAA

ümit: flooding is all he can do

yüksel: ÜMIT YOU ASSHOLE I GRADUATED! YOU DIDN'T EVEN FINISH SCHOOL YOU BUM AROUND THE STREETS YOU SONOFABITCH YOU CAN GO TO ISTANBUL AND CLEAN SHOES YOU UGLY PARASITE HA HA WITH YOUR LOOKS YOU'LL NEVER HAVE A GIRLFRIEND PÜHAH[2]

This lively chat dialogue gives a small impression of my workday. I lurk, I compress recorded chat-logfiles, until something like the above text emerges. Sometimes I'm astonished that all this is original. Sometimes I ask myself: how come I want to even spread such conversations?

Before, my workdays looked differently. I typed, I linked, I searched in scripts for forgotten semicolons for many beautiful hours, I tried to guess the hexadecimal codes of monitor-colors for fun and was right more and more often. And now? Döners, bio-trash and sons-of-bitches?

What happened?

Or better: What is constantly happening?

Readers stubbornly try to print out my hypertexts—with poor results. Readers want to know from me whether they have seen and read everything of a certain text, and if not, how they can control that. Readers ask for my preferred way of reading of the story. Readers ask for structural maps, for sketches, early versions, for secret information that must exist for which they would be forever grateful.

To be honest: I am suspicious of them. These are not normal readers. They don't sit in front of the screen in the evening for a nice read. They have something different in mind, much less naïve: they are reworking their own works.

This leads me to three painful questions:

1. How many readers of net literature do I know?
2. How many of them read without writing about it themselves?
3. Are there something like normal readers in the net, i.e. people who are neither looking for contacts nor writing themselves, people for whom the

> encounter with net literature is valuable in itself? In order not to have to answer I do what I always do in these cases, I google: In this case: "Death of the Author" (in German)—3130 hits. "Death of the Reader" (in German)—14 hits. I am pondering these numbers for a long time. Aren't these two deaths connected with each other? Doesn't the disappearance of the author conclusively lead to the disappearance of the reader? How humiliating for the reader that only 14 people in the German net cry about his loss. If there is talk about his disappearance at all, then rather in a Buddhist manner with the motto "liberation of the reader." (Google: "liberation of the reader" (in German)—94 hits, "liberation of the author" (in German)—15 hits, "liberation of the reader"—72 hits, "liberation of the author"—11 hits). As if the traditional reader in his cozy wing chair had had such a lamentable existence that he can be only too glad to disappear.

This is how I interpret the numbers: the common reader is not really appreciated in the net. And I am collecting more evidence: the net goads even those rare epicurean hypertext-readers whom I know into the urge to communicate and this craving shortens the actual reading time.

And I? If I surf without communicating I feel asocial. Shouldn't I at least say "HI" if I land on a non-commercial site? Is it rude to just look and read in the net? As rude maybe as reading at a party?

Why am I thinking of a party just now?

One sits in front of a computer. Or is one just hanging out at the computer? OK, but usually unlike going to a party one has a concrete goal in the net and then . . . hmmm . . . one ends up with people with similar intentions, people who catch you with bizarre domains and breathtaking search words. There you hang out for a while. And then, after a while, their whole linked group of friends arrives. An entertaining small group has found each other. The intention? Later. First now there is . . . oh well, somehow party, as always in the net. Some more guests come by with intentions and plans. They certainly all want to know ones opinion. There are thousands, millions, billions, trilli . . . gulp! How relaxing it would be now to meet a couple of quiet readers. Who want nothing at all from you. Whom you just hear breathing and clicking on. Wouldn't it be pleasant to feel part of a big, silent community? Shouldn't therefore as many people as possible learn to continually remain silent in the net? So that we don't feel rude if we just look and read?

I swallow—once, twice, then I think about changing my profession. After all, the strengthening of the reader's role is of a top-level interest for an author. I should be a good example and retrain as a net-reader.

I am happy with this idea. Without further hesitation I am moving to the center of the typing maniacs—a chat channel. Ending up at the following lines:

hüpfer:	are we alone here? rolli
rT -rollitom:	nope. I think they are listening in. And laughing their heads off about me and you........... *gggggg .... gluttons *lllllooooooooooooollllllllll
hüpfer:	oh come on, reading is always interesting here, don't need to read a novel? Laughs
rT -rollitom:	book, paper, what's that? I only read DD in certain places *gfff

I feel I'm on the right track and internally nod at the silent community of readers. As I'm reading on, astonished, reading on, bored, I am pondering whether here a reader isn't something like a listener and whether an author-reader should not be an especially good and professional reader and likewise listener as well, and whether the particularly good listeners whom I know do not also do other things than listen. On the contrary. Don't the best of the particularly good listeners help the speakers find their real reckless phrases with pointed provocations and weird behavior?

Yes. Then: I have to be provocative, much more provocative.

A little later I'm chatting as a *dumbwaitress*.

cloud ex_undea:	hey u hey u yeah u hey speak no no say something your profile is crap as well
dumbwaitress:	dumbwaitress bows
cloud ex_undea:	oh how old hey u
dumbwaitress:	dumbwaitress sizes up cloud dumbwaitress frowns
cloud ex_undea:	what kinda girl are u wot u want
dumbwaitress:	dumbwaitress raises a brow
cloud ex_undea:	no no
dumbwaitress:	dumbwaitress shakes head

cloud ex_undea: well lets talk regular
really
no no
-.-"
pleez
pleez
pleez
or shall i go[3]

I continue my silence and in the meantime harbor more silent ideas. I could, for example, just simply repeat the phrases of the previous speakers or change them a little stylistically or continue to grill them with a few words.

The more variations of silence I can think of, the clearer it becomes to me that almost everything will work, provided I keep to my own strict rules. Namely that the stricter I keep to my ploys, the more attention I'll get. It even seems to me more and more that the real work is to suppress every spontaneous idea, response, to reply in set phrases and to come across with the varied vocabulary of a bot. That galls the public but gets me the most reading time.

I see myself only as a kind of language-machine in the chat, as a catalyst—and I record the reactions. I read them. Because I want to uphold this deserted place of the reader, the public therefore can plea, beg, implore as much as they want, they won't get anything out of me, I won't answer any more questions. I'm reading.

The net-author who I once was and will be again, however, wants to bring back the read and collated, condensed and partly fictitiously expanded text into the net—as an audio installation.

This is how it will look: an empty white screen will open. Behind it, invisible for the website visitor, are hidden six fields of different sizes, sensitive to the mouse. When the visitor is approaching one of these fields with his mouse, the hidden dialogue will be heard—at first low, getting louder the closer the pointer approaches the center of the field, reversely softer as soon as one leaves the field. If the fields are visited several times the dialogue continues where it was interrupted. All texts run in loops.

Thereby the listener samples his own "radio"-play. Like a silent, blind guest at a party he moves from one group to the next. In the best-case scenario he will recognize himself in his own role as the silent protagonist and will enjoy being a politely silent reader.

Notes

1. *ich sterbe gleich, schatz*. Dir. Manfred Kerklau. Perf. Gabriele Brüning. Sounddesign Detlef Piepke. Première: Theater im Pumpenhaus, Münster. 29 Oct. 2004.
2. *i'm dying honey:* continuation of dialogue 1; chat room West coast:

ümit:	flooding is all he can do
yüksel:	ÜMIT YOU SONOFABITCH YOU COPY SHORT PHRASES YOURSELF THEN YOU PASTE BUT YOU CANT FLOOD THATS WHY YOURE PIS SED YOU ASS-FUCKED ASS-FACE YOU CANT RE SIST ME YOU STEAL MY WORDS? AND MY SAYINGS? SHAME ON YOU YOU UGLIN
ümit:	looool yüksel, dont you know how to fight?
yüksel:	ÜMIT YOU ASSHOLE I GRADUATED! YOU DIDNT EVEN FINISH SCHOOL YOU BUM AROUND THE STREETS YOU SONOFABITCH YOU CAN GO TO ISTANBUL AND CLEAN SHOES YOU UGLY PARASITE HA HA WITH YOUR LOOK YOU WILL NEVER HAVE A GIRLFRIEND PÜHAH
ümit:	I talk normally with them
yüksel:	ÜMIT YOU SONOFABITCH WITHOUT FATHER . . . WHOSE FATHER I HAVE SOLD YOU BASTARD CHILD WERE FOUND IN ORGANIC WASTE AS MAGGOT AND YOU WERE BRED AS A MONSTER YOU SONOFABITCH WITH YOUR DO NER
ümit:	without flooding he can't handle it, he makes a fool of himself
yüksel:	I NEVER MAKE A FOOL OF MYSELF because i have no friends here
ümit:	im crying already.)
yüksel:	I NEVER MAKE A FOOL OF MYSELF because I have no friends here
ümit:	im superior to him in all things :) he cant cope with that haha
yüksel:	I NEVER MAKE A FOOL OF MYSELF because I have no friends here

jungguy13: *(whispers to she-lurker)* hi mouse, do you want to chat press press 54545454545454545454545554545545454444

yüksel: im gonna continue!!
BECAUSE I HAVE NO FRIENDS ON THE NET

My86Salsa: who has no friends press 555555555555555

yüksel: I HAVE NO FRIENDS ON THE NET
NOT IN THIS CHANNEL
im only here TO DISS ÜMIT
BECAUSE HE JUST THINKS HE IS KING

XxILHAN05xX: FUCK YOURE ALL FUCKED UP

ümit: haha

jungguy13: hi

My86Salsa: yüksel?

yüksel: ÜMIT YOU ASSHOLE I GRADUATED! YOU DIDNT EVEN FINISH SCHOOL YOU BUM AROUND THE STREETS YOU SONOFABITCH YOU CAN GO TO ISTANBUL AND CLEAN SHOES YOU UGLY PARASITE HA HA WITH YOUR LOOK YOU WILL NEVER HAVE A GIRLFRIEND PÜHAH

ümit: everybody knows how i look,
i dont hide like you
i look normal :)
not like pig :D

yüksel: FERHAT HAS FRIENDS
BUT DOESNT LIKE YOU
BECAUSE YOURE A LIAR

ümit: ferhat is also ugly, ferhat sucks up to you
because youre chat friends :)

yüksel: flood!

jungguy13: who wants to chat 12-15 press 6565656565656565656565

ümit: if we had met for real, you would be an outsider because youre so ugly

yüksel: AAA AAAAAAAA AAAAAAAA AAAAAAA AAAAAAAAA AAAAAAAA AAAAAAAAAA AAAAAAA AAAAAAAAA AAAAAAAAAA AAAAAA AAAAAAAAAAAAAA AAAAAAAAAAAAAAA AAAAAAAAAAAAAAAAAAAAAAAAAAAAAAAAAAA AAAAAAAAAAAAAA AAAAAAAAAAA

ümit: yüksel, nobody can stand next to you because your nose pu shes them away, it is so crooked:)

yüksel:	ÜMIT YOU SONOFABITCH WITHOUT FATHER . . . WHOSE FATHER I HAVE SOLD YOU BASTARD CHILD WERE FOUND IN ORGANIC WASTE AS MAGGOT AND YOU WERE BRED AS A MONSTER YOU SONOFABITCH WITH YOUR DONER
ümit:	you have gargamel nose

3. *i'm dying honey:* continuation of dialogue 3; chat room Dream-Island:

c4f-prinz:	hello how can i/we help you?
cloud ex_undea:	hey prinz d u know her?
c4f-prinz:	cloud whats ur prob
cloud ex_undea:	she dont speak man
c4f-prinz:	cloud . . . the monitoring is not an official chat room, it only serves the purpose to find the navigators faster. please leave this room after having expressed your problem so that other chatters get the opportunity to get help as well. dumbwaitress . . . as it seems you donot have a problem and you do not leave the room, you will now be expelled from the monitoring.
dumbwaitress:	dumbwaitress tumbles rubs eyes, nose sniffles knocks
operator:	Channel Erotic is full! Change not possible
dumbwaitress:	dumbwaitress knocks
operator:	Channel Erotic2 is full! Change not possible
dumbwaitress:	dumbwaitress knocks
operator:	You will enter the channel Dream-Island
dumbwaitress:	dumbwaitress smiles vaguely
wtüte:	go get me some coffee @waitress gg
printed matter:	what do we need a brain for? jellyfish dont have one, dont need one either. they just hover thru the sea, like heyyyyy, hoooo . . . i used to live like that also . . . and then i left school . . . *gg
wtüte:	waitress . . . wheres that coffee??? confound@domestics!
printed matter:	waitress?
dumbwaitress:	dumbwaitress scurries towards cold coffee

wtüte:	why cold???
	want it warm . . . cheekwarm please
schwester2000:	*(whispers to dumbwaitress)* where u from??
dumbwaitress:	dumbwaitress vaguely points into the distance
operator:	you have been invited into tellmemore by schwester2000; to get there enter /go tellmemore; or /sg
dumbwaitress:	dumbwaitress lowers eyes
printed matter:	?????? confused is @really waitress
wtüte:	i told u matter . . . u just have to listen right?
schwester2000:	*(whispers to dumbwaitress)* sorry bout the attack
printed matter:	is she some jeanny or gini or what that thing from the bottles called
wtüte:	dunno matter . . . wait i rub her. . . . ggg
printed matter:	löööööl
	do we have wishes???
	enthusiastically jumping from one foot to the other
wtüte:	yep matter . . . but u need your own waitress . . . remember . . . like in (life of brian) monty python . . . only one waitress each . . . gg
printed matter:	oahhhhh blast! . . . stonekick
dumbwaitress:	dumbwaitress gets back with sludge in cup
wtüte:	sludge? what sludge?
	receives cup
	slurps@coffee . . .
dumbwaitress:	dumbwaitress watches curiously
wtüte:	welllllllll@waitress . . . might be bit sweeter n bit less hot . . . other hand u r still learning
schwester2000:	*(whispers to dumbwaitress)* dont like to meet?? im a decent girl. hello??
dumbwaitress:	dumbwaitress lifts eyebrow
printed matter:	do u lend her out sometime
wtüte:	nope
printed matter:	come on . . .
	only oooonce
dumbwaitress:	dumbwaitress lifts eyebrow dangerously high
schwester2000:	smiles
printed matter:	waitress?????
wtüte:	ahm . . . are u going to be rebellious waitress???
	obey!!! talk with voice like thunder
printed matter:	WAIIIIII TREEEEEEESSSS

wtüte:	shes mine@matter . . . gg
printed matter:	cooome chick chick chick
wtüte:	away matter
wtüte:	*(whispers to dumbwaitress)* smile . . . great you play along . . . hello i am WTüte
printed matter:	you are too rude to her and besides you cant pay the non-wage labor costs
schwester2000:	*(whispers to dumbwaitress)* just 5 minutes
wtüte:	hey . . . shes a waitress!! those are used@rude
printed matter:	that right waitress? talks with lovely voice
dumbwaitress:	dumbwaitress strides to and fro
schwester2000:	try restart @dumbwaitress
operator:	you have been invited into thisismylastinvitation by schwester2000; to get there enter /go thisismylastinvitation; or /sg
dumbwaitress:	dumbwaitress fingers for napkin
schwester2000:	*(whispers to dumbwaitress)* hmm . . . have we met before??
dumbwaitress:	dumbwaitress is horribly clumsy
wtüte:	no prob waitress . . . we will learn that together . . .
dumbwaitress:	dumbwaitress hits full cup with elbow
wtüte:	aaaarghhh . . .
dumbwaitress:	dumbwaitress looks like nothing happened
schwester2000:	meditation . . . smiles
printed matter:	did she take a vow of silence . . . ?? then i will step on her foot . . . like in life of brian
schwester2000:	*(whispers to dumbwaitress)* say something—smiles
schwester2000:	if i dont hear anything else from you
wtüte:	loool@matter@steps on foot
dumbwaitress:	dumbwaitress waves napkin
wtüte:	hm hm hm . . . good domestics are so rare these days . . .
dumbwaitress:	dumbwaitress wipes on printed matter
printed matter:	on meaaaaaaaaa? what are you wiping on me?
dumbwaitress:	dumbwaitress looks skeptically at smeary printed matter
printed matter:	looks skeptically back
dumbwaitress:	dumbwaitress picks up printed matter gingerly dumbwaitress shakes head
printed matter:	eyyyyyyyy . . . let me down
wtüte:	no . . . this is not my printed matter . . . throw him away gg

schwester2000:	*(whispers to dumbwaitress)* shall i invite us once more??
dumbwaitress:	dumbwaitress wraps boa constricta around her neck
schwester2000:	*(whispers to dumbwaitress)* special girl u r or guy perhaps
schwester2000:	smiles
dumbwaitress:	dumbwaitress struggles for air. mpfffft
schwester2000:	laughs
printed matter:	you are all assholes . . .
dumbwaitress:	dumbwaitress gathers up her skirts
printed matter:	ill stay with u vfg
schwester2000:	*(whispers to dumbwaitress)* so . . . no chance???
wtüte:	*(whispers to dumbwaitress)* coffees all gone!!
schwester2000:	*(whispers to dumbwaitress)* a pity, really . . . well . . . where u from??
dumbwaitress:	dumbwaitress waves goodbye with napkin
printed matter:	you stay waitress
schwester2000:	*(whispers to dumbwaitress)* stay here waitress just stay here waitress
schwester2000:	hey
schwester2000:	*(whispers to dumbwaitress)* just stay one minute or 2 hours i beg you loudly to whisper back at me for once
dumbwaitress:	dumbwaitress whispers incomprehensibly and rustles away
schwester2000:	hey

Translated by Brigitte Pichon and Dorian Rudnytsky

Thomas Kamphusmann

Another ABC

A Model for "Augmented Business-Communication"

Abstract:

Communication doubtlessly has become the major driving force in the development of actual business. Without an efficient inner- and inter-organizational communication no knowledge-driven business can survive. In contrast to the traditional resources (raw material, capital, machines, manpower) communication is not yet considered as a resource in its own right. The reason for this disparity between the importance of and the attentiveness to the subject of communication can be seen in the complexity of the topic that hinders a straightforward modeling and optimization as it can be achieved e.g. in production processes.

This essay briefly sketches an approach to a comprehensive model of business related communication that lays the basis for methods to elicit, model, analyze and last but not least to design business communication on technical, procedural and structural levels.

After suggesting some background thoughts this article introduces the setting and gives an overview over the relevant perspectives on business communication as well as over the significant theories concerned with the subject. It then proposes a model by means of delineating the central concepts of that model. It closes with a prospect on the work that still has to be done in order to make use of these fundamental concepts.

1. Background

For the longest time communication has been a more or less unconscious means that human beings have used for purposes they considered as "outside" of themselves. Communication was considered identical to the transmission of natural language and as such nothing more than a cluster of sentences; communication thus could not be handled in the same way they handled tools. Language and thus communication was more likely part of their brain, heart or body than part of their tools, workshops or factories.

Changes occurred through a series of "revolutionary" inventions—writing, book printing, synchronous transmission, or automated modification. They were followed by appropriate and more-or-less scientific reflections like contemplation of textual structures, intertextuality, cognitive processes etc. Nowadays they encompass individual, aesthetic and social impacts of "literary machines."

Literature and bibliographies on this subject fill libraries. Just to mention a few examples, I want to name Leroi-Gourhan's *Gesture and Speech* for anthropological aspects, Goody's *The Logic of Writing and the Organization of Society* for the studies on the formation of early societies by means of the written text, Giesecke's *Der Buchdruck in der frühen Neuzeit* for the "Gutenberg revolution," an enormous amount of semiotic studies like those by Eco (e.g. *Il Segno)* as well as research of telecommunications based on Shannon's *The Mathematical Theory of Communication* and even including work on artificial intelligence, e.g. Hofstadter's *Gödel, Escher, Bach,* Kurzweil's *The Age of Intelligent Machines* or, of special importance for my considerations, *Understanding Computers and Cognition* by Winograd and Flores. This covers only a very small range of the scholarly research into communication.

2. Introduction, Motivation and Scenario

In my opinion this situation doesn't justify neglecting the topic. With the advent of the so-called "knowledge societies" individual and small-group communication has become one of the most productive "tools." Parallel to the goods-oriented notion of value chains that addresses mainly the primary sector and the activity-oriented notion of business processes addressing mainly the second and partly the third sector, we are in need of a communication-oriented notion of information-flows covering the knowledge-intense parts of the services sector. Studies of IDC et al. prove this need clearly and quantitatively (Feldman). This missing notion does not only have to focus on "information supply" but has to take into account communicative processes as a whole. Otherwise the needed support (from the workers' perspective) for improving quality of work by reducing error-prone tedious work while enhancing concentration on the central tasks can not be achieved; nor can the needed growth in efficiency (from the company's perspective) be realized by coordinating and thus reducing redundant work thereby fostering a higher usage of given intellectual resources.

To achieve these two major aims a rational and open communication on communication has to be set in motion, especially in companies that increasingly depend on the use of their intellectual resources. This meta-communication must involve the persons affected in a higher degree than every other enterprise-resource-planning; otherwise the necessary compliance is not likely to be achieved. It therefore must not only be an open communication implying that every aspect of business related aspects can be addressed, but also a transparent meta-communication. Transparency is primarily grounded on a common understanding of the terms and concepts used and their relations to each other. This is not at all a trivial task. Such meta-communication thus needs a "model" of communication, whether explicitly or implicitly. This model neither can be presupposed as "naturally given" nor does it have to be formal or mathematically proven. But in order to conduct such a meta-communication that supports the sketched goals the model has to consider business communication in all facets that are relevant in relation to the goals.

This article will sketch and discuss such a model in section 4. Before that, in section 3, it will give a brief overview over the theoretical foundations, the questions that are to be answered by the model and its architecture. Yet prior to that a fabled example from ComPro Ltd. will show what this is all about. Let's hear what Mr. Prolead says about the internal communication at ComPro:

> Normally we meet for the regular meetings of our department on Monday at 9am at a conference room. Sometimes somebody from other departments such as marketing or from the IT attend the meeting or parts of it. First an agenda is set up but it's no problem to add more points during the meeting. Sometimes this disturbs the meeting because the focus of the discussion becomes unclear. The head of department normally conducts the meeting—even if there is nothing to conduct because he is the only one who speaks. Mostly the meeting gets minuted but mostly the quality of the minutes is very poor so that somebody who wouldn't attend the meeting better asks a colleague. Well, normally the meeting is a "must" but there are two or three colleagues that regularly don't attend it.
>
> There is a wide variety of topics, ranging from administrative themes and resource planning over customers, projects to strategic issues. More personal topics such as further education or personal objectives are rarely discussed there but in bilateral talks. Some of the topics are very constant such as projects and dates. But even the weekly reports on the projects are seldom interesting for the others. Some project-leaders just say "in-time'n'budget," others torture with technical details on the cable-infrastructure of the clients. And for the dates: Do you think that I can remember the weekly schedule for nearly 20 colleagues?
>
> The most interesting discussions in my opinion arise from problems within projects that lead to development of our competencies and offers. Often they arise from talks about projects or customers. These discussions sometimes extend to hours and result in new ideas that are sketched on flip-charts. Some of them were framed and ended on

the "wall of fame," if you've time left I can show you and tell more on some of them.
Once a quarter of a year one of the Chief-Officers attend our regular meeting. Even if the agenda doesn't differ from the regular ones the atmosphere is completely different. The project reports are fairly prepared and even the dates are not only mentioned like "Wed, 10 to 12 at Customer A" but more like "Well prepared and promising meeting with the CEO of Customer A, scheduled Wed, 10 to 12 but might extend in the afternoon."

At this point we leave fictitious Mr. Prolead's report on regular meetings at ComPro Ltd. and return to more general thoughts. These are concerned with the appropriate theories for a systematic analysis of the above scenario, the questions that employees and managers might have facing the described situation and the layers on which analysis and support should be located.

On the other side, but not going hand in hand with the scholarly research, is the commercial side of communication. Today we are facing a ubiquitous trend towards a rationally planned and targeted usage of communication for political and economic purposes. This not only addresses areas of mass-communication like "public communication" or "marketing" but increasingly also the communication of individuals and small groups. In contrast to the former these attempts to support, guide or control communication touch or enter the private sphere of workers, sometimes hidden and without their prior consent.

3. Perspectives, Questions and Theories

To develop a comprising model for the description, analysis and design of supporting mechanisms of business related communication we start with a differentiation of relevant perspectives on the communication in question. Remembering the "interview" we can distinguish at least four of them:

1. A *structural* perspective in which we can distinguish "regular meetings of our department" from "bilateral talks." These structures for communication are not necessarily formalized but sometimes they occur in terms of standing orders. These determine, formalized or not, necessary and optional roles, admitted types of utterances and other presentation forms, topics and often location, time, as well as necessary and optional attendees.
2. A *procedural* perspective discussing the necessary tasks before, during and after communication such as "setting up the agenda" and "taking down

and distributing the minutes." In this perspective we will identify typical processes such as "presentation of project results" or "discussion and decision on alternative solutions."

3. An *individual* perspective in which the personal preferences, the prerequisites and deficiencies as well as the individual positioning within different environments can be addressed. Obviously the difference between "in-time'n'budget" and "details on the cable-infrastructure" within the same type of communication (project-report) is strongly linked to the individuality of the speaker in conjunction with the persons present.
4. A *semiotic* perspective that deals with the content, its topics and their relations as well as with its technical presentation and the transmission as messages. "Discussions on projects and customers" and the "sketching of new ideas" as well as the notion of "technical details" point directly to this perspective.

It is obvious that all four perspectives are heavily interdependent—which in other words says that an all-embracing model of communication needs all of them. Roughly speaking these interdependencies are determined by subordination. As sketched in fig. 1 the structural perspective frames the other three, the procedural perspective guides the individual and the semiotic one, whereas the individual perspective uses the semiotic one in a specific way. It must be said that the lower levels, namely the semiotic one, of course have a formative influence on the higher ones but that these processes are far slower and belong to a diachronic point of view.

Apart from the more theoretical justification of these perspectives they are useful to guide the process of explication, modeling and analysis of business communication. Only in very rare cases will all of the perspectives be necessary for a targeted analysis. In most cases preliminary knowledge of the problems to be considered can be assumed, leading the elicitation and analysis to one of the perspectives as the primary access to further investigation.

With the goal of a systematic analysis of business related communication—based on a guided elicitation and modeling and aiming toward designing efficient communication—and with the four perspectives in mind it is possible to denote some key questions that will be answered if the elicitation, modeling and analysis succeed on all levels of perspective.

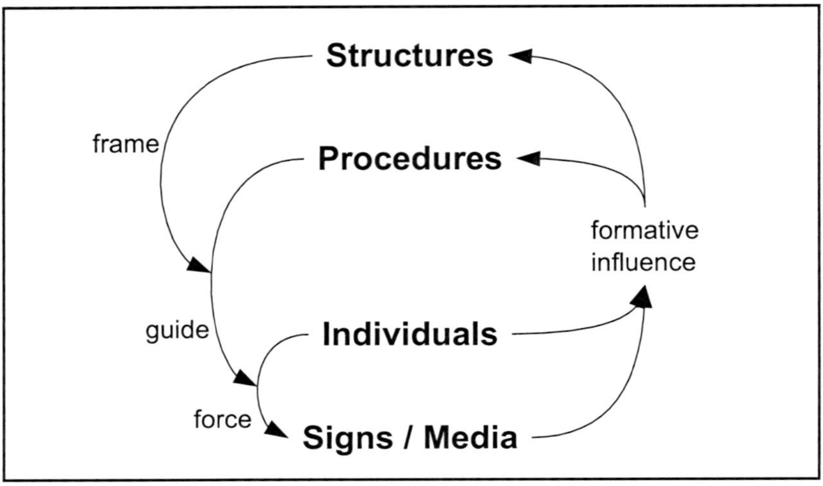

Fig. 1. Interdependencies between the relevant perspectives
on business communication.

- For the structural perspective questions concerning the appropriateness and the design of established forms of communication like different types of meetings and talks as well as less formal norms and habits have to be answered. These communicative forms and norms are seldom subject to and rarely the result of a systematic development. With the efforts in mind that are invested in regular meetings and all the other discussions and the complaints about ineffective, lengthy, useless, unfocused, redundant . . . meetings it is obvious that a systematic design of these forms and habits of communication would support existing resources more efficiently—and especially the highly paid ones.
- Questions regarding the *procedural* perspective on communication show substantial similarities to business process-modeling and -analysis. This is not surprising since business communication processes are by definition related to or even identical with business processes. This is highly true in service- and even more so in knowledge-oriented businesses. Thus the questions about used resources, latency, roles, dependencies and surplus value are in general relevant to communication processes as well. Differences occur when we take a closer look at the specific attributes that describe communication processes in contrast to business processes in general.

- Questions concerning the *individual* perspective on business communication clearly invade the privacy of the employees. But, this does not necessarily hinder a rational discourse about efficient practices. Questions concerning the personal organization of communicative work such as "answering e-mails in two blocks a day" or "preferring to read longer reports as printouts" as well as support in finding the relevant documents (relative to the personal knowledge) for preparing a discussion in time are examples for an individual but still rational perspective on communicative work.
- For the *semiotic* perspective we will be confronted with questions about the internal and external structure of documents as well as with questions about the transmission of messages. Thus, this perspective covers the textual form and the content side as well as the low-level, transmission oriented aspects of communication. Within this perspective questions about the segmentation and recombination of texts, the semantic clustering and relations as well as the appropriate media of storage, transmission and modification have to be posed and answered.

All these questions are not completely new and none of them have less than a few answers. Therefore a look at relevant theories that can help answering these questions in this particular perspective regarding enhancing business communication will conclude this section.

Looking at theories and methods that are relevant for the description and analysis of business communication could fill one's (and others') lifetime. Therefore the following selections are certainly not incomplete but sparse, they nevertheless serve as an overview for more detailed and future work.

- Judging the broad range of *sociological* theories regarding their usability for our aim we firstly find the differentiation between macro- and micro-theories. Neither of them are sufficient for the full explanation of business related communication. The structural perspective on the one hand demands a view of social structures that is covered by macro-theory, while on the other hand sociological views that point to micro-theories are necessary when looking to the dependent individual activities. Despite the basic work of Giddens *(The Constitution of Society)* and Symbolic Interactionism (cf. Kim) we find few operational starting points for the modeling of targeted communication. This criticism holds strongly for the macro-theories, above all Luhmann and Habermas that don't focus on the analysis of concrete parts of social subsystems. But also the descriptive and analytical power of micro-theories like Rational-Choice-Theory or Game-Theory is lacking regarding the phenomena sketched in the scenario.

- A lot of work (e.g. Scheer, Rejswoud et al., Medina-Mora et al., Goldkuhl and many others) has been done to model business processes as well as their environments, dependencies and resources. These *procedural* models aim at optimizing internal value chains in minimizing the usage of resources of all kinds, reducing latency and improving quality especially of the so called core-processes. Yet none of these consider communication as a core process in its own right. On the other hand there are a few attempts at modeling all the facets of communication, resulting in unmanageable models (e.g. "KODA," cf. Dämmig et al.), models that don't show the relation of communication and the other processes ("KSA," cf. Krallmann) or in attempts that don't show any practical relevance (Integrated Process- and Communication Modeling on top of the "transactive memory systems" approach, cf. Remus). This variety of theories has led to approaches that integrate by designing meta-models and derive transformations from these. Kethers' work is certainly inspiring in judging the selected theories (even though she left out some of the most powerful ones like DEMO and BAT) but it doesn't extend the underlying methods of process-modeling to the needs of business communication (Kethers, *Multi-Perspective Modeling and Analysis of Cooperation Processes).*
- For the *individual* perspective psychological explanations wouldn't help much in a setting where privacy is of high value and where violating privacy would lead to seriously reduced commitment to any measures that try to change the way of communication within the company. Therefore we stay within the area of personal configurable settings (or "preferences") supporting individual work. This can be achieved by a wide variety of measures like demand-driven information supply supported by text mining features (cf. Kamphusmann, "Anwendungen für Wissensarbeiter"; *Text-Mining*) as well as by context-sensitive functionality of integrated operational, informational and communicative software-systems. The latter point to current work in end-user development. Adopting these starting points to systems that have the potential to guide inter-human communication will take a lot of attempts and empirical investigation.
- For the *semiotic* perspective we can and have to make use of structuralist theories on all levels of textual complexity as well as of information studies. Evidently this covers findings in intertextuality (e.g. Lachmann) as well as "core-semiotics" (e.g. Saussure), research on "syntactic structures" (e.g. Chomsky) and the use of speech act theory (e.g. Searle) (and its applications like Winograd and Flores) that give a sound base for analyzing utterances in their respective context. This (in a broad sense semiotic) base is accompanied by a technical counterpart that deals with the transmis-

sion of content that goes back to Shannon's *The Mathematical Theory of Communication* and its applications like "Prediction and Entropy of Printed English." It has to be stated a second time that this overview on relevant theories and methods is sparse and has to be extended substantially into all directions. Some work has been launched at Fraunhofer ISST (e.g. in the "Interval" (= 'Internet and Value Chains') Project, founded by the German Ministry of Science) as well as in the doctoral thesis of Guy Vollmer (to appear in 2006) leading to first practical results (e.g. Vollmer) and preliminary theoretical foundations (e.g. Kamphusmann, "Zeichen in Wissensmanagementsystemen") as well as drafts on software architecture (e.g. Kamphusmann and Gerhard).

A mere combination of theses approaches might lead to an encompassing model but will probably not be consistent enough to give a sound base for systematic elicitation, modeling and optimization of business communication. Even if enhanced with a meta-model covering all the concepts of the underlying models as in Kethers's dissertation there will be redundancies as well as gaps, especially for the semiotic perspective. Therefore, in the following I will create a new model for the description and analysis of business communication—of course bearing the aforementioned models in mind.

4. Entities, Concepts and Models

It would take far too much time and space to develop the whole communication model in detail. For this reason this chapter introduces only the main concepts for each perspective as well as their relations and their main attributes. Another reason is that the model is not totally developed yet: notably missing is sufficient proof in real world projects. But the consideration of recent projects concerning a project (with a real estate company as well as another concerned with communication in software engineering) suggests the suitability of the following model. There are some design issues that should be mentioned in advance of a more detailed view of the model itself. For the sake of practical relevance the development of the model had to ensure a maximal proximity to real life situations. This was ensured by an iterative method switching back and forth between the conceptual level and the instance level. Some of the minor important concepts are still lacking a precise definition and therefore are represented by a single term (on the conceptual level) that only has one attribute for the instance level: "description," which is a single term as well. In these cases common sense replaces the missing precise definition. The model is defined in terms of ontology, using

Protégé as the modeling tool. According to the overall aim of a communication model of high practical relevance the central concepts should be operational to a high degree in order to contribute to an easily viable and precise elicitation. Therefore we prefer concepts that are of low complexity and that resemble everyday experience of the workers in question.

The following paragraphs are headed by visualizations of the central concepts that are explained afterwards. For the sake of readability these screenshots only show connections between the concepts and suppress types, attributes and other details.

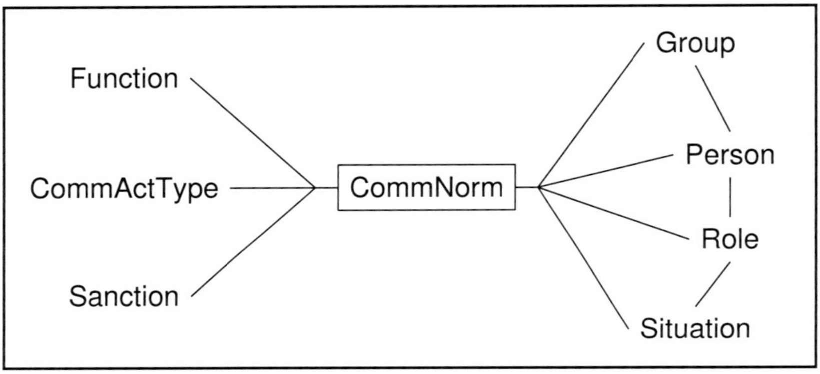

Fig. 2. "Communicative Norm."

For the structural perspective we consider "communicative norms" as the central concept. Such norms can be considered as obligations that—in certain situations—bind certain persons (determined by their individuality or by their role within the given situation) or groups to a certain communicative behavior. The commitment thereby can arise from different sources. It can either be a personal commitment that is indifferent to the role or to the membership within a group. Such commitments may arise from history within the company or from personal experience. There are two other sources of such commitments: Firstly we can name adopted roles or memberships in a group that in turn plays a decisive role (such as being superiors who are not allowed to use information from personal files in group discussions). Secondly there can be "anonymous" commitments stemming from a common understanding of a relevant group of arbitrary size and complexity, ranging from "western society" to department and below e.g. demanding to answer technical questions using an appropriate vocabulary. To model the requested or forbidden communicative behavior I am using the typology of communicative acts as introduced in the paragraph on "communicative acts"; for the modeling of situations see the respective paragraph below.

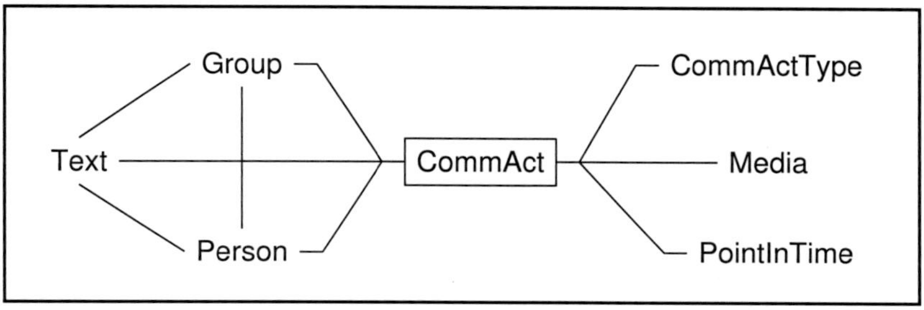

Fig. 3. Communicative Act.

For the *procedural* perspective a "communicative act" as the smallest activity-oriented entity will serve as the central concept. This concept takes the transport of a single message (or "text") by a (not necessarily technical) communication medium as its core. With this the transmission of a single message is chosen as the atomic entity. This is used to construct more complex communicative processes by timely ordered and logically related sequences of such transmissions. Yet this purely technical view on communication is enhanced by some additional attributes characterizing the communicative act according to the needs of an encompassing analysis of business communication. In particular, I am introducing a typology that resembles the illocutionary forces of Searle. But this typology differs methodologically from SAT in two important ways. First, the types are types of the speech itself and are not drawn from the intended activity of the addressee. Second, the typology is far less generic than Searle's but is oriented towards common sense and common naming of communicative acts on different levels of genericity. In short it is somewhere in-between Speech Act Theory, business process modeling and dialogue-grammars. Apart from this, I am introducing some basic roles as well as a simple concept of groups. With these we denote not only sender and receiver of some arbitrary text as with Shannon but also the differentiation between addressee and receiver as well as the creator of the transmitted text. In addition groups as potential senders, addressees and receivers of the transmitted data are established as well as introducing a "semantic" hook with the notion of the "theme" of a text. Together with the notion of references from one communicative act to others that it refers to and from one text to others we can build up sequences in time (chains of acts) and intertextual structures (nets of texts). It seems that these types of references are sufficient to model the low-level structures of communication seen from the processing point of view.

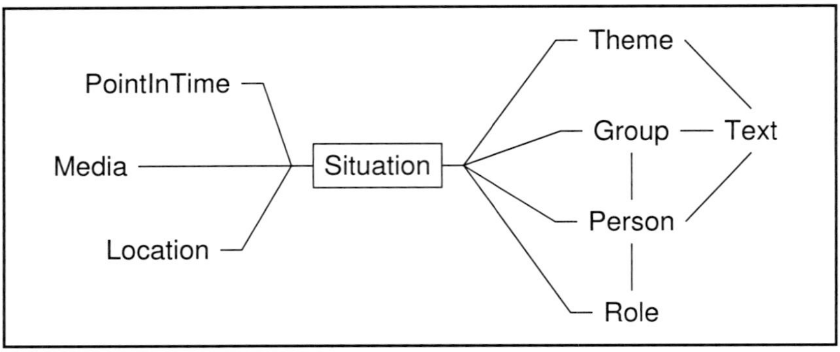

Fig. 4. Situation.

For the *individual* perspective "situation" will be the concept around which the secondary concepts are arranged. This choice is not a natural one since it incorporates not only personally bound attributes (that are often summarized as "preferences") but also such descriptions as "taken roles" or "medial infrastructure" etc. Yet these seemingly impersonal attributes determine largely the possible activities of the workers as well as their actual behavior. For this reason the seemingly over-individual concept "situation" is taken as the main concept for the individual perspective regarding communication. Modeling situations from a communication point of view offers a wide range of alternatives what to include and what to leave out. Admitting that this article is preliminary in many points we take the relation of necessary, facultative and forbidden roles, the instantiation of roles by certain people, the spatial and temporal as well as the thematic and technical characteristics as constitutive for a situation. Apparently the affected person(s), relevant documents and the activated communicative norms in most cases further specify situations. Examples from business situations are "regular departmental meetings," "project meetings" as well as "customer telephone calls," "demonstration" etc. Thus from an individual point of view situations are highly dynamic and characterized by necessary, facultative and forbidden relations to a wide variety of instances. The less important parts consist of the necessary relations ("attributes" and their values). This makes situations hard to define in a traditional way and hard to recognize by automatic means—even if the affected persons can name them precisely and spontaneously. This must be seen as a reflection of the fact that in practice this concept is rarely defined explicitly in existing models of organizations and their processes.

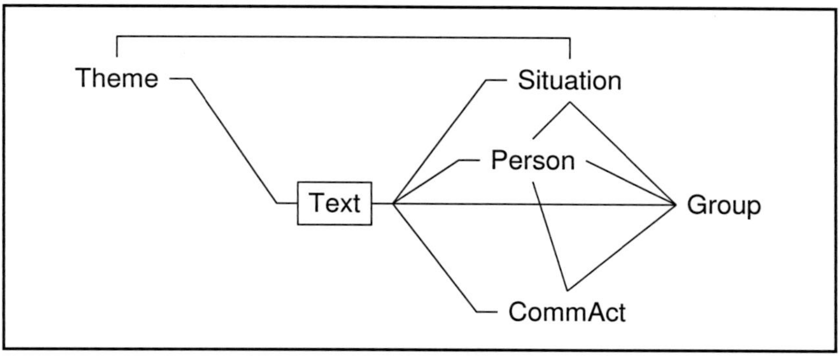

Fig. 5. Text.

For the *semiotic* perspective the "text" is the primary concept—even if it is a term that bears countless years of discussions around its definition and use. Primarily we take every identifiable significant as a possible text and introduce four main relations for our purposes. There should be a creator, let it be a single person or a group, who can be named. Secondly we only deal with texts "in action" that is to say we only regard texts used in communicative acts. Thereby we export the transport of texts to the communicative act and reduce redundancy if the text is used several times (as in forwarded e-mails or other copies). A generic relation to an arbitrary representation (let it be keywords, entries in a thesaurus or an ontology) of its theme is necessary to enable testing the appropriateness on a semantic level. A recursive relation to other instances of "text" allows building up intertextual networks. It is obvious that the latter two relations have to be attributed by themselves since there are different possible relations to themes (e.g. "introduction to something," "overview on latest developments," "canonical source") and different types of relations to other texts (e.g. "cites text," "builds upon," "contradicts"). Since the necessary types are specific to certain domains they are missing in the generic model presented here. It should be stated that the relations to communicative acts, themes and other texts loosely resemble the linguistic levels of pragmatics, semantics and syntax.

5. Prospect

Modeling business communication for its own sake might be interesting because it foresees problems that are also potentially relevant to other areas of modeling, such as the question how to handle concepts that only have very few or even no necessary attributes. Yet this attempt as a whole aims at an analysis and augmentation of business communication, thus making the communication-model into the central concept for the elicitation, notation and analysis of business-related communication and its problems. Furthermore it should give us strong indications for the design of supporting computer systems. From this point to that end there are apparently several steps that are to be taken:

- Enhancing the model to a version that has no "loose ends" (e.g. concepts that are represented only by its names), checking the theories more in depth for each concept and proving that the model is not self-contradictory.
- Developing methods and supporting tools like questionnaires and log-file analysis for eliciting and acquiring descriptions of factual communication.
- Using these methods in pilot projects with potential users from industrial environments, checking the appropriateness of the model and the methods and tools of elicitation as well as revising them.
- Developing appropriate methods and tools for visualization and modeling the factual communication using the described model (which in fact is the meta-model for the factual models).
- Analyzing the models of factual communication to derive typologies, best practices and constraints for the concepts and their relations and revising again the (meta-)model as well as the elicitation tools.
- Evaluating common problems and needs from the analysis of the factual models that can serve as requirements for software systems that enhance business communication.
- Constructing a software platform that serves as the base for all required functions concerned with business communications, especially for all rule-based analysis and modification of inner- and inter-organizational communication.
- Implementation of appropriate software systems using the platform and integration of these into standard communication systems like e-mail, computer-integrated telephony including voice-over-IP, fax and ERP-systems.

This working program will take some time and effort, although some of the tasks should be performed parallel to each other. This is particularly true for the interdependency of the more theoretical work on the model, the methodological work on elicitation, acquisition and notation and the practical work within the industrial pilot-projects.

Works Cited

Andreas Abecker, Knut Hinkelmann, Heiko Maus and Heinz J. Müller, eds. *Geschäftsprozessorientiertes Wissensmanagement.* Berlin: Springer, 2002.

Chomsky, Noam. *Syntactic Structures.* 's-Gravenhage: Mouton, 1957.

Dämmig, Ingo, Uwe Hess, and Claudius Borgmann. "Kommunikationsdiagnose (KODA): Einstiegsmethode und -werkzeug in das praktische Wissensmanagement." *Geschäftsprozessorientiertes Wissensmanagement.* Ed. Andreas Abecker, Knut Hinkelmann, Heiko Maus, and Heinz J. Müller. Berlin: Springer, 2002. 123-158.

Eco, Umberto. *Il segno.* Milan: Istituto Editoriale Internazionale, 1973.

Feldman, Susan. "Special IDC Report. Enterprise Search Technology: Information Disasters and the High Cost of Not Finding Information." *Portals Magazine.* December 2003. 7 May 2006 <http://www.portalsmag.com/articles/default.asp?ArticleID=5286>.

Giddens, Anthony. *The Constitution of Society: Outline of the Theory of Structuration.* Cambridge: Polity, 1984.

Giesecke, Michael. *Der Buchdruck in der frühen Neuzeit: Eine historische Fallstudie über die Durchsetzung neuer Informations- und Kommunikationstechnologien.* Frankfurt am Main: Suhrkamp, 1991.

Goldkuhl, Göran. "Generic Business Frameworks and Action Modelling." *Proceedings of the First International Workshop on Communication Modeling: The Language/Action Perspectice.* Ed. Frank Dignum, Jan Dietz, Egon Verharen, and Hans Weigand. Berlin: Springer, 1996. 8 Jun 2006 <http://ewic.bcs.org/conferences/1996/comms/papers/paper8.htm>.

Goody, Jack. *The Logic of Writing and the Organization of Society.* Cambridge: Cambridge UP, 1986.

Hofstadter, Douglas R. *Gödel, Escher, Bach: An Eternal Golden Braid.* New York: Vintage Books, 1980.

Kamphusmann, Thomas. "Anwendungen für Wissensarbeiter: Inhaltsbezogene Recherche und Informationsangebote." *Report Informationslogistik: Informationen just-in-time*. Ed. Wolfgang Deiters and Carsten Lienemann. Düsseldorf: Symposion, 2001. 82-93.

———. "Zeichen in Wissensmanagementsystemen." *Informatik 2001: Wirtschaft und Wissenschaft in der Network Economy—Visionen und Wirklichkeit.* Ed. Kurt Bauknecht, Wilfried Brauer, and Thomas Mück. Vienna: OCG, 2001. 743-747.

———. *Text-Mining: Eine praktische Marktübersicht.* Düsseldorf: Symposion, 2002.

Kamphusmann, Thomas, and Michael Gerhard. "CommunicAID: Kommunikationsunterstützende Systeme." *Zukünfte des Computers.* Ed. Claus Pias. Berlin: diaphanes, 2005. 157-170.

Kethers, Stefanie. "Multi-Perspective Modeling and Analysis of Cooperation Processes." Diss. U of Aachen, 2000. 8 Jun 2006 <http://deposit.ddb.de/cgi-bin/dokserv?idn=962847658>.

Kim, Tae Won. *G. Simmel, G.H. Mead und der symbolische Interaktionismus: Geistesgeschichtliche Zusammenhänge, soziologische Systematik.* Würzburg: Ergon, 1999.

Krallmann, Herrmann. "Die Kommunikationsstrukturanalyse—Voraussetzung für das papierarme Büro." *GI—16. Jahrestagung II, Berlin, 6.-10. Oktober 1986, Proceedings.* Ed. Günter Hommel and Sigram Schindler. Berlin: Springer, 1986.

Kurzweil, Raymond. *The Age of Intelligent Machines.* Cambridge, MA: MIT P, 1990.

Lachmann, Renate. "Intertextualität als Sinnkonstitution: Andrej Belyjs 'Petersburg' und die fremden Texte." *Poetica* 15 (1983): 66-107.

Leroi-Gourhan, André. *Gesture and Speech.* Cambridge, MA: MIT P, 1993.

Medina-Mora, Raul, Terry Winograd, Rodrigo Flores, and Fernando Flores. "The Action Workflow Approach to Workflow Management Technology." *Proceedings of the 4th Conf. Computer-Supported Cooperative Work, 31. Oct.-4. Nov. 1992.* Ed. Jon Turner and Robert Kraut. Toronto: ACM, 1992. 281-288.

Remus, Ulrich. "Integrierte Prozess- und Kommunikationsmodellierung als Ausgangspunkt für die Verbesserung von wissensintensiven Geschäftsprozessen." *Geschäftsprozessorientiertes Wissensmanagement.* Ed. Andreas Abecker, Knut Hinkelmann, Heiko Maus, and Heinz J. Müller. Berlin: Springer, 2002. 91-122.

Reijswoud, Victor E. van, Hans B. Mulder, and Jan L. Dietz. "Communicative Action Based Business Process and Information Systems Modelling with DEMO." *Information Systems Journal* 9.2 (1999): 117-138.

Saussure, Ferdinand de. *Course in General Linguistics.* London: Peter Owen, 1959.

Scheer, August W. *ARIS: Vom Geschäftsprozeß zum Anwendungssystem.* 3rd ed. Berlin: Springer, 1998.

Searle, John. *Speech Acts: An Essay in the Philosophy of Language.* London: Cambridge UP, 1969.

Shannon, Claude E. *The Mathematical Theory of Communication.* Urbana, IL: U of Illinois P, 1949.

———. "Prediction and Entropy of Printed English." *The Bell System Technical Journal* 30 (1951): 50-64.

Vollmer, Guy, and Katrin Gassner. "Quality improvement of email communication in work groups and organizations by reflection." *Proceedings of the 2005 international ACM SIGGROUP conference on Supporting group work.* New York: ACM, 2005. 124-127.

Winograd, Terry, and Carlos Fernando Flores. *Understanding Computers and Cognition: A New Foundation for Design.* Norwood, NJ: Ablex, 1985.

Contributors

Jean-Pierre Balpe was Head of the Hypermedia Department and of the *Paragraphe* laboratory at the University Paris 8 until September 2005. He has published both scholarly and technical books. His most recent theoretical book is *Contextes de l'art numérique* (2000). Since 1985 he has created or participated in exhibitions of art and new technologies. His last work *Interdit au moins de seize ans* (2006) was a generative play. He now manages an international poetry festival near Paris.

Susanne Berkenheger is a net artist and journalist. Her best known works are the hyperfictions *Zeit für die Bombe* (1997), *Hilfe!* (1999), *Die Schwimmmeisterin* (2002) and the English relaunch *The Bubble Bath* (2005). Her most recent art work is the mobile art service www.zentralinstanz.de (in collaboration with Klaus Ungerer). For more information see <http://www.berkenheger.de>.

Friedrich W. Block is Director of the Brückner-Kühner-Foundation and the Kunsttempel in Kassel, Germany. He is the curator of numerous exhibitions and has also worked as an artist. Since 1992 he has been responsible for the *p0es1s* project on digital poetry and since 2000 for the *Kasseler Komik-Kolloquium.* His research focuses on contemporary literature, language art and humor. Block is co-editor of the *Kulturen des Komischen* series. His books include *Beobachtung des "ICH"* (1999) and *p0es1s: The Aesthetics of Digital Poetry* (2004, co-editor). He has recently edited *Komik—Medien—Gender* (2006).

Philippe Bootz is Assistant Professor of Information and Communication at the University of Paris 8. He is also head of publication of the review *Regards croisés/Crossed looks* and publisher of the review of e-poetry *alire* (since 1989). Most recent significant papers in e-poetry in 2006 are contributions to the conference LUDOVIA and to the reviews *Rilune, Formules, DOC(K)S.* His next authored hyper-book *Les Basiques: la littérature numérique* is due to be published on-line in 2006 by LEONARDO.

Laura Borràs Castanyer is Professor of Literature at the Languages & Cultures Department at the Universitat Oberta de Catalunya (UOC). She also teaches Comparative Literature and Literary Theory at the Universitat de Barcelona. She is Director of the *Hermeneia: Literary Studies and Digital Technologies* research group and leader of the project *Texto, cibertexto, hipertexto: literatura érgódica y práctica críti-*

ca en el paradigma digital of the Spanish Government. She is author of several articles related to the subject and of the books *Textualidades electrónicas: Nuevos escenarios para la literatura* (2005) and *Testualità elettroniche* (2006).

Markku Eskelinen is an independent scholar and experimental writer of ergodic prose and critical essays. According to *The Review of Contemporary Fiction* he is "easily the most iconoclastic figure on the Finnish literary scene." Currently he is writing a two-volume history of Finnish literature due to be published in 2009. He is also one of the editors of *Game Studies: The international journal of computer game research,* and with Raine Koskimaa, he edits *Cybertext Yearbooks.*

Frank Furtwängler is a physicist and scholar of Literature, Media and Art; he is currently a fellow at the Ph.D. scholarship programme *Bild—Körper—Medium: Eine anthropologische Perspektive* at the Hochschule für Gestaltung Karlsruhe. A selection of recent publications include "Computerspiele am Rande des metakommunikativen Zusammenbruchs," "Mensch-Maschine Computerspiel: Über eine notwendige Paartherapie in der Medienrealität" (both 2006) and several others on media and video games forthcoming in 2006 and 2007.

Peter Gendolla is Professor of Literature, Art, New Media and Technologies at the University of Siegen. He also is Head of the Human Collaborative Research Center *Media Upheavals* and directs the sub-project *Literature in Nets/Net Literature.* With Georg Braungart and Fotis Jannidis he is editing the *Jahrbuch für Computerphilologie* and *Forum Computerphilologie.* He is currently completing a book on simulation and he is also co-editor of *Wissensprozesse in der Netzwerkgesellschaft* (2004), *Formen interaktiver Medienkunst* (2001) and *Bildschirm-Medien-Theorien* (2001). For more information see <http://www.likumed.uni-siegen.de>.

Loss Pequeño Glazier is a poet, professor of Media Study, and Founder and Director of the Electronic Poetry Center <http://epc.buffalo.edu>, Dept. of Media Study, State University of New York, Buffalo. He is the author of the digitally informed poetry collection *Anatman, Pumpkin Seed, Algorithm* (2003), the award-winning *Digital Poetics: The Making of E-Poetries* (2002), and other works. Recent performances of "Baila," his digital poem for dancers, have occurred in London and Buffalo. He is organizer and director of E-Poetry: An International Digital Poetry Festival, the first and one of the most celebrated digital poetry series in the field. His work has been shown at various museums and galleries, including the Kulturforum, Berlin, the Royal Festival Hall, London, and the Guggenheim, New York. Selected digital projects and other works are available on his EPC author page <http://epc.buffalo.edu/authors/glazier>.

Fotis Jannidis is Professor of German Literature at the Technical University of Darmstadt. He is co-editor of the *Jahrbuch für Computerphilologie,* the *Journal of Literary Theory* and the series *Narratologia* and *Revisionen: Grundbegriffe der Literaturtheorie.* Main areas of research: Narratology, especially character, and the history of German narratives; concepts of literary theory like that of "author," digital editions. He is working on a book on *Computerphilologie* (due 2007), and is co-editing a volume on the concept "literature" (also due in 2007). For more information see <http://www.jannidis.de>.

Thomas Kamphusmann is a Postdoctoral Research Fellow at the Fraunhofer Institute for Software- and Systems Research. His focus lies on the technical and process-related support of business communication. His most recent publication (together with Guy Vollmer) on that topic is "Erhebung, Modellierung, Analyse und Gestaltung innerbetrieblicher Kommunikation." (2006). His recent books include *Literatur auf dem Rechner* and *Text-Mining* (both 2002).

Mela Kocher is a Postdoctoral Research Fellow at the Swiss Institute for Children's and Youth Media in Zurich, Switzerland. She lectures on digital literature and popular culture at the German Department of the University of Zurich and at the Institute for Media Studies of the University of Basel. Her most recent book is *Folge dem Pixelkaninchen! Ästhetik und Narrativität digitaler Spiele* (2007).

Marie-Laure Ryan, a native of Geneva (Switzerland), is an independent scholar based in Colorado. She is the author of *Possible Worlds, Artificial Intelligence and Narrative Theory* (1991), of *Narrative as Virtual Reality: Immersion and Interactivity in Literature and Electronic Media* (2001), and of *Avatars of Story* (2006). She is also the editor of *Cyberspace Textuality* (1999) and *Narrative Across Media* (2004), as well as the co-editor, with David Herman and Manfred Jahn, of the *Routledge Encyclopedia of Narrative* (2004).

Jörgen Schäfer is a Postdoctoral Research Fellow at the Human Collaborative Research Center *Media Upheavals* at the University of Siegen. He is author and editor of several books and articles on avant-garde and contemporary literature, net literature and media theory. His most recent books include *Exquisite Dada: A Comprehensive Bibliography* (2005), *Wissensprozesse in der Netzwerkgesellschaft* (2004, co-editor) and *Pop-Literatur* (2003, co-editor).

Roberto Simanowski is Assistant Professor of German Studies and Digital Aesthetics at Brown University and editor of *dichtung-digital.org,* an online journal for digital aesthetics. He is author and editor of several books and articles on literature and culture in the 18th century, nationalism, the literary salon, media theory and digital culture. His most recent books include *Transmedialität: Studien zu paraliterarischen Verfahren* (2006, co-editor) and *Interfictions: Vom Schreiben im Netz* (2002). His forthcoming book *Ästhetik des Spektakels* will be published in 2007.

Noah Wardrip-Fruin is a digital media writer, artist, and scholar. His writing/art has been presented by galleries, arts festivals, scientific conferences, DVD magazines, VR Caves, and the Whitney and Guggenheim museums. He is the author of the forthcoming monograph *Expressive Processing* and has recently edited three books: *The New Media Reader* (2003, with Nick Montfort); *First Person: New Media as Story, Performance, and Game* (2004, with Pat Harrigan); and *Second Person: Role-Playing and Story in Games and Playable Media* (2006, also with Harrigan), all published by the MIT Press. He is currently Assistant Professor in the Communication Department at the University of California, San Diego.